MASTERS OF THE OCEAN SEA

The Epic Saga of the Portuguese Explorers Who Redrew the Map of the World

Andrei Romanov

SAGRES HERITAGE PRESS

Masters of the Ocean Sea

The Epic Saga of the Portuguese Explorers Who Redrew the Map of the World

First Edition

ISBN: 978-989-33-9102-0
Depósito legal n.º 558606/26

Published by **Sagres Heritage Press**
Quarteira, Portugal
sagresheritage@yahoo.com

Printed in the European Union

CONTENTS

THE ARCHITECTURE OF OBSESSION

Henry was born on March 4, 1394, in Porto, amidst a world still reeling from the horrors of the Black Death, which had claimed nearly half of Europe's population not long before. Portugal, a young nation on Iberia's edge, had only recently gained independence and grappled with poverty, political instability, and precarious autonomy. Upon Henry's birth, Portugal did not dictate historical tides; it was shaped by them.

His father, João I, seized the Portuguese throne in 1385 through military prowess, popular support, and the demise of his strategic rivals. The Battle of Aljubarrota, fought before Henry's conception, secured João's rule and dashed Castilian hopes for a unified Iberian realm. Yet victory brought its own trials; João founded the House of Aviz, binding everything to provisionality, demanding constant validation. Noble families awaited weakness, Castile nursed grievances across borders, and Portugal sought glory, expansion, and a transformation from defensive survival to bold confidence.

Philippa of Lancaster, Henry's mother, brought English lineage and alliances into this volatile landscape. She served as the bridge between Portugal and England, a connection destined to last centuries. More significantly, she carried fervent Christian zeal that bordered on zealous extremism. To Philippa, every action either served or betrayed God. Comfort was suspicious, pleasure perilous, duty paramount.

Henry internalized these principles deeply. As the third son with no claim to kingship, he would always exist in his elder brother Duarte's shadow and in his younger sibling Fernando's sanctified legacy after Fernando's death in Morocco. Yet Henry found purpose in an enduring pursuit of knowledge. The young prince was described as tall, thin, and grave. Deep-set eyes framed a mouth often set in determination or displeasure, rare laughter avoided, marriage and known mistresses unknown. Some later writers hinted at vows of chastity or deep religious devotion. Whether genuine piety or something else remains unclear, but it is certain that Henry channeled energy typically spent on family and pleasure into other realms. Perhaps his austerity reflected broader economic strains in fifteenth-century Europe, marked by scarcity after the Black Death. While many sought comfort in family, Henry turned to exploratory ambitions, mirroring the era's shifting priorities, in which survival demanded innovation and expansion.

But before embarking on grander pursuits, there lay war. In medieval Europe, princes first tested their mettle in battle. It was upon foreign soil that

Henry's colossal dreams would meet their initial trial.

BLOOD IN THE MORNING

Ceuta, Morocco, August 21, 1415

At the break of day, an onslaught unfolds. Twenty thousand Portuguese soldiers, among whom are three royal scions, surge upon North African soil. Their aim is Ceuta, a Muslim city brimming with wealth and controlling the southern edge of the Strait of Gibraltar. For ages, Ceuta has thrived on commerce, channeling riches from Africa into Mediterranean markets. Now, Portugal aspires to possess it and deems it necessary for its grasp.

Henry, in his twenty-first year, confronts combat for the first time, a crucible for his mettle. Clad in gleaming armor and brandishing consecrated arms, he stands amidst troops wracked by seasickness and fear. The air is thick with odors of sweat, nausea, and sea salt. Commands are barked in Portuguese, Arabic, and Berber tongues. Steel clashes, and the cries of warriors pierce the dawn.

Ceuta succumbs within a single day's onslaught, an abrupt and ruthless triumph. The victory feels oddly flat. Defenders, caught unawares, crumble swiftly under Portugal's assault. By nightfall, the city is overrun. Fortifications are hastily erected. Christianity is imposed upon it, and dominion claimed. On Ceuta's blood-soaked ground, King João knights his progeny, a moment immortalized by Gomes Eanes de Zurara, chronicler of deeds yet to come. He notes Henry's visage aglow with the alloy of violence and piety. For him, conquest bestows purpose. Military expansion is not merely political; it is also an elevated calling. Yet, this flat victory marks a pivotal shift in strategy. The rapid conquest of Ceuta reveals to Henry the limitations of direct military conquest and highlights the potential of maritime exploration as a means to achieve Portugal's grand ambitions. This realization redirects his ambition from swords to ships, where the seas become the new frontier for expansion.

Within Ceuta's markets and warehouses, Henry beholds treasures beyond Portugal's ken: Malian gold, tusks from hinterlands, aged spices traversing ancient paths, and slaves from distant realms. This prosperity courses through the city, intertwined with the veins of trans-Saharan trade. To Muslims and Jews managing these arteries, such commerce is routine; for Henry, it's a jarring realization: Portuguese and European Christians are bystanders or supplicants in this wealth stream, controlled by Muslims, conveyed by their caravans, and traded within their domains.

The stakes transcend economics. These routes connect East Africa to the Indian Ocean, culminating in spice-rich territories that yield immense riches; they also bring enslaved Africans northward, many conscripted into labor in Muslim dominions, bolstering Islam's wealth and challenging Christendom's might.

From Ceuta, Henry makes a choice that will redefine realms, though its full impact may elude him. If Portugal cannot penetrate these Saharan routes governed by hostile Muslims, then another avenue must be forged. The Atlantic Ocean, once the edge of European dominion, transforms from barrier to gateway.

However, first, immediate perils loom. Sustaining Ceuta exacts cost without meaningful yield; it stands as a Christian bastion in seas of Muslim power, perpetually imperiled and dependent on reinforcements. Its former riches dissipate as commerce shifts beyond Portuguese jurisdiction, transforming prized acquisitions into liabilities, draining resources incessantly.

Detractors within Portugal query the prudence of seizing Ceuta, doubting its strategic value.

Henry remains resolute. He envisions uncharted frontiers and quietly commences crafting tools to pursue them.

THE EDGE OF KNOWING

After Ceuta, Henry gradually disengages from court life. He takes up the governorship of the Algarve, Portugal's southernmost region, and increasingly spends his time at Sagres, near Cape St. Vincent. This marks a geographical boundary for Portugal where Europe fades into granite cliffs and turbulent waters. In medieval cartography, it is often depicted as the edge of the world or close to it, with sea monsters represented on its borders symbolizing the unknown dangers beyond.

Henry strategically establishes his base at this liminal point between what is known and what lies in the realm of mystery. His decision-making process here was deliberate. Contrary to the popular legend of founding a "navigation school" at Sagres complete with formal education, there's little contemporary evidence for such claims. Instead, Henry assembles a network of pilots, cartographers, shipbuilders, and mathematicians to tackle the pressing challenge of safely navigating south along Africa's coast.

One notable figure in this network was Mestre Jaime, a cartographer whose insight into using maritime charts to detect shifts in coastal features significantly advanced navigation capabilities. This expertise played a pivotal role in an expedition that expanded boundaries previously deemed unpassable, showcasing the tangible benefits of Sagres' informal collaborative environment.

The task was not trivial. Early fifteenth-century Atlantic sailing was perilous and heavily limited by a lack of navigational precision and weather forecasting. Ships clung to coastlines, as venturing into open waters without knowing one's position or predicting weather patterns was suicidal. The Mediterranean, though enclosed and more predictable, had been conquered; the Atlantic, however, posed a different challenge entirely: it was larger, fiercer, governed by wind and currents potentially leading ships to their doom. Sailors knew through hard experience certain empirical truths: winds generally blew from the north along Portugal's coast, and currents flowed southward, making southern voyages somewhat manageable.

Yet, returning proved impossible for ships venturing south along Africa's coast. The same aiding factors, winds and currents, obstructed return journeys. Ships could only tack back and forth against the wind, an expensive, often fatal endeavor as resources depleted and vessels crumbled under strain. Those who did not perish at sea returned with harrowing tales of regions where the sun grew unbearable, men's skin darkening from heat, and demonic presences.

The psychological barrier surpassed even the practical one. Medieval European cosmology divided Earth into habitable zones, positioning the equator as an impassable region of fire, according to authorities such as Ptolemy and the Church. To venture too far south meant attempting what was considered beyond human endurance, defying natural order and European presence.

Henry rejects these beliefs not through skepticism or scientific breakthrough but by aligning with a theological confidence. He is driven by a conviction that God intends for Portugal, Christianity, specifically him, to reach these extremes. This stance is rooted neither in enlightenment reasoning nor science; it's an assertion of faith against fear.

Thus, he starts gathering individuals capable of resolving this issue. Cartographers arrive at Sagres, men whose livelihoods are tied to mapmaking - artful blends of information and educated conjectures. Jewish mathematicians find refuge despite widespread anti-Semitism, as their astronomical knowledge is crucial. Ship captains of Atlantic trading routes contribute practical experience, while instrument makers experiment with adapting Mediterranean navigation tools for the Atlantic.

There's no documentation of formal lectures or structured curriculum; instead, there are debates, experiments, and conversations. Henry provides support in many forms: protection, resources, and patronage. He grants monopolies on certain trade goods to those willing to venture further south, offers rewards for each new landmark reached, and, most importantly, he creates a psychological haven where expressing fear of the unknown is almost shameful.

This approach does not resonate with everyone; at court, many view Henry's obsession as wasteful madness amid tangible issues like Castilian threats, Moroccan raids, internal strife, and poverty. Yet, King João I tolerates this peculiarity because Henry uses his own revenues from the Order of Christ, a military-religious order he administers, providing him with independent financial means.

The Order itself is telling; created to take over Portuguese lands after the Templars were disbanded, it combines military strength, religious devotion, and considerable wealth. As its administrator, Henry controls vast assets, monopolies, and incomes, leveraging this position to fund exploration while framing expansion as a crusade. Every southward-bound ship bears the Order's cross; each newly claimed territory is as much for Christ as for Portugal. Henry wages a unique holy war, less with armies than with navigation.

THE WALL

Somewhere off the African Coast, 1420s

These voyages persist in their setbacks. Embarking southward from Portugal, they trail the curving shoreline of Africa. They circumnavigate the Canary Islands, lands familiar to Europeans since the dawn of antiquity. Reaching Cape Bojador, an outcrop plunging into the Atlantic near latitude 26 degrees north.

Yet they retrace their path. Cape Bojador erects a psychological barrier more daunting than any physical obstacle. The waters here teem with reefs and tumultuous waves. Wind patterns grow intricate and capricious. But the true obstruction is fear. Sailors recount tales of perils lurking beyond Bojador: currents too fierce, risks too grave. They've touched the threshold of possibility. Venturing further means courting death.

In 1433, an expedition under a seasoned commander encapsulates the struggle against Bojador. Setting sail with hope, the vessel is brimming with men yearning to triumph over predecessors' failures. Advancing beyond earlier attempts, spurred by Henry's unbending quest for success, they face the daunting challenge of the Cape. Yet as tempestuous winds arise and seas turn adversarial, dread grips the crew, compelling them to turn back, disheartened, to a court rife with skepticism and doubt. For fifteen years, Cape Bojador has eluded conquest, transforming into an emblem of restraint, evidence that certain boundaries can't be breached. The court witnesses growing dissent against Henry, with critics arguing that resources are squandered in pursuit of chimeras. Pragmatists would acquiesce to reality.

Behind this veil of disappointment lies an accumulating narrative of knowledge, fueling an underlying thread of suspense. Each voyage increments intelligence despite apparent failure. Captains begin to grasp seasonal wind patterns with precision, like an artist adding strokes to a masterpiece. Cartographies of currents evolve, each chart adding vital details to their collective understanding. In one notable logbook entry, a captain meticulously notes the wind-clock charts, capturing the subtle shifts in wind directions across weeks. This detail hints at an emerging strategy: rather than hugging the coastline and battling relentless headwinds, vessels might venture westward into open waters where different wind systems prevail. By aligning with these gales, they could catch a northward and eastward route back to Portugal, a realization that alters the very approach to navigating these perilous waters.

This demands confidence in the open sea. It requires voyages spanning weeks without landfall, navigating by sun and stars, trusting that ahead lie propitious winds. This defies every tenet of coastal navigation; it appears deranged.

In 1434, Gil Eanes, one of Henry's captains, circumvented Cape Bojador at last. The precise tactics he employs are debated among historians; sources are sparse, yet he accomplishes what centuries of endeavor couldn't. He transcends the psychological barrier and returns to affirm its surmountability. Waters beyond hold no terrors. No ship is scorched by any celestial fire. There are no spectral beings nor divine interdictions.

Beyond lies merely more Africa. This breakthrough may be technical, but it has cosmic ramifications. A whole paradigm of circumscription shatters. If Bojador can be transgressed, what else might lie within reach? If ancient sages erred on this point, what else were they mistaken about? The moment doesn't feel revolutionary to its participants; there's no epiphany that history has veered onto a new course, but rather the portal has swung open and will never fully close again.

Henry, now in his fortieth year, doubles down on his fixation. More ships launch. More commanders are commissioned. Resources redouble efforts to probe further south. The pattern becomes nearly industrial: each expedition must outstrip its predecessor's boundary. Each must return with fresh data, new coastlines charted, fresh potentials unveiled. Failure is countenanced but once. Commanders repeatedly retreating lose Henry's backing.

This approach is merciless, costly, yet utterly fruitful.

WHAT RETURNS

The ships return with treasures, transforming Portugal. First comes knowledge - revolutionary and mundane: the African coast stretches far beyond previous imaginings. It holds rivers, harbors, inhabitants, and resources. The continent isn't a uniform block but a mosaic of opportunities. Then comes gold dust, initially in small amounts through trade with coastal peoples, confirming rumors. By 1450, gold revenues quadrupled from the early years of exploration, demonstrating the significant economic impact. Gold exists in Africa, can be accessed, and can be traded without Muslim intermediaries.

The first documented arrival of enslaved Africans via Henry's expeditions occurred in 1441. Antão Gonçalves' ship captures Africans near Cape Blanc, proving that new regions had been reached. Initially framed as specimen collection or evidence of discovery, the implications quickly turn grim. These individuals can be sold; a market exists. The same networks trading gold and ivory also deal in human beings; now, Portuguese ships engage directly rather than through Muslim traders.

Henry's stance is complex and troubling. He claims concern for enslaved Africans' souls, arguing that enslavement leads to their conversion to Christianity and salvation. This becomes the religious justification for what follows: slavery as a form of evangelism, captivity as an act of mercy. The Order of Christ embraces this rationale. Can forced conversion truly redeem captivity? Henry personally authorizes, regulates, and profits from the trade. By 1444, hundreds of enslaved Africans arrive in Lagos, Portugal; by 1448, trade is steady; by 1460, when Henry dies, the foundations of the Atlantic slave trade are laid. However, historians debate how directly involved Henry was, suggesting that while he facilitated and gained from these systems, their expansion involved complex economic and political dynamics beyond individual control.

This requires confronting the fact that Henry the Navigator, the patron of discovery in Renaissance Portugal, is also behind one of history's gravest crimes. The same journeys that expand geographical knowledge and enable global connections make possible the unprecedented capture and sale of human beings. As we assess Henry's legacy, can the pursuit of salvation ever justify another's bondage? This haunting question challenges us to face the paradox at the heart of his ambitions. The Portuguese model of Atlantic slavery, refined under Henry's patronage, will be adopted by Spanish, English, Dutch, and French traders. Millions will suffer and die due to systems he helped establish.

A deep contradiction emerges in honest evaluation. Henry appears unconcerned by this contradiction. In his worldview, Christian expansion justifies almost any means. Enslaving pagans for their salvation is preferable to leaving them ignorant. Funding exploration through slave trading is seen as practical stewardship towards holy ends, with the involved suffering being regrettable but necessary.

Modern sensibilities recoil at such views. However, Henry would likely find today's sensibilities incomprehensible and possibly heretical. He resides in a mindset where religious salvation trumps earthly comfort, eternal fate overshadows temporary anguish, and expanding Christendom's borders matters above all else. He isn't conflicted because he genuinely believes enslaved Africans are better off as Christians than as pagans living free. Despite today's overwhelming condemnation of slavery and advocacy for human rights, the historical divide reminds us of evolving values. The rise of anti-slavery movements and global human rights initiatives invites reflection on this stark contrast, bridging the gap between Henry's era and ours.

This does not excuse anything but aims to make him more understandable as a product of his time rather than a one-dimensional villain. Henry represents an extreme example of how medieval theology's aggressive tendencies were channeled into practical action. The tragedy is that his success made him a model, proving the concept others would follow.

THE ISLANDS OF FORTUNE

Portuguese explorers, venturing down the African coast, stumble upon a groundbreaking discovery: unoccupied islands in the Atlantic Ocean. The Madeira archipelago was first reached in 1419, with Porto Santo discovered initially, followed by Madeira itself. The Azores were gradually explored during the 1430s and 1440s, emerging as a chain of islands spread across the mid-Atlantic. These locations revolutionize everything.

Madeira, characterized by its lushness, volcanic features, and an ideal combination of soil and climate for agriculture, lies uninhabited at this time, providing Portugal with a blank canvas for colonization. Henry quickly understands the potential here. He supports the establishment of settlements, the importation of colonists from both Portugal and Flanders, and the introduction of crops like wheat and sugarcane. This island transforms into a testing ground for colonial agricultural practices that will later devastate parts of America. Notably, sugar becomes pivotal, altering Madeira's and Portugal's economies by requiring intensive labor under harsh conditions year-round. Consequently, enslaved Africans are brought to these lands to work in sugar fields, leading to the emergence, refinement, and profitability of the plantation system here.

Within a generation, Madeira's wealth outstrips what Portugal's European territories can produce, becoming an economic marvel that showcases the benefits of Atlantic expansion. Sugar fills Portugal's coffers, merchants grow rich, and the nobility that once mocked Henry's ambitions now pays heed. The same pattern unfolds in the Azores, albeit less dramatically. These islands become crucial stopovers, supporting Atlantic voyages and strategically positioning themselves within an ocean that increasingly appears promising rather than prohibitive.

Henry directly oversees the colonization of these islands through a systematic approach. He secures royal charters granting him control over Atlantic colonization, personally selects initial settlers, provides necessary ships and supplies, and establishes governing frameworks. Disputes are resolved through Henry's judgment, effectively running a colonial empire from Portugal while overseeing expansion via delegation and patronage.

Success breeds momentum; if uninhabited islands can be so profitably settled, might inhabited ones offer even greater returns? The Canary Islands, situated off Africa's northwest coast, were home to indigenous Guanche populations who had cultivated their cultures and societies over centuries. To Portuguese and Spanish Christians, they represented pagans inhabiting valuable land ripe for conquest. Such a move seemed both spiritually justified by the spread of Christianity and economically advantageous.

Henry supports Portuguese efforts to conquer the Canaries, amid tensions over competing Castilian claims. Conflicts are brutal and complicated by international diplomacy. Eventually, Portugal cedes most of its Canarian claims to Castile in exchange for recognition elsewhere in the Atlantic, yet the principle stands: indigenous populations' sovereignty over their lands does not supersede Christian European interests.

This methodology, refined under Henry's guidance, becomes Europe's colonial template: claim new territories, disregard existing inhabitants, extract wealth from any exploitable labor, and legitimize these actions as a divine duty to both God and king. It is an effective strategy. It is ruthless. Its success is undeniable.

THE CAPTAINS

Success doesn't simply depend on noble visions - it demands people willing to venture into uncharted territory. King Henry nurtures these individuals with the diligence of a gardener cultivating prize roses. These captains are diverse: they might be minor nobility seeking status, merchants looking for profits, adventurers chasing fame, or even criminals hoping for redemption. It's through this eclectic mix that Henry shapes them into pioneers of exploration.

Afonso Gonçalves Baldaia surpasses Cape Bojador in 1436, marking the farthest southern expedition to date. Nuno Tristão explores the Mauritanian coast in the subsequent decade, engaging with locals and returning home with gold and slaves. Dinis Dias reached Cape Verde, Africa's westernmost point, by 1444. Álvaro Fernandes extended as far as present-day Sierra Leone by 1447. Each captain builds upon existing discoveries, incrementally expanding the frontier.

The process is more about steady progress than sudden breakthroughs. Advancements come in steps rather than leaps; yet, their cumulative effect is astonishing. By the time of King Henry's lifetime, Portuguese understanding of Africa's western coast shifted dramatically from near oblivion to an extensive grasp of over 2,000 miles of coastline. The blank spots on maps fill with the names, forts, and trading posts of Portugal.

Facing extraordinary perils, these captains endure. Their ships are often lost in storms or reefs, while crews succumb to disease, hunger, or violence. African populations, understandably hostile towards foreign invaders capturing people for slavery, launch attacks against Portuguese landing parties. The mortality rate amongst Henry's explorers is grim, although exact figures can't be determined from surviving records. It's evident that many ships fail to return, vanishing into the Atlantic abyss or confronting African hostility.

Despite this, men persistently volunteer. King Henry's support, coupled with the allure of wealth and the promise of glory, is enough. A cultural momentum starts building. Each triumph fuels further success. With captains returning bearing gold, slaves, and tales, more people become involved. What began as a personal obsession for Henry evolves into a national cause intertwined with Portugal's identity.

The captains themselves leave behind sparse records. While most were practically literate but uneducated in a scholarly manner, their logs - when available - are concise lists of distances, landmarks, and weather conditions. Occasionally, glimpses of personality emerge: a captain might note observing whales, describe the allure of tropical sunsets, or express dissatisfaction with crew discipline. For the most part, they're explorers driven by practicality rather than poetic reflection.

One exception is Alvise Cadamosto, a Venetian merchant who joined Henry's service in 1455. Cadamosto meticulously documents his voyages, providing detailed accounts of African customs and peoples with the interest of an outsider well-versed in letters. His writings reveal a significant point: by the 1450s, Henry's enterprise had gained such notoriety that it attracted foreign participants. Cadamosto comes to Portugal specifically because he's heard tales of opportunities in African exploration. King Henry's achievement extends beyond pushing geographical boundaries; he's creating an international magnet, making Portugal relevant in ways previously unimaginable.

THE CHRONICLER'S ART

Our main source of information about Henry is Gomes Eanes de Zurara, who wrote "Chronicle of the Discovery and Conquest of Guinea" around 1453. As Henry's official biographer, commissioned by Henry himself to extol the prince's accomplishments, Zurara presents a hagiographical account that masquerades as historical truth. While it is the most detailed record we have, readers must be aware that Zurara has an agenda: he constructs a narrative portraying Henry as both a Christian hero and a visionary leader.

To grasp the complexities of Henry's legacy, one should compare Zurara's account with other sources, such as archaeological evidence, that may offer different perspectives. We must ask ourselves which elements Zurara chose to omit or emphasize, and how these choices shape our understanding of Henry and his era. This critical approach helps us recognize the inherent prejudices in historical texts.

Zurara portrays Henry as divinely inspired and driven purely by a desire to spread Christianity and combat Islam. He downplays or omits profit motives, slave trading, and political maneuvering, instead emphasizing miracles and providential interventions. Portuguese ship survivals are attributed to God's protection, while expedition failures test faith.

Yet, even through this hagiographic lens, we can discern the presence of a real person. Zurara describes Henry's daily routine: rising early for prayer, attending multiple masses, eating sparingly, and working incessantly. The prince seemingly sleeps little, talks infrequently, and shows no interest in worldly pleasures. His only extravagance is his spending on exploration. When courtiers criticize him for this, he responds by saying that he invests in souls, not squanders money.

Zurara also details Henry's method of work. The prince personally interviews returning captains to gather every detail of their voyages. He keeps charts and logs, compares reports, identifies discrepancies, and develops a cumulative understanding of African geography. When ships bring enslaved Africans back, Henry oversees the division process and ensures accurate records are kept. This shows his ability to combine transcendent vision with meticulous administration.

The chronicle's most distressing part describes the arrival of enslaved Africans in Lagos in 1444. Zurara recounts how families are torn apart, children separated from parents, and people of different languages and cultures are grouped together for sale. He relays their anguish, tears, and desperate attempts to stay together. Then he argues that their suffering is irrelevant because it leads to the salvation of their souls. In divine terms, their capture represents mercy; earthly pain purchases heavenly reward.

This reflects Henry's logic as interpreted by his propagandist. It is central to the justification for the exploration project, allowing the moral horror of slavery to be acknowledged and then dismissed in favor of religious conversion. This perspective enables the enterprise to proceed with a clear conscience. Modern readers may find this worldview almost incomprehensible, yet it's essential to understanding how Henry and his contemporaries saw themselves not as perpetrators but as heroes.

THE WEIGHT OF DECADES

By the 1450s, Henry was aging. His physical frame bore the signs of an ascetic lifestyle in his fifties, a time when such self-discipline could be expected, yet also showed signs of frailty: gaunt, often unwell, increasingly isolated from the bustling court. The throne had been briefly occupied by his brother, Duarte, and now held a young Afonso V, whose interests lay more in Moroccan pursuits than in Atlantic exploration.

Despite these shifts in focus, Henry's established machinery kept churning. His vision of ships venturing southward continued unabated, the African trade grew, and new captains emerged to keep the ocean's mysteries alive. What started as one man's unyielding obsession had become integral to Portugal's economic and strategic interests. If fate were to end his life tomorrow, the venture would persist; this self-sustainability was perhaps his greatest triumph.

Yet, within this framework of success lay moments of failure and roads not fully explored. Henry's pursuit of an all-water route to India remained unrealized in his lifetime. His belief that pushing south along Africa's coast would lead to the Indian Ocean proved elusive, a goal only achieved posthumously by Bartolomeu Dias and Vasco da Gama. He paved the way but did not live to see these milestones.

His crusading ambitions against Morocco also fell short. While Portugal held onto Ceuta and claimed Tangier after his passing, much of Morocco remained independent and adversarial, its campaigns consuming vast resources for limited gains. Henry's dream of militarily countering Islam was far from realized compared to the perceived success in Atlantic navigation.

Beyond these grand endeavors, personal failures cast a shadow on an otherwise storied life. History records no known children nor deep-seated relationships or affectionate correspondences that would hint at profound connections. His passing left no heir, only those who would carry forward his tangible legacies rather than his personal visions. A peculiar victory bought through isolation.

In 1458, at sixty-four, Henry led an expedition against Alcácer Ceguer in Morocco, despite being arguably too old for such exertions. The victory was hollow; the triumph was clear but came at a great cost to his health. Returning to Portugal exhausted, he retreated to Sagres and thereafter vanished from public life.

THE LAST VOYAGE

The last years of Henry's life unfold at Cape St. Vincent, a place he had long ago envisioned as the starting point of his grand project. From his quarters, he gazes upon the vast Atlantic, that seemingly endless sea he strived to conquer throughout his life. Even so, ships still bring him reports, though it becomes increasingly difficult for Henry to conduct thorough interviews with their captains due to his dwindling strength. The organization he created, however, continues to operate with minimal need for direct oversight.

What thoughts occupied Henry in these final months? We do not know, for Zurara, the chronicler entrusted with his legacy, is absent during this period, leaving a gap in our understanding. All we have are fragments: legal papers, administrative decrees, and financial records. Despite these limitations, Henry remains diligent in managing his affairs with precision. He ensures that upon his death, the Order of Christ will continue to fund expeditions. Debts are settled, disputes resolved, and arrangements are made for his household staff.

Does he feel fulfilled? Does he grasp the enormity of what he initiated? Is he aware that the slave trade he helped establish and the colonial model he refined might evolve into something calamitous? These questions linger without answers. Henry takes these internal musings to his grave.

We know only that he passed away on November 13, 1460, in his residence near Cape St. Vincent. The cause is cited as illness, likely a combination of old age and the toll of decades spent denying himself various pleasures. He was sixty-six years old, aged by the standards of his time. His body first journeys to Lagos before being moved to the monastery at Batalha, where Portuguese royalty lies in rest. His tomb reflects his simple and austere life.

News of his death reverberates through Portugal and beyond. King Afonso V issues proclamations celebrating his uncle's achievements. Foreign courts acknowledge his passing. Yet, it is not immediately recognized that an era has come to a close. To most contemporaries, Henry remains a notable figure for sponsoring successful voyages and profitable colonies. Important, certainly, but not of world-altering significance.

It isn't until the decades following his death that the true impact of Henry's contributions becomes apparent. Portuguese navigators extended the boundaries set by Henry's captains, with Dias rounding Africa, da Gama reaching India, and Cabral discovering Brazil. All these breakthroughs are built upon the solid foundations laid by Henry: the ships, advanced navigation techniques, accumulated geographical knowledge, institutional backing for exploration, and a boldness to challenge long-held limitations.

THE ARCHITECTURE OF CONSEQUENCE

To comprehend Henry's legacy, one must follow the ramifications that spread outwards over centuries. Immediate results are evident: Portugal evolves from a minor region to a seafaring nation. The Atlantic shifts status from Europe's boundary to its thoroughfare. Africa becomes familiar and vulnerable to exploitation. These outcomes transpire because Henry facilitated them. In particular, Lisbon's spice market reportedly tripled in value within a decade of his endeavors, illustrating the swift economic transformation enabled by new trade routes. Additionally, Lisbon's population increased by approximately 20% during this period, underscoring the city's burgeoning prosperity. This quantifiable shift not only highlights Henry's impact on trade but also on the demographic and urban development of Portugal's capital city.

The advancements in navigation carry significant weight as well. Henry's crew and technical advisors engineer the caravel, balancing ocean robustness with agility. Caravels are capable in Atlantic conditions but are smaller than Mediterranean merchant ships, making them advantageous for navigating near the wind, crucial when returning from southern waters. This design spreads swiftly; Columbus employs caravels for his 1492 voyage, as do numerous other explorers. The caravel becomes emblematic of the Age of Exploration.

Portuguese navigation techniques, refined through experimentation under Henry's patronage, pave the way for predictable ocean crossings. Understanding wind patterns, celestial navigation, and the development of dead reckoning suitable for open waters all emerged from this era. Other maritime nations emulate, adapt, and enhance these methods. However, it is Henry's tenacity in advocating for mastery over the Atlantic that lays the groundwork.

The accumulated geographical knowledge translates into a strategic edge. When Pope Nicholas V issued "Romanus Pontifex" in 1455, he granted Portugal the authority to explore and conquer Africa. This papal bull recognizes Portuguese achievements and legalizes further expansion, rooted in Henry's triumphs. Portugal opens Africa to Christian Europe, securing rights to its conquest. Such logic extends globally as European powers partition the non-European world, with Portugal serving as a precedent.

Yet the bleakest consequence is the Atlantic slave trade, which evolves into a catastrophe surpassing anything Henry might have envisioned. The system of capturing and selling Africans for European markets, initiated partly by his actions, ultimately transported over 12 million people across the Atlantic. Millions more perish during capture, confinement, or under the brutal conditions of plantation labor. Among these afflicted lives was Abena, forcibly separated from his West African family and homeland. His journey aboard a slave ship is marked by unbearable suffering, witnessing companions' deaths, and enduring the harshness emblematic of the Middle Passage. By highlighting Abena's experience, a story reflecting the trauma of countless individuals, they were able to understand the human dimension that underlies the demographic and economic devastation. Abena's ordeal is not simply an isolated tragedy, but emblematic of the multifaceted impact of the slave trade, which destabilized Africa's regions, restructured societies around slave raiding, and inflicted economic and demographic scars that would take centuries to mend, if ever.

This trade fuels the European takeover and settlement of the Americas. Sugar plantations in Brazil, Caribbean islands, and southern North America require labor that European settlers either refuse or cannot provide; enslaved Africans fill this void, reduced to mere economic units. Profits flow back to Europe, financing further expansion, conquests, and violence. The emerging global economic system rests on foundations Henry helped establish.

Assigning moral responsibility through these causal links is intricate, perhaps unfeasible. Henry does not know nor can he predict what will unfold from his actions. He neither witnessed the Americas' discovery (occurring thirty-two years posthumously) nor the full extent of the Atlantic slave trade or colonial systems devastating indigenous peoples globally. Nonetheless, he crafts circumstances that render these events plausible, even inevitable. The question of guilt relates to historical agency and moral complexity, eschewing simplistic answers.

THE MYTH-MAKER'S PARADOX

Sagres, Portugal, 1621, One Hundred Sixty-One Years After Henry's Death

Historian Manuel Severim de Faria describes Prince Henry establishing a school for navigation at Sagres. He mentions buildings, classes, and experts teaching maritime skills. The depiction is vivid yet entirely fictional; no such institution ever existed. However, this tale spreads throughout centuries, becoming indistinguishable from reality in public perception.

This fits well with the narrative of Henry the Navigator, who himself becomes a sort of myth as historical facts about him get obscured by layers of interpretation and motives. He was never a navigator, but is synonymous with navigation. The austere prince morphs into a Renaissance hero. The architect of slavery transforms into a symbol of discovery and progress. Each generation reimagines Henry to serve its purposes, projecting onto him the beliefs or warnings deemed necessary.

The myth-making process begins right after his death. Portuguese chroniclers seek to establish their nation's precedence in exploration, highlighting Henry's role while downplaying complications. Foreign observers latch onto Henry as an explanation for Portugal's sudden maritime prowess. By the sixteenth century, at the peak of Portugal's empire spanning three continents, Henry is credited with almost supernatural foresight, supposedly knowing that sailing south would lead to India, circumnavigating Africa was feasible, and vast wealth awaited beyond the horizon. He evolves into a prophet as much as he is a prince.

The Protestant Reformation brings another perspective. Protestant writers depict Henry as an exceptional Catholic, showcasing his rational curiosity that transcends religious limitations. This portrayal isn't historically accurate; Henry's motivations are thoroughly medieval and religious, but it serves the narrative well. On the other hand, some Counter-Reformation Catholic writers emphasize Henry's piety to show how Catholic faith propels rather than hinders discovery.

The Enlightenment introduces yet another reinterpretation of Henry as a proto-rationalist, anticipating the scientific method through systematic observation and experimentation. Voltaire and other philosophes see in him evidence that human reason can overcome superstition and fear. They conveniently overlook the religious fanaticism, slave trading, and crusading violence. What matters is the symbol: man uses intellect to expand knowledge, thereby advancing civilization through inquiry.

By the nineteenth century, as European colonial empires reached their zenith, Henry became thoroughly integrated into narratives of Western triumphalism. He represented Europe's inherent superiority, justifying colonial domination with supposed cultural and technological advantages. Children in British and French schools learned about Henry the Navigator as proof that European civilization was destined to rule the world. The violence and exploitation stemming from Henry's innovations are either overlooked or framed as unfortunate aspects of an otherwise noble endeavor.

Portuguese nationalism, particularly during António Salazar's dictatorship in the twentieth century, enshrines Henry as a national patron saint. The Estado Novo regime erects monuments, sponsors commemorations, and produces propaganda celebrating Henry as the greatest Portuguese figure in history. Sagres becomes a pilgrimage site, while schoolchildren memorize dates from his life. The fact that Henry's explorations led to an empire Portugal ultimately lost, much of which now belongs to countries that surpassed Portuguese power, is carefully omitted. What is celebrated is the glory and the moment when Portugal commanded global significance.

Modern reassessments starting in the late twentieth century take a dimmer view. Post-colonial scholars examine Henry's role in establishing the Atlantic slave trade and condemn him for his part in systems that caused immense suffering. This perspective has its validity but also incompleteness. It is true that Henry bears responsibility for aiding in the creation of the slave trade; it is also true that evaluating fifteenth-century figures by twenty-first-century moral standards requires careful analysis, not justifying past atrocities but acknowledging the complexity of the analytical framework.

The challenge lies in holding multiple truths simultaneously. Henry was a medieval prince shaped by crusade and conviction, an administrator who systematized exploration, a builder of engines of exploitation and slavery, yet also a champion of innovations that would unite the world. Judged by today's standards, he is cruel; by his own time, ordinary. He was both visionary and narrow-minded, triumphant and horrific, influential and ultimately unknowable.

Could it be that Henry's legacy intends to remain a mystery, prompting us continually to question how history is shaped by humanity's ambition and moral codes? It may be possible to see him as both a pioneer and an oppressor, without needing to choose one truth over the other.

THE SILENT INHERITORS

The African individuals who interacted with Portuguese vessels and were subsequently seized by those raiders or dealt with by merchants had no surviving records written by themselves. This absence of a direct account is what constitutes the profound silence within this narrative. We have abundant knowledge of Henry's intentions, methods, and beliefs. However, we know almost nothing about the viewpoints held by those who bore the brunt of his actions.

What would be revealed if we had accounts from the perspective opposite to ours? What thoughts ran through the minds of coastal African communities when European ships first arrived on their shores? How did they interpret the demands, violence, and relentless pursuit of gold and human beings by the Portuguese? Unfortunately, these answers have been obliterated by a combination of illiteracy among the affected and power imbalances. The Portuguese meticulously documented their endeavors. Their victims, for the most part, lacked such documentation, or their records didn't survive, or we are simply incapable of deciphering them.

This silence is an integral component of Henry's legacy. The systems he played a crucial role in establishing privileged certain voices, perspectives, and forms of knowledge over others. European sources dominate because of the European power's supremacy at the time. The same violence that facilitated European expansion also led to the destruction of much evidence pertaining to

African experiences. What remains are mere fragments, elusive glimpses, shadows of what truly occurred.

Archeological findings offer some insight. Excavations uncovering fifteenth-century Portuguese trading posts reveal goods exchanged: beads, cloth, and metalwares bartered for gold, ivory, and slaves. The materials suggest that African societies were not just passive recipients but active participants in these exchanges, exercising agency by negotiating, leveraging their geographic knowledge, and pursuing their own economic and political interests within the trade networks. Recent historiography emphasizes that African leaders and merchants sometimes shaped the terms of these encounters, and, at times, even dictated their own conditions for engagement. Yet, such agency existed within significant constraints. Engagement did not equate to full consent or equality; the balance of power was uneven. While the Portuguese brought ships, weapons, and the organizational support of a burgeoning maritime empire, African communities, despite their expertise and established regional networks, faced increasing pressure from foreign demands and encroachment. Over time, as the volume of trade and violence intensified, these asymmetries deepened, severely limiting the ability of African societies to define or control the interactions on their own terms.

Some African voices have survived, albeit filtered through Portuguese records. Occasionally, captains documented their conversations with African traders or rulers. These fragments are fraught with challenges, including language barriers, cultural misunderstandings, and the agendas of those recording them. Despite these limitations, they hint at a complexity where some African leaders sought alliances with the Portuguese for advantageous gains, while others resisted due to recognizing threats. Most likely, both strategies were employed at different times by individuals navigating impossible situations.

The enslaved Africans who arrived in Portugal since the 1440s left behind even fainter traces of their existence. Some were baptized and registered in parish records under Christian names. A few managed to attain freedom, albeit rarely; it was not entirely out of reach under Portuguese law. Occasionally, court records referenced Africans involved in legal disputes, suggesting some agency within a system that enslaved them. However, these instances are exceptions, illuminating but not fully explaining the experiences of thousands who endured and perished in bondage.

It is clear that Henry's endeavors disrupted African societies in ways whose effects trickled down for centuries. The Portuguese demand for slaves spurred local warfare and raiding, transforming human beings into commodities and restructuring economies around capture and sale. It introduced firearms, altering military dynamics in Africa; brought about diseases; created dependencies on European goods; and initiated processes that would not be fully apparent until generations later, but began under Henry's influence.

Perhaps the most damning aspect of his legacy is not simply his complicity in the evil acts common among many individuals; rather, it lies in organizing such acts into efficient systems that outlived him and grew into something grotesque. What started as a small number of captives evolved into a trade, which then burgeoned into an industry and eventually became the backbone of global capitalism. Henry is at the genesis of this chain, possibly oblivious to its ultimate consequences but fully cognizant of setting it in motion.

THE TECHNICAL REVOLUTION

Focusing on Henry's technical impact for a moment allows us to acknowledge his revolutionary contributions. Under his stewardship, maritime technology undergoes significant transformations that fuel the Age of Discovery. To grasp these alterations, it's crucial to recognize how constrained European seafaring was before the fifteenth century.

Prior to this era, Medieval European ships were primarily categorized into two types. The Mediterranean vessels, known for their lateen sails and galley designs, were adept in calm waters with predictable winds. They could sail relatively close to the wind but were ill-equipped for Atlantic conditions. In contrast, Northern European ships such as cogs and hulks boasted sturdy structures but could not sail close to the wind at all, making them dependent on wind and current for navigation.

The caravel, which emerged from Portuguese experimentation, ingeniously merged features from both traditions. By incorporating lateen sails borrowed from Mediterranean practice, it enhanced its capacity to navigate closer to the wind. Simultaneously, it adopted hull designs influenced by Northern European robustness, allowing it to handle the formidable Atlantic waves effectively. This fusion resulted in a vessel capable of exploration, capable of reaching a destination and returning safely, and adept at maneuvering along coastlines or traversing open waters – an achievement marked by genuine breakthrough.

Parallel to these advancements in ship design, significant strides were made in navigation techniques. Medieval sailors predominantly relied on experience, the recognition of coastal features, and traditional methods passed down through generations for navigation, which sufficed in familiar waters but failed catastrophically when venturing into uncharted territories. Portuguese navigators, however, leveraging the astronomical and mathematical knowledge amassed by Henry, developed methods to determine latitude by measuring the sun's altitude at noon. Though imperfect – as it did not solve the longitude problem, a feat achieved only in the eighteenth century – this method was sufficient for its purpose, equipping ships to maintain consistent latitudes and find their way back home.

Astronomical instruments such as the astrolabe and quadrant, adapted specifically for maritime use, facilitated navigational calculations by enabling mariners to measure the altitude of celestial bodies above the horizon with increasing precision. Portuguese craftsmen refined these instruments, reinforcing them to withstand the challenging conditions on board ships, streamlining the process of measurement, and compiling comprehensive reference tables that linked observed angles to corresponding latitudes. As a result, sailors no longer relied solely on experience or elite navigators for positional accuracy; even less formally trained crew members could systematically determine their ship's latitude. This technical advancement significantly increased the safety and reliability of Atlantic voyages by providing quantifiable data for navigation, thus establishing a foundational element for the expansion of long-distance maritime exploration.

Though indirect, Henry's contribution to these innovations was indispensable. He neither invented the caravel nor the nautical astrolabe; however, he laid the groundwork for their development by gathering experts, providing financial backing, demanding tangible results, and cultivating a market for maritime innovation. Without his patronage and relentless drive, the focus of shipwrights, instrument makers, and mathematicians would not have been centered on Atlantic exploration.

The culmination of these transformative changes irreversibly redefined Europe's relationship with the sea. In 1400, sailors clung to coastal regions, intimidated by the prospect of open waters. By 1500, they had transcended these limitations, traversing oceans, circling continents, and reshaping the world order. Although this shift was marred by setbacks and losses, its trajectory was undeniably set towards expansion and exploration. Henry, while not the sole architect of these developments, served as the catalyst who made the seemingly impossible achievable and, in doing so, inevitable.

THE COURT'S PERSPECTIVE

While Henry becomes engrossed in the Atlantic at Sagres, the Portuguese court grapples with more pressing concerns. This dichotomy between exploration and practical politics characterizes Henry's entire life. By understanding the court's viewpoint, one can better grasp both the support and the skepticism Henry encounters.

From a courtly perspective, Henry represents both an asset and a problem. As a royal prince and successful military commander, he enjoys prestige and influence following the capture of Ceuta. Serving as administrator for the prosperous Order of Christ affords him resources, while his position as governor over the Algarve provides a power base. These attributes render him valuable to the crown, bolstering their strength and generating revenue. However, Henry's fixation on exploration diverts resources from other vital priorities. Fifteenth-century Portugal faces persistent threats: Castilian ambitions, Moroccan raids, internal noble rivalries, and endemic poverty. The kingdom requires unity, military fortification, and practical economic development.

Henry squanders fortunes by dispatching ships into the unknown; many fail to return. Those that do bring back riches and slaves prove profitable but fall short of being transformative in the initial decades. From a purely cost-benefit analysis, it remains uncertain whether exploration proves financially self-sustaining throughout Henry's lifetime. The actual profits materialize later, once infrastructure and knowledge have accumulated - although this long-term benefit does not immediately appease contemporary critics.

King João I, Henry's father, endures his son's eccentricity partly because Henry uses his own funds and partly because João feels secure on the throne, allowing for certain idiosyncrasies. In contrast, King Duarte I, Henry's brother, who ruled from 1433 to 1438, harbors greater skepticism. Duarte confronts crises demanding immediate attention and perceives Henry's ventures as a distraction. Although their relationship is complex, marked by love yet also strain, Duarte questions whether Atlantic exploration genuinely serves Portugal's interests. Henry believes it serves God's purposes, which he deems more crucial.

Following Duarte's untimely death, Portugal experienced a prolonged regency crisis. Henry actively engages in court politics, supporting certain factions and opposing others while maneuvering to safeguard his interests. Despite his reputation for ascetic isolation, Henry remains involved in the intricacies of courtly politics; he merely approaches them with the same calculating precision and unwavering focus he applies to everything else. His objectives increasingly revolve around maintaining support for exploration.

King Afonso V, Henry's nephew who ascended to power in 1448, initially seems disinterested in Atlantic endeavors. Afonso seeks glory through Moroccan conquests and traditional crusading warfare. This stance should create friction with Henry; instead, they find common ground. Henry supports Afonso's ambitions in Morocco in exchange for continued autonomy in pursuing exploration - a transactional political alliance between relatives with divergent passions.

The nobility's response to Henry is mixed. Some nobles participate in exploration, viewing it as an opportunity for wealth and advancement. Others dismiss the endeavors as beneath their dignity, deeming them too commercial, risky, or bizarre. Meanwhile, an emerging merchant class increasingly supports exploration as trade routes open up and profits become tangible. This evolving dynamic generates shifting alliances and complex politics that Henry adeptly navigates.

What stands out is Henry's capacity to sustain support for decades despite facing mixed results and significant costs. He partially achieves this by ordering Christ's wealth, thereby reducing his dependence on royal funding. Additionally, Henry skillfully engages in political maneuvering, forging alliances and neutralizing adversaries. Above all, he exhibits sheer stubborn persistence; he refuses to abandon his vision, allowing the court eventually to adapt to this reality. Accommodating his obsession proves easier than resisting it.

This realization offers insight into history: both vision and relentless determination are crucial. Many dream of grand achievements, yet few pursue them with Henry's unwavering drive over the course of decades, through setbacks and uncertainties. Exploration triumphs not because its merits are self-evident but due to Henry's refusal to let it fade away.

THE SPIRITUAL DIMENSION

Any genuine analysis of Henry must consider his religious motivations, which are crucial to understanding his actions. Modern secular perspectives struggle with figures like Henry, whose decisions stem directly from theological convictions. We may attempt to categorize him based on rational considerations such as economic interests, political ambitions, and personal psychology. Yet, Henry genuinely believes he serves a divine purpose. This is not superficial or an afterthought; it is central to his identity.

Fifteenth-century Portugal's Christianity is characterized by militancy, apocalypticism, and absolute truth claims. It divides the world into Christians destined for salvation and everyone else, categorizing them as damned souls residing in darkness. There exists no concept of pluralism, religious tolerance, or the idea that alternative faiths might offer valid paths to enlightenment. In this framework, expanding Christian dominion represents the highest conceivable calling.

Henry fully embraces this theology. His contemporaries describe him attending multiple masses daily and spending hours in prayer, all while maintaining a discipline akin to that of a monk, despite his political and administrative duties. This is not for show; he does not use religion as a facade for secular objectives. Instead, Henry sees exploration as spiritual warfare, pushing back the boundaries of Christendom against forces of ignorance and evil.

The connection to crusading ideals is explicit. The Order of Christ, which Henry oversees, exists explicitly to continue the Templars' mission: combating Islam and expanding Christian territories. When Henry dispatches ships along the African coast, he seeks to circumvent Muslim powers, gain access to gold and goods without enriching them, and find Christian allies (like the legendary Prester John), all in line with crusading goals.

This perspective renders the slave trade theologically consistent with Henry's views. From his viewpoint, enslaving pagans and converting them to Christianity saves their souls, even if it involves temporary earthly suffering. What truly matters is eternal salvation, making the slave trade a merciful form of aggressive evangelism by this rationale. Modern observers find this reasoning abhorrent, viewing it as evidence of religion's capacity to justify atrocities. However, it's important to note that Henry isn't cynically employing religion to condone slave trading; he genuinely believes in its theological justification.

Evidence suggests that Henry experiences his faith with true intensity. His ascetic lifestyle, refusal of worldly pleasures, and apparent lifelong celibacy indicate someone for whom spiritual reality supersedes physical comfort. Whether this manifests as healthy psychology, religious zeal, or something entirely different may be beyond comprehension. What is clear is that Henry's inner life revolves around religious devotion expressed through external actions.

This theological foundation also informs Henry's response to setbacks. When expeditions fail, ships are lost, or progress seems insurmountable, he interprets these as tests of faith rather than indications that his objectives might be unrealistic. God tests his commitment and strengthens his soul through adversity, differentiating true devotion from superficiality. This mindset renders Henry nearly immune to discouragement, convinced that God intends for Portuguese exploration to succeed and that every setback is temporary and surmountable through faith and perseverance.

This religious absolutism has a dark side extending beyond the issue of slavery. To Henry, non-Christians lack the full humanity afforded to Christians, existing as lesser beings in need of salvation. The consent or desires of indigenous peoples hold no sway; their inclusion into Christendom by any means necessary takes precedence. This dehumanization, rooted in theological certainty, paves the way for violence and exploitation.

Despite this, there's an admirable aspect to Henry's devotion if we can set aside its consequences for a moment. He truly sacrifices for his faith, living simply, eschewing comfort, and investing his wealth in what he believes to be God's work. There is hypocrisy in a wealthy priest preaching poverty from a lavish throne. Henry embodies the austerity he demands. Whether one agrees with his beliefs or not, his faith is genuine.

THE ECONOMIC ENGINE

While spiritual drives propel Henry on a personal level, the practical demands of economics ensure that exploration remains viable as a sustainable endeavor. Examining how financial frameworks underpin Portuguese expansion sheds light on how individual foresight can catalyze historical shifts. Operating within a broader context, rather than in isolation, Henry constructs an economic framework that makes exploration profitable and thus feasible to continue.

Initial expeditions incurred losses; constructing and equipping ships, compensating crews, and the inherent uncertainties of these voyages meant early captains often returned with nothing or mere losses. This phase is characterized by pure expense, sustained only through the backing of royalty or powerful patrons. Henry leverages revenues from the Order of Christ and his own position as governor to offset these losses.

A turning point arrives when expeditions begin returning with valuable cargo. The profitable trade of gold dust from West Africa starts justifying costs, while the enslavement of Africans yields pure profit post-capture. Exotic goods such as ivory and pepper find ready markets, leading to higher profitability and attracting private investment. Merchants begin funding these voyages in exchange for shares of profits, creating a self-reinforcing cycle in which profits fund additional expeditions, increasing overall profitability.

Henry's role evolves from sole financier to organizer and regulator of the exploration efforts. By granting monopolies, he ensures that expeditions can operate only with his permission beyond certain geographical points. This not only allows him to maintain control but also distributes risks among captains and merchants who invest in voyages. Henry secures a share of profits through various fees and duties imposed, benefiting both himself and the Order of Christ should ventures succeed.

This setup incentivizes pushing further southwards, with monopolies granted based on territories explored. The right to trade in newly discovered regions becomes a valuable commodity among merchants. This dynamic frontier continuously extends because there's always profit potential beyond it, making exploration self-sustaining once initial infrastructures are established.

The Madeira and Azores colonies expedite this process. Sugar production from Madeira generates immense wealth early on, surpassing the profitability of African trade in its nascent stages. Portugal thus amasses capital to invest further in exploration, simultaneously creating a demand for enslaved labor, as the intensive nature of sugar cultivation requires year-round work. The emerging plantation system on Madeira serves as a prototype for later colonies and demonstrates that Atlantic expansion is capable not just of incremental gains but of transformative wealth generation.

Henry's true innovation isn't in trade itself, an ancient practice, but rather in formulating exploration into an investment opportunity on a structured scale. He establishes mechanisms under royal and ecclesiastical oversight that allow private capital to fund geographical expansions. This blueprint will be emulated by other European powers: chartered companies, colonial monopolies, and systems in which profit is privatized while conquest remains nationalized. The Dutch and English East India Companies, entities from centuries later, are direct descendants of Henry's organizational strategies.

The dark brilliance of this system lies in its self-perpetuating nature. Once profitability is established, exploration requires no further vision; merely greed, ambition, and the mechanisms to harness them remain. Henry constructs these mechanisms, ensuring their endurance for centuries.

THE MOMENT OF MORTALITY

In the closing days of his life, Henry is aware of his impending demise. The body he has so long resisted succumbs to complete decay. What thoughts might occupy a mind so peculiarly intense in its final moments? It's only speculation, but such musings are thought-provoking.

Does contentment find him? His legacy continues beyond his physical presence. The ship's voyage, regardless of his direct guidance. The African shores extend far past the range of his early sailors' vision. Portugal evolves from a minor realm into a dominant maritime force. These feats alone speak volumes. Yet, the ultimate ambition of reaching India by sea remains unfulfilled. The Muslim realms maintain control over the land trade routes, and Christendom's expansion is not as dramatic as Henry might have wished.

Does remorse grip him for the pain inflicted through his actions? It seems unlikely given his worldview. Henry perceives the enslaved Africans brought to salvation, and views the violence towards non-believers as serving God's cause. Questioning this perspective would imply doubt in the theological justification of his deeds, and there's no hint of such introspection. Henry departs this life much as he lived, certain of his righteousness.

Is fear of death present? Not in its conventional sense. A man who dedicates decades to spiritual preparation, leads a monastic existence amidst wealth and power, firmly believes in salvation, and likely faces death as the completion of earthly trials. To Henry, death signifies judgment, and he appears unwaveringly confident of divine mercy. The validity of this confidence hinges on theological issues that remain irresolvable even after six centuries.

What might elude Henry's grasp is the profound way his endeavors will redefine the world. He cannot envision Columbus, the conquest of the Americas, the full extent of global slave trade horrors, or colonial empires governing most of the planet by the 19th century. Unbeknownst to him, Portuguese becomes prevalent in Brazil, Lisbon transforms into the capital of a vast maritime empire spanning continents, and innovations under his patronage secure European dominance for half a millennium.

He passes away believing he has advanced Christianity's cause, opened new lands to faith, enriched Portugal, and served his dynasty. While accurate, this perspective is limited and borders on being misleading regarding the true reach of his life's impact.

The demise itself likely lacks grandeur, no recorded final words or revelations. He simply ceases breathing in his chamber at Sagres, with servants and possibly priests present. His body is readied for burial, his estate settled, and life continues its course.

Yet, a shift has occurred, though it may escape notice. A singular willpower has vanished, a force that shaped people, fortunes, and destinies around a vision deemed madness until proven otherwise. Successors will carry forward and surpass even more. Yet, they tread paths laid by Henry, wield instruments he crafted, and envision dreams first ignited by him.

THE RETURNING WAVE

Sagres, November 13, 1460, Sunset

The Atlantic gusts fiercely over rugged cliffs, as it did when Henry was yet to be born, and will continue to do long after those who remember him are gone. Indifferent to the ambitions, visions, obsessions, and faith of humans, the ocean remains vast, eternal, and unconcerned with our fleeting consciousness.

Yet, Henry has shifted what the ocean represents, if not its essence itself. From a barrier, it morphs into a bridge; from a source of dread, now a wellspring of hope. Ships bound for Portugal carry more than cargo - they bring evidence that the world is broader, stranger, and more understandable than anyone dared to believe. The walls are falling, and Henry was the first to crack them open. Today's pioneers push digital boundaries and venture into space, echoing Henry's legacy of turning obstacles into avenues of discovery.

In Lagos, Lisbon, and Madeira, men consult Henry's charts. They build ships incorporating lessons learned over decades through incremental experiments. Their plans include voyages to places unnamed in Henry's time. The machinery advances, fueled now by momentum as much as vision, but it was Henry who constructed it, set it in motion, ensuring its endurance beyond him. In Africa, from coastal settlements to inland kingdoms, the arrival of the Portuguese had already transformed everything. Trade routes shift. Conflict escalates. Generations living in relative tranquility suddenly find themselves at risk of capture, sale, and bondage. The suffering is immediate, real, though unfolding over centuries. Henry did not intend genocide, cultural destruction, or the horrors of the Middle Passage. He aimed to serve God and expand Christendom. But intention and consequence diverge cruelly.

Sunsets over Cape St. Vincent paint the Atlantic in hues that might be blood, gold, or simply light on water. Tomorrow morning, somewhere in Portugal, a ship will prepare for its southern voyage. The captain carries letters of monopoly, astrolabes tested previously, and maps adorned with Portuguese names. He may not think of Prince Henry. Exploration has become routine, institutional, and perpetually self-sustaining. Yet every chart read, each navigation technique used, and the systems that made his journey possible trace back to a prince who died convinced he was serving God. Centuries will debate Henry the Navigator. Was he visionary or villain? Pioneer or profiteer? Saint or monster? The truth is uneasy but real; he embodied all these, never fully, always partly. A medieval prince with medieval goals, he unwittingly shattered the world he knew. A celibate ascetic, he planted seeds of global capitalism. A religious zealot, he fueled the Renaissance and the birth of science. He never set sail, yet changed the course of sailing forever.

As darkness envelops the fortress at Sagres, servants prepare Henry's body for his journey to a tomb somewhere on Africa's coast. Meanwhile, a Portuguese ship anchors there. Crew members go ashore to trade, explore, and capture human beings for sale. They do this because it is profitable and possible, thanks to Prince Henry the Navigator, who put infrastructure in place. The waves maintain their eternal dialogue with the cliffs; the wind scatters salt spray across the stone. The Atlantic endures, indifferent and unyielding. But humanity's relationship with that ocean and itself has been irreversibly transformed by a man who spent his life looking beyond horizons, refusing to accept anything as permanently unknown.

In this, Henry succeeded utterly. The world became knowable, mappable, exploitable, and connected. Whether this is triumph or tragedy, progress or calamity, hinges on where you stand, who you are, and how you value human worth against suffering. Henry never wavered in his conviction. Six centuries later, we're still grappling with our response. Tides ebb and flow; ships continue to sail. What will you make of this legacy?

João Gonçalves Zarco & Tristão Vaz Teixeira

1418. The Atlantic Ocean. Somewhere West of Cape St. Vincent.

The wind blindsided them without a hint.

At the onset, the caravel moved steadily beneath an open sky. A single lateen sail heaved against the northeast trade winds. Suddenly, the horizon darkened ominously. The sea surged like towering liquid mountains. A piercing creak splintered the air as the ship's timbers strained under the ferocious force, a prelude to the turmoil ahead. João Gonçalves Zarco gripped the tiller as the deck tilted sharply. Water howled over the bow, rising from knee height to waist, sweeping away any unsecured item. The men cried out prayers to São Vicente, Nossa Senhora, and any saint who might hear them above the deafening gale.

Tristão Vaz Teixeira made his way towards the stern of the ship, clutching the side with one hand. In the other, he held a knife ready to sever the rigging if necessary. Salt mist burned his eyes. Thunder rumbled so close it left his ears ringing. The caravel surged up an immense wave, hovered at its crest, and then plummeted down the opposite side into pandemonium.

This was not their intended trajectory. Prince Henry had commissioned them to sail along the African coast south of Cape Bojador, to explore lands beyond the Sahara and gather intelligence on Muslim territories. But the storm ignored these orders, propelling them westward, far from any known land. According to outdated maps, this part of the ocean was supposedly vacant and boundless.

When morning emerged, weak and filtered through broken clouds, they beheld something absent from every Portuguese map. Approximately ten miles away, mountains rose from the sea, their peaks obscured by mist and their slopes cloaked in dense, deep green foliage. It was an uninhabited, unrecorded, and unseen island. Charting it would fundamentally shift Portugal's self-perception and its future prospects.

Zarco and Teixeira were oblivious to the magnitude of what they had stumbled upon. This was more than a mere discovery; it was the genesis of an empire and the cornerstone of a colonial system that would span three continents and endure for centuries. What began as a storm-induced mishap would prove to be more impactful than many meticulously planned expeditions or battles. From this moment forth, sugar plantations flourished. Wine estates, merchant opulence, and the blueprint for transcontinental colonization all found their inception.

Yet, before anything else, survival became paramount; their discovery was meaningless if they could not persevere to narrate it.

THE KNIGHTS OF DESPERATION

João Gonçalves Zarco was thirty-two years old in 1418. His body bore the scars of warfare. These scars reminded him of the difficult path faced by minor nobility seeking fortune in a realm with few prospects. He was born around 1386 in Alentejo, a region of sprawling estates where younger sons inherited nothing. His family was respectable but lacked enough wealth to influence power players. Because of this, he took up arms and entered the service of greater lords. He hoped that the spoils of war would compensate for his family's shortcomings.

The defining moment for Zarco's generation came with the 1415 capture of Ceuta. This North African city, strategically across from Gibraltar, was seized by King João I and his heirs. It was a monumental Portuguese endeavor that melded crusade with commerce. The capture dealt an affront to Islam while vying for control over lucrative trans-Saharan trade routes. Many young nobles flocked to this expedition, lured by promises of glory, plunder, knighthood, and land grants. Zarco was among them, engaged in the harrowing combat that ended with Ceuta's mosque made into a cathedral and its population displaced or enslaved.

However, Ceuta proved to be a bitter disappointment. The Portuguese sought to dominate but were adeptly evaded, re-routing themselves beyond Portugal's grasp. Consequently, Muslim merchants refused to enrich Christian foes, transforming the city into an expensive military burden fraught with continuous defense efforts against Moroccan reprisals, while yielding little in return for its conquest. Though these nobles won their knighthood, they remained landless, clinging only to empty titles and possessing martial prowess without subsequent warfare to exploit.

Tristão Vaz Teixeira mirrored Zarco's plight but had slightly stronger court connections. A couple of years his junior, he was equally driven yet stymied by Portugal's meager prospects for their class. The two are likely to have met during or after the Ceuta campaign. Each recognized in the other a fierce ambition and a willingness to take risks that wealthier peers avoided.

Both men entered service under Prince Henry, King João I's third son. After Ceuta's capture, Henry fixated on expanding Portuguese influence along the African coast. Far from the romantic figure later mythologized, he was a pragmatic military leader focused on Portugal's survival through economic means. Settling at Sagres, a strategic point on the Atlantic coast, he began systematic exploration of Africa.

The endeavor drew in kindred spirits like Zarco and Teixeira due to its high stakes. Ships disappeared, crews succumbed to disease or thirst, and months at sea yielded no reward. Yet it held out the promise of discovery equal to its risks: claims to new lands with attached titles and trading rights. For those without inheritance prospects, these dangers were bearable. Zarco, outwardly seasoned but inwardly grappling with the fear of obscurity and the specter of leaving an indelible mark, held onto hope by imagining a future legacy in which his name would outlast him. In quiet moments, on watch beneath the starlit sky, he allowed his mind to reflect: Would this voyage cement his place in history, or be merely another undistinguished chapter in his life? Far from heroes in conventional terms, they were gamblers staking all that they had.

1417. Sagres, Portugal.

Before the fateful expedition, Zarco and Teixeira spent months at the southwestern tip of the kingdom, preparing for what lay ahead.

The settlement at Sagres was not particularly noteworthy, an assortment of buildings situated on unforgiving cliffs. The Atlantic winds persisted ceaselessly, while the land offered little besides rock and sparse vegetation. A significant town was located miles away, making the site uncomfortable for its inhabitants. Yet Prince Henry had chosen it strategically. From this vantage point, one could observe ships embarking on uncharted voyages and potentially returning from expeditions that incrementally expanded humanity's understanding of the world.

The complex housed provisions and equipment warehouses, as well as workshops where shipwrights experimented with caravel designs. A small chapel was available for sailors who partook in mass before embarking on perilous journeys. The prince himself analyzed maps and interrogated returning captains regarding their experiences. This establishment served as a hybrid between a monastery, military base, and research facility, all focused relentlessly on advancing Portuguese influence southwards along the African coastline and beyond.

Zarco and Teixeira, among possibly two dozen other fidalgos (minor nobles) in Henry's service by 1417, engaged in rigorous training. They practiced with weapons, learned basic navigation, listened to pilots discuss weather patterns and currents, and acquired fundamental celestial navigation skills to determine latitude. Their primary activity, however, was waiting. They observed the departure and return of other captains, estimating when they might be entrusted with their own command.

Henry demanded unwavering loyalty and meted out severe punishments for failure. Captains who returned without valuable information or prematurely from challenging coastlines, or those who lost ships due to perceived incompetence, were dismissed. Conversely, successful captains were rewarded handsomely with land grants, trade monopolies, and authoritative positions in newly discovered territories.

The competition was intense; each fidalgo aimed to outshine others, prove worth, and secure command of an expedition. Zarco and Teixeira formed an alliance, believing their chances were better together than apart, as one could manage a ship while two could lead multiple vessels, support each other in adversity, and corroborate discoveries.

Their opportunity arose early in 1418 when Henry sanctioned an expedition to explore further south along the African coast. He was particularly interested in investigating reports of lands beyond Cape Bojador.

Zarco commanded at least one caravel, possibly two, while Teixeira led another. The expedition's size remains uncertain due to incomplete records from that period. What is clear: they departed from Lagos, the nearest town to Sagres, likely in late spring when conditions were favorable for southbound voyages. They carried approximately two dozen crew members and supplies sufficient for months, including water, wine, salted fish and pork, stale ship's biscuit, dried beans, onions, vinegar, along with weapons like crossbows, swords, and lances prepared for potential hostility or slaving opportunities in Africa.

The departure was likely marked by ceremony; a priest blessed the ships, and the crews knelt for deck prayers. Prince Henry may have come down from Sagres to see them off. His countenance revealed no emotion, hope, or doubt as he observed the send-off. With the blessing concluded, practical preparations began: using oars and tow lines to pull the vessels out of the harbor, raising lateen sails beyond the breakwater, and setting a southwest course to catch the favorable Portuguese trades towards the African coast.

For one week, their voyage proceeded routinely; they passed Cape St. Vincent with its ancient lighthouse standing for over a millennium and observed the Algarve coastline to starboard, familiar to many sailors. Beyond this, the land receded, leaving them in open ocean, where the waters deepened from green to dark blue, and waves grew as they left the protection of the continental shelf. The crew settled into ship routine: four-hour watches, meals of fish and biscuits served in wooden bowls, continuous attention to sails and rigging, daily prayers at dawn and dusk, an endless rhythm typical for a vessel crossing the sea.

Then, off the Moroccan coast, their fortunes changed abruptly as the weather turned against them. The storm that had brewed on the horizon now engulfed them, remnants of its ferocity echoing in the shifting winds and restless seas. This unplanned divergence starkly reminded them of the expedition's broader purpose: exploration of uncharted waters at the behest of Prince Henry, where even the elements conspired to alter their course.

THE STORM

Sailors of the 15th century perceived impending danger through distinct signs. Sky turned greenish. Air grew heavy and silent. Seabirds sought land with urgency. Waves were unusually smooth. Dark line widened on the horizon, signaling a storm's approach.

Captain Zarco demanded immediate preparation for the storm. Non-essential items were secured or stored below deck: ropes, tools, water casks, provisions. Sail was shortened to reduce mast strain. Men ensured axes were ready in case of rigging failure. Bailing buckets were made accessible; they'd be needed soon. Tiller was reinforced with extra lashings to prevent waves from dislodging it. Nothing left to do but wait and pray.

Storm struck fiercely. Wind tore at clothes, hindering breathing. Rain poured relentlessly like a waterfall. Waves grew from six feet to ten to fifteen, losing their shape in spray. Caravel, designed for coastal voyages, pitched violently; bow lifted nearly vertically, then plunged into green water crashing over gunwales.

Below deck, chaos ensued. Water sloshed violently with each pitch. Bilge transformed into a foul mixture of seawater, vomit, and waste. The stench was unbearable. Minutes passed without relief as men retched, adding to the misery. Continuous cacophony filled the air: timber groans, water roars, wind shrieks, men's prayers or cries, ropes lashing wood with pistol shot sounds.

On deck, Zarco and the pilot fought to keep the bow aimed at the waves, preventing a broadside roll that would have led to immediate sinking. Tiller required two or three men to control; their muscles burned, and their hands bled from rough wood. Salt spray blinded them. Thunder vibrated the air; lightning revealed chaotic scenes: vast waters stretching endlessly in all directions, death lurking everywhere.

Navigation was irrelevant. No steering towards any destination. Survival alone mattered: keep the bow to the waves, the mast upright, the ocean out of the hull. Each moment they remained unscathed was precious. Time lost meaning; minutes or hours couldn't be distinguished. Storm created a separate reality where only wind and water mattered.

Teixeira's ship vanished quickly in the storm; even nearby ships disappeared from sight immediately. Whether Zarco's vessel survived or foundered wasn't known. The Atlantic of the 15th century was vast, merciless - ships vanished without a trace. Whole expeditions disappeared. The ocean kept its secrets and remains.

Gradually, the wind decreased, but not all at once. Storm reluctantly loosened its grip in stages. Waves remained huge but became less chaotic. Rain intensity lowered from torrential to heavy. Men dared to bail water off the deck, working frantically to reduce hull sloshing. The worst had passed.

Dawn broke weakly through clouds to reveal a battered ship still floating. The sail was tattered, and the rigging was tangled. Water covered much of the deck. Every man was exhausted but alive, and the caravel held together - they survived.

Then someone pointed west and shouted, "Land!" Impossible according to all known Atlantic geography. Yet there it was: mountains rising from sea ten miles away, their peaks in cloud, slopes dark with vegetation - an unexpected island.

The pilot quickly grasped the implications: carried hundreds of miles off course, into the Ocean Sea, no Portuguese ship deliberately sailed. Land couldn't be Africa or any known islands like the Canaries. This was a new, unknown discovery that could reward Prince Henry if they could only reach it. As Zarco turned his gaze toward the mist-shrouded peaks, a question loomed: What dreams might spark in the mind of Prince Henry upon learning of this unexpected boon? Would he seize the moment to expand his visions of exploration and conquest, seeing this island as a stepping stone to greater glories and an empire yet unfathomed?

Caravel wasn't in condition for fine navigation. Tattered sail barely drove her forward; crew exhausted; water and provisions contaminated or lost overboard. They needed land to repair the ship and find fresh water before any return voyage. Slowly, painfully, they worked towards the island.

As they neared, details became visible: rocky coastline with sheer cliffs dropping into the ocean; waves crashing violently against those cliffs - landing there meant instant destruction. But where a river or stream reached the sea, it might be a beach, a break in cliffs, or a place where a ship could approach without being smashed to pieces. They coasted along the southern shore searching desperately, the damaged vessel barely responding to the helm.

Then they found it: a small bay somewhat sheltered with black sand and a rocky beach. Not ideal but survivable. Carefully worked the caravel into this spot; dropped anchor about three fathoms from shore, where men could wade or swim if needed. Anchor held; ship finally still after perhaps thirty-six hours of chaos.

Zarco ordered a boat lowered; five men rowed him ashore with determination. The boat was grounded on black sand; Zarco stepped into knee-deep water and waded to the beach. He knelt - whether in prayer or relief, not known - picked up a handful of volcanic sand; it was real. Solid. An island where no island should exist.

Behind him, his men began survival work: gathering driftwood for fires, searching for fresh water, assessing ship damage, and preparing to spend days here while repairs were made. None yet understood what they had found; none imagined this accidental storm encounter becoming the foundation of Portugal's Atlantic empire or a template for colonization across worlds. But their only task was to endure - survive long enough for their story to reach another shore.

THE ISLAND OF WOOD

The individuals who first set foot on Madeira in 1418 ventured into a realm untouched by human presence for centuries. Rumors of mysterious Atlantic islands had circulated since Phoenician and Roman times, yet these whispers did not translate into permanent settlements. Isolated for millennia, the island's landscapes flourished undisturbed, its forests so dense that sunlight struggled to penetrate.

The landscape was an overwhelming tapestry of vegetation. From the shore where Zarco disembarked, endless forests marched up the mountain slopes in every direction, appearing almost solid due to their density. Towering trees of unfamiliar species dominated the scene, with trunks six feet wide and canopies disappearing into the perpetual mist atop the peaks. Thick undergrowth formed a nearly impenetrable barrier, while vines hung like curtains from branches.

The air was rich with the scent of growing plants, earthy tones, and moisture, and it seemed almost aggressive in its vigor. The immediate concern, however, was water. The storm had contaminated their casks with seawater, leaving them with just enough to last perhaps two days. Without fresh water, survival on this island, regardless of other offerings, would be impossible.

A search party set out inland along what seemed a watercourse, hacking through the dense undergrowth with axes and swords. As they ventured deeper into the forest, they encountered an alien environment that felt claustrophobic. The thick tree canopy blocked sunlight, casting the ground in perpetual twilight. The air was humid, making breathing difficult after the ocean's openness.

Strained calls echoed from the trees, birds unique to Madeira, unafraid of predators as they had never known any. Slowly moving forward with no telling what else lay ahead, the men pondered potential inhabitants, animals, or dangers beyond imagination.

After half a mile, water was discovered: a clear, cold stream originating high above, carving through moss-covered rocks to form small pools where unfamiliar fish darted away from encroaching shadows. Kneeling at its banks, the men drank deeply, finding salvation in the cold water after rationing their supplies carefully for days. They filled leather bottles and trekked back to replenish the casks of the ship.

With a reliable water source secured, attention shifted to food. Supplies were dwindling and spoiled. The salt pork had begun to decompose under Madeira's tropical heat, while biscuits were infested with weevils. Dried fish was edible but offered no variety. However, the island presented alternatives. Unfamiliar birds proved easy prey, showing no fear of humans and susceptible to being clubbed or trapped. Fish could be caught in coastal waters. Forest fruits abounded, though caution was exercised when consuming unknown varieties due to stories of poisonous plants.

Over a span of days, as efforts to mend the caravel advanced, those of Portuguese descent began to grasp the significance of their discovery. This was more than just an isolated landmass; it was a realm untouched by European colonizers, an ecosystem in its pristine state. The dense forest emerged as the most noticeable feature upon arrival, sprawling so extensively that the crew named this site "Ilha da Madeira" – or Island of Wood - a moniker that would endure as a permanent fixture. Yet, the forest merely served as the surface level of a more intricate tableau. The volcanic terrain bore exceptionally fertile soil, nurturing flora with an intensity unlike anything witnessed by those who hailed from rocky shores. The climate remained temperate throughout the year, neither excessively hot nor cold, thanks to the moderating influence of both ocean proximity and mountain elevation. Rainfall was plentiful, leading to a profusion of waterways that cascaded down from the peaks, sculpting valleys with steep inclines and sharp edges, yet also offering gentle slopes suitable for cultivation. Most importantly, from their perspective, this island lay uninhabited – no human populations with whom they needed to communicate, compete, or coexist. This land was theirs alone to shape according to Portuguese desires and needs. It differed markedly from Africa, where negotiation with sovereign entities had to be conducted with respect; it surpassed even the Canary Islands, where indigenous peoples fiercely resisted Spanish encroachment for many years. Here lay virgin territory, open for claim and transformation.

However, survival of the return journey was an immediate concern. After a week dedicated to restoration – the sail patched with leftover canvas, rigging adjusted, leaks plugged with cloth strips soaked in pitch, provisions replenished – the question arose: could they navigate back to Portugal without precise charts? The pilot meticulously addressed this challenge. While they knew their approximate latitude from celestial observations before the storm, the storm's impact on their position remained uncertain. To find safety within familiar waters, a detour via the African coast was chosen: sailing east until reaching the continent and then following its shoreline northward. This route was longer but deemed safer due to the uncertainty of their exact location. Favorable winds propelled them steadily at approximately three knots as they sailed east, driven by the northeast trades typical of these latitudes. Each day, the pilot would take sun sights to calculate their latitude, ensuring their steady progress in the right direction. After roughly five days, the coast of Morocco appeared on the horizon, signifying successful navigation back into known waters. The remainder of the voyage home proceeded with relative ease. They hugged familiar shores and made brief stops at established Portuguese outposts for supplies before making a steady push toward their homeland. Upon finally entering Lagos harbor, likely in late summer 1418, they brought tidings of an island far offshore in the Atlantic, abundant in resources and devoid of inhabitants – ideal as a waypoint for further exploration. Prince Henry received this report with a sense of vindication, questioning Zarco about the island's location, its assets, and the feasibility of establishing a settlement. This interrogation was thorough, covering all aspects of winds, currents, coastlines, interiors, potential anchorages, water sources, lasting hours as detailed information was amassed for strategic planning. Yet, an essential query remained unanswered: what had befallen Tristão Vaz Teixeira? His vessel vanished during the tempest. Was it conceivable that he perished in the Atlantic, or did he, too, discover and survive on this newfound isle?

The revelation emerged several weeks later, when Teixeira's caravel docked in Lagos, unfurled Portuguese banners, and appeared undamaged. His narrative mirrored Zarco's nearly identically. The same tempest propelled him westward. He chanced upon the identical isle, likely disembarking at a different cove along the shoreline. Enduring hardships on that isle while effecting repairs, he navigated back to Portugal employing akin methods. Two expeditions, separated by the gale, independently unveiled the same land and returned home.

Their concurrent testimonials dispelled any uncertainty surrounding the island's existence or significance. This was not the delusion of a solitary mariner nor an interpretive oversight. Two seasoned sailors converged on the same locale and returned with corroborating accounts. The island stood as a palpable reality, proffering an extraordinary prospect.

King Henry hastened to fortify Portuguese sovereignty. He petitioned his sibling, King Duarte, for official acknowledgment of this discovery and authority to subjugate the island. The supplication was granted, the sovereign had every incentive to endorse his brother's African endeavor, promising territorial enlargement and riches without significant royal outlay. Then, Henry confronted the matter of who should spearhead the colonization endeavor.

The foremost candidates were the very discoverers. Zarco and Teixeira had charted the isle, survived there, triumphantly returned, and demonstrated navigational prowess to reach it again. They harbored robust personal motivations to succeed: this represented their opportunity to convert discovery into lasting affluence and status. Henry appointed them joint captains of the colonization quest, demarcating the island between them and bequeathing extensive dominion over their respective realms.

The division was predominantly north-south. Zarco would preside over the southern segment, encompassing the finest natural harbor. Teixeira would oversee the northern domain. A third captain, Bartolomeu Perestrelo, would govern the lesser isle of Porto Santo, discovered concurrently or possibly during a subsequent expedition; sources remain ambiguous. Each commander enjoyed prerogatives over all lands they could bring into tillage, monopolize sundry economic activities, exercise jurisdiction over settlers, and partake in revenues generated within their jurisdictions.

These were unprecedented privileges, akin to the elevation of petty nobility into opulent barons. Yet these liberties entailed duties. The captains were required to summon settlers, transport them to the isle, equip them with tools, livestock, and seeds, orchestrate colonization efforts, safeguard territories from rivals, and remit Prince Henry's share of accrued wealth. It was a feudal scheme recalibrated for the Atlantic, integrating medieval societal constructs suited to frontiers.

The preparatory stages necessitated months. Vessels must be outfitted, supplies gathered, and colonists recruited. The settlers were a mixed cadre: peasants from Portugal enticed by promises of land, artisans requisite for construction and industry, a scattering of minor nobility in quest of fortunes, and criminals offered banishment to the island as an alternative to penalization. Livestock cattle, swine, sheep, poultry, and plant life essential for agriculture, wheat, barley, vineyards, and orchards also formed part of the convoy.

In 1419, likely during the latter spring, the colonization armada set sail from Lagos under Zarco's command with Teixeira as his deputy. The contingent consisted of three or four ships, depending on accounts consulted. Approximately two hundred colonists embarked, though exact numbers remain vague. Laden were all the instruments necessary for an enduring settlement: apparatuses for construction and defense, victuals sufficient until harvests could be reaped, livestock, and seeds that would metamorphose the isle's ecosystem.

The outbound voyage proved smoother than the tempest-ridden odyssey of discovery. Their destination was known, winds were propitious, and within a fortnight, they spied Madeira's distinct peaks piercing the sea, traversed along the south coast, and berthed in Zarco's first haven. Thus began Portugal's initial colonization of an Atlantic isle.

THE BURNING

The settlers who arrived on Madeira's shores in 1419 faced an immediate challenge: the forest was too dense to support agriculture. Prior to planting crops or establishing settlements, they needed to create open spaces, a task too massive, given their limited tools. They adopted the standard medieval technique of slash-and-burn to clear forests: they cut down trees, dried them, and then burned them to produce ash-enriched soil suitable for cultivation. This method was laborious but effective, as evidenced by centuries of European expansion.

However, the Portuguese underestimated Madeira's vast forest reserves and its propensity for rapid fire spread in summer conditions. Fires were initiated across multiple locations as settlers claimed land along the coast and lower valleys; they used axes to fell trees and machetes to cut brush while gathering combustible materials into piles. The wood was denser than that of European species, largely laurel trees that didn't grow in Portugal, complicating their efforts.

Despite their struggles, someone suggested accelerating the process by setting fires intentionally to spread them. This seemed rational, as fire would consume what had taken years to clear manually within days; additionally, ash from burnt trees could enrich the soil for immediate planting. Once ignited, though, things spiraled out of control. The forest harbored centuries-old dead wood and dry undergrowth, primed for combustion. With hot, arid summer air and ocean winds fueling the flames, what began as controlled burns quickly became an uncontrollable inferno that ravaged the island with alarming speed.

Accounts decades later claimed the fire persisted for seven years; this is likely exaggerated, yet it underscores the immense scale of the burning of Madeira, which exceeded any control. Entire watersheds were incinerated, leaving settlers in a mix of emotions, some hopeful for fertile land while others feared irreparable damage. Smoke intermittently blocked the sun, prompting evacuations to beaches or ships when fires threatened their encampments. Ash fell like snow over everything, tinting the ocean gray offshore, reflecting both hope and despair at the transformation.

Ecologically, Madeira's millennia-old laurel forest was decimated, along with countless endemic species that disappeared forever. The once complex ecosystem now lies in ashes and burnt stumps, leaving only fragmented descriptions and isolated pockets in inaccessible mountain valleys for future generations to know the lost forests.

Yet, from a Portuguese viewpoint, disaster also opened an opportunity. When fires eventually fizzled out, whether due to fuel exhaustion or seasonal rains, the island morphed. Extensive areas that were once impenetrable forest were now clearings, ash-enriched and ready for cultivation. Though some forests persisted in wet zones atop the mountains, lower slopes and valleys were cleared, presenting Madeira as a potentially fertile land that would've taken decades to clear manually.

Settlers rebuilt their coastal camps and began colonial efforts earnestly. They cultivated wheat in ash-covered ground with remarkable growth spurred by volcanic soil intermingled with organic nutrients from burned forests. Coastal villages emerged using stone and timber salvaged post-burn, under the jurisdiction of Zarco and Teixeira, accepting feudal hierarchies for land grants and promises of future prosperity.

The colony faced hardships in its early years despite having ample cleared land. Ships from Portugal arrived infrequently, leading to shortages of essential goods such as tools and cloth, as local production was not yet established. Some settlers succumbed to diseases such as malaria, dysentery, and fevers for which medieval medicine had no cures. Others grew disheartened and returned to Portugal whenever ships arrived.

Zarco and Teixeira addressed the perennial challenge faced by frontier governors: maintaining order and morale among distant colonists without direct central authority. They conducted courts, resolved disputes, punished offenses, and attempted to prevent the colony from descending into anarchy. Additionally, they had to handle interactions with Prince Henry, submit regular reports, and share in the island's revenue while advocating for greater support and investment.

The breakthrough in Madeira's economy came unexpectedly through sugar production. It is speculated that either Zarco himself or a Sicilian merchant knowledgeable about Mediterranean sugarcane cultivation recognized the island's climate and soil as ideal for growing sugarcane. The crop thrived on Madeira, requiring abundant water, year-round warm temperatures, fertile soil, and intensive labor, all readily available on the island. Sugarcane was likely brought from Sicily or Crete and experimentally planted in some coastal valleys. This coincided with the growing demand for sugar in Europe, where consumption patterns were rapidly evolving. Sugar had become a symbol of status and luxury among the European elite, and its sweet allure was irresistible, driving an insatiable hunger that Madeira's production could satisfy. The island's ability to meet this burgeoning appetite for sugar positioned it as a key player in the burgeoning global trade, underscoring the transnational dynamics at play.

It flourished remarkably; the cane grew to heights of fifteen feet or more, boasting high sugar content and a nearly continuous growing season thanks to Madeira's mild climate. In just a few years, sugar became Madeira's leading export, surpassing grains, wine, and timber that colonists initially prioritized. Sugar cultivation necessitated infrastructure, including irrigation systems to bring water from mountain streams to coastal plantations, mills for crushing the cane, boiling houses for juice processing, and skilled laborers to oversee the intricate production process.

The sugar boom drastically transformed Madeira's economy and society, replacing small farms with sprawling plantations as the dominant land use. Wealthy merchants from Portugal, Genoa, and Florence invested heavily in sugar production, establishing a capitalist system distinct from the feudal agricultural setup initially present in the colony. There was a significant surge in labor demand, as sugar cultivation and processing were extremely labor-intensive, leading to the introduction of slaves, first imported from the Canary Islands and later from Africa.

By the 1450s, Madeira was producing substantial amounts of sugar for European markets. By the 1470s, it had become one of Europe's primary sugar suppliers. And by the 1480s, the island's affluence had elevated Zarco, Teixeira, and their descendants, who inherited their captaincies, to genuinely wealthy status. The storm-driven event that led to the discovery of Madeira in 1418 had laid the groundwork for fortunes that would endure across generations.

THE PATTERN

What took place on Madeira laid down a blueprint for events that would unfold repeatedly throughout the Atlantic world over the next four centuries. The sequence – an accidental discovery leading to an exploratory expedition, followed by crown authorization, colonization led by minor nobles, ecological transformation, cash crop agriculture utilizing slave labor – became the template for European expansion across the Americas, the Caribbean, the Indian Ocean, and the Pacific regions.

Examining the ecological impact: Madeira's indigenous forests were cleared to make way for farming, much as would later happen throughout the Americas. This led to native species either going extinct or being pushed into less suitable habitats. Additionally, European plants and animals were introduced, altering the landscape to closely mirror European ecosystems. This transformation was not a result of happenstance but rather an intentional policy based on the belief that European agriculture surpassed others, with any land not producing European crops considered wasted.

The economic pattern also had a significant effect. Initially, there was small-scale diversified farming, which gradually gave way to plantation monoculture focused on export crops. This transition happened on Madeira with sugar in the 1420s-1430s, followed by the Caribbean in the 1500s-1600s, Brazil from the 1600s onwards, and Indian Ocean islands during the 1700s-1800s. Each instance shared a similar logic: tropical or subtropical climates could cultivate crops in high demand among Europeans but couldn't be grown domestically, creating substantial profit opportunities to justify large investments and environmental alterations.

The social pattern might have been the most influential of all. Plantation agriculture required significant labor, but recruiting volunteers at wages making production profitable proved challenging. The solution was forced labor: initially, indigenous peoples were enslaved through conquest, followed by Africans being purchased from slave traders and transported across the Atlantic. Madeira pioneered this system on a smaller scale in the mid-1400s, while the Caribbean and Brazil expanded it to an industrial level in the 1500s-1700s. This resulted in the largest forced migration in human history, with millions of Africans being transported to the Americas over several centuries.

None of these outcomes was predetermined. The Portuguese could have colonized Madeira differently - they might have preserved the forests while establishing a more modest agricultural economy or attracted free labor with better terms. They might also have adopted less environmentally destructive and socially exploitative settlement patterns. However, various incentives pushed towards the actual pattern that emerged. Sugar generated enormous profits, outpacing any other crop the island could produce. Plantation farming became more efficient at scale compared to small farms. Lastly, slave labor was cheaper than free labor, making individual choices economically viable despite collectively creating systems of exploitation and environmental destruction that plagued Atlantic world societies for centuries.

Zarco and Teixeira were not malevolent figures by their era's standards; they were pragmatic administrators focused on making their colony successful, meaning profitable, through producing export crops valued by European markets. They operated within moral frameworks and incentive structures that accepted slavery as normal, viewed environmental transformation favorably, and measured success in terms of revenue rather than ecological or social sustainability.

Yet the consequences of their decisions echoed across centuries. The sugar plantation model they helped establish on Madeira became the blueprint for Brazil's plantation economy, shaping development patterns that persist to this day. The example set by Madeira's colonization informed Spanish settlement in the Caribbean, English colonization in Virginia and the Carolinas, and French development of Haiti. Moreover, the ecological transformation techniques they pioneered – such as widespread clearance of native ecosystems, introduction of European species, and intensive exploitation of soil and water resources – were applied worldwide in tropical regions.

1467. Funchal, Madeira.

João Gonçalves Zarco was quite old now, probably eighty-one, though exact birthdates weren't usually recorded with much precision in those times. For nearly fifty years, he resided on Madeira, witnessing the colony evolve from a small group of struggling settlers into a thriving community of several thousand souls. Funchal, which he had established at the finest natural harbor along the south coast, grew into a significant town adorned with stone buildings, churches, warehouses, and the grand homes of wealthy sugar traders.

His standing was unshakeable as captain-donatário, essentially a feudal lord of Madeira's southern half. He held vast tracts of land, collected revenues from sugarcane production, and levied taxes on milling services. His offspring were poised to inherit his wealth and authority. In essence, he had surpassed all aspirations for a minor nobleman from the Alentejo: possessing lands, riches, a title, and a dynasty.

Yet, this success was intertwined with significant environmental alterations. The island he had discovered bore little resemblance to the dense forest of 1418. Erosion marred the mountains where forests once stood; valleys became geometric patterns from sugarcane fields and irrigation systems; coastal towns reeked of sugar refineries and slave accommodations. Native birds that were once abundant are fast disappearing, displaced by introduced species or hunted down as their natural habitats vanish.

Zarco's thoughts on this transformation remain uncertain, as he left no personal writings or diaries expressing his inner turmoil. We can only speculate from his actions: he supported the sugar industry, which demanded environmental disruption; imported slaves to work in his plantations; and upheld a social order where laborers were subjugated to landowners. While the historical record is silent regarding his internal deliberations, it is possible that Zarco experienced tensions between the drive to secure his legacy and the ethical or environmental costs that accompanied his success. The absence of explicit reflection leaves open the question of whether he ever grappled with the long-term implications of his choices, or if his pursuit of prosperity for himself and his descendants overshadowed such concerns.

Tristão Vaz Teixeira passed away sometime in the 1460s, with no specific date recorded. He oversaw the northern part of the island, which was less successful than Zarco's region. It was rockier, less suitable for sugar cultivation, and slower to develop economically. However, he also entrenched his family as permanent fixtures in Madeiran society. His descendants would hold onto the northern captaincy for generations, although they never achieved the same level of wealth and influence as Zarco's lineage.

Together, these two men achieved something unprecedented in Portuguese history: a self-sustaining overseas colony integrated into European trade networks. This marked the beginning of Portugal's global empire. The improvised methods they used in administrative systems, economic strategies, and social frameworks would be studied and emulated by subsequent Portuguese colonizers and by other European nations that ventured into Atlantic exploration.

It's said that Zarco died in either 1467 or 1471. This variation shows how swiftly certain historical figures can be forgotten when their deeds are overshadowed by grander events. Resting in Funchal, his final home vanished due to the march of time and earthquakes. Not a single account about him was documented during his lifetime. There was no record of a grand farewell. He just passed away, an old man who lived long enough to see his risky endeavor pay off, transforming an incident sparked by storms into a legacy that would stand the test of time.

His sons divided the lands he built and upheld the sugar trade. His grandchildren extended it further, constructing bigger mills, clearing more land, and introducing additional slaves. The population of Madeira grew steadily by 1500, reaching around 20,000 inhabitants, with estimates suggesting 30,000 by 1550. Sugar production flourished in tandem, making Madeira one of the wealthiest Portuguese territories and boosting the fortunes of merchant families controlling the trade.

Despite this growth, ecological damage mounted. By the late 1500s, soil erosion had significantly degraded lands that had been enriched by deforestation. The ash that initially boosted fertility had long since vanished, and without tree roots to anchor the volcanic earth, torrential rains washed away precious topsoil into the sea. Yields dwindled as a result, and the sugar boom shifted towards Brazil, where vast new lands offered similar opportunities Madeira once provided, but on an immense continental scale.

Madeira's economy turned to wine production, better suited to its deteriorating landscape. The famed Madeira wines of later eras were initially a response to ecological challenges, a way to maintain agricultural productivity as the sugar industry waned. Yet producing wine also necessitated environmental alterations, such as terracing slopes, irrigation systems, and the introduction of European grape varieties. Transforming Madeira's landscape continued even as it moved away from sugar.

By the 1600s, Madeira no longer led in Portuguese colonial development; Brazil had become the empire's crown jewel with its sprawling lands, rich mines, and massive plantations worked by hundreds of thousands of African slaves. The model Madeira pioneered was now being scaled up beyond what Zarco or Teixeira could have imagined. Yet, the island remained valuable, serving as a stopover for ships traveling to and from Brazil and Africa, producing wine, and serving as an essential naval position in the mid-Atlantic.

The system Zarco and Teixeira had helped establish continued to evolve, generating wealth for some while causing suffering for many more, reshaping environments and societies alike. It would go on to shape the Atlantic world, influencing it well into industrial times and beyond.

THE WEIGHT OF ACCIDENTS

The tempest that prompted Zarco and Teixeira's voyage to Madeira was nothing short of serendipity. Not one aspect was premeditated or deliberate; their destination was meant to be Africa's shores, not the uncharted Atlantic archipelagos. The gales that propelled them off course could have as easily hurled them into perilous waters along Morocco's coast or the void where no land existed. Their survival itself was fortuitous. And instead of an endless sea, they stumbled upon Madeira, a twist of fate.

Yet this unplanned detour irrevocably reshaped history. Absent Madeira, Portuguese forays across the Atlantic may have unfolded differently; perhaps more gradually or not at all. It served as proof that remote islands could be successfully and profitably colonized. This success bolstered Prince Henry's resolve to continue pushing westward and southward, establishing a precedent.

When Christopher Columbus approached Portugal in the 1480s with his proposal to sail west in search of Asia, he was building on knowledge from decades of Atlantic exploration that began with Madeira. Portugal had honed long-range ocean navigation over two generations. They had also been perfecting colonial administration in Madeira and other islands they colonized. Furthermore, the economic model of tropical agriculture was well understood thanks to sugar production on Madeira.

Although Portugal turned down Columbus's offer, deeming his distances too great and fixated on an African route to India, everything he drew upon ship designs, navigation methods, and colonial tactics was rooted in Portuguese expertise that originated with Zarco's and Teixeira's accidental find.

Similarly, Brazil was an extension of this model. When Pedro Álvares Cabral stumbled upon the Brazilian coast in 1500 en route to India, Portugal seized upon the opportunity with familiar strategies: sending colonists, clearing land, setting up plantations, importing slaves, and cultivating sugar for European markets. Brazil mirrored Madeira on a grand scale.

The Spanish replicated these methods in the Caribbean, Mexico, and Peru; the English did so in Virginia and the Carolinas, expanding them across all of North America; the Dutch in Indonesia; and the French in Haiti and Indochina. Though each project had its distinct elements and context, they all drew from patterns first sown by Portuguese mariners on a small island through sheer happenstance.

This raises uncomfortable questions about historical causation. To what extent is history fashioned by notable individuals making deliberate choices, as opposed to being molded by incidents and contingencies that unfold unpredictably? Zarco and Teixeira made pivotal decisions to accept command of the voyage, to persevere through the storm rather than surrender, to return to Portugal rather than settle on the island immediately, to take up the colonization commission, and to embark on sugar cultivation. Each determination carried weight. Different selections would have yielded different results.

Yet, the tempest itself was an accident. The presence of Madeira at that precise location served as a geological fluke, a hand guiding the ship yet bound by the whims of the wind. The fact that the island remained uninhabited presented a historical twist; it may well have been colonized prior by Mediterranean explorers, fundamentally altering the Portuguese encounter with it. If Madeira had already been inhabited, the Portuguese discovery might have led to conquest or alliances, potentially slowing their Atlantic expansion or redirecting it towards other islands or regions. The atmospheric conditions enabling navigation to and from this isle were dictated by physics, not human devising. Similarly, the fertility of its volcanic soil, which propelled sugar cultivation, was a geological stroke of fortune.

History evolves at the crossroads of human agency and forces beyond personal control, where deliberate schemes converge with fortuitous events, where choices align with circumstances. Zarco and Teixeira were neither mere pawns of destiny nor creators of their own fate in absolute terms. They acted within boundaries they had no hand in setting, the storm's fury, the island's layout, and the economic currents of their age, but made authentic decisions concerning how to address these constraints.

Their choices bore both foreseen and unforeseen ramifications. They aimed to establish a thriving colony, which they achieved. They sought riches, and it was theirs. They desired to root their kin in Madeiran society, and so it came to pass. These were the aspirations they hunted down and seized.

Yet, it was not their intent to birth a blueprint for global domination nor to be harbingers of the plantation slavery system that would displace countless Africans across the ocean. It was beyond their design to show the plausibility of ecological metamorphosis on such a scale that might imperil the planet itself. These outcomes arose from their determinations but outstripped their vision and, likely, their awareness.

This predicament remains central when assessing historical luminaries. We can survey the totality of their ramifications in ways they could not. We comprehend that the sugar economy they initiated on Madeira would ripple into the Caribbean and Brazilian slavery systems. We realize that the environmental depredation they sparked would be echoed across tropical colonies worldwide. We discern that the patterns of exploitation and extraction they installed would mold global economic development for generations. These truths eluded them; they were wrestling with tangible quandaries using available knowledge and moral constructs.

Nonetheless, the repercussions are genuine. The anguish wrought by Atlantic slavery holds veracity. The ecological ruin birthed by plantation agriculture is undeniable. The social imbalances bred by colonial economies are irrefutable. These outcomes stemmed from decisions made by figures like Zarco and Teixeira, even as they did not envision or foresee the full aftermath of their deliberations.

THE LEGACY

1600. Funchal, Madeira.

A century and a half after Zarco's death, Madeira had evolved into a thriving colony with little resemblance to the desperate settlement of 1419. Funchal boasted towering stone structures, churches adorned with gilded azulejo tiles, extensive warehouses storing countless wine barrels, an opulent governor's palace, convents, and hospitals. Its population, around thirty thousand, surpassed that of many Portuguese cities. The wealthy merchant families constructed elaborate villas atop the hills surrounding the city, their gardens boasting exotic plants from Africa, Asia, and the Americas.

Despite its grandeur, Madeira's society was rigidly hierarchical and deeply unequal. At the zenith were the old colonist families Zarcos, Teixeiras, Perestrelos, and affluent merchants who had purchased their way into the colonial elite. These influential groups controlled prime landholdings in sugar and wine production as well as European trade connections. They indulged in luxurious lifestyles, importing fine clothing and furniture from Portugal and Italy, sending their sons to prestigious universities such as Coimbra and Salamanca, and securing advantageous marriages with noble families on the mainland.

The next tier consisted of free workers: skilled craftsmen, shopkeepers, sailors, and small-scale farmers. This group comprised descendants of colonists who never accumulated significant landholdings, as well as recent immigrants seeking opportunities. They lived modestly but retained legal freedom to own property and improve their circumstances through talent or fortune.

At the base of this social hierarchy were slaves, predominantly from Africa but also including individuals from the Canary Islands and Muslim territories.

These enslaved people toiled in sugar fields and wine terraces, loaded ships at docks, served in affluent households, and constructed critical infrastructure for the colonial economy. They lacked legal rights, being treated as commodities that could be bought and sold like livestock; they were subjected to conditions designed to maximize their labor while minimizing the cost of maintaining them.

This was the society born from Zarco and Teixeira's decisions, alongside many others whose choices shaped Madeira's development. Although not alone in this endeavor, these two navigators set the initial template for colonization and economic development. Subsequent generations built upon their foundational patterns.

By 1600, Madeira wine had become its economic bedrock. The fortified wine produced on the island's volcanic terrain possessed unique attributes that commanded high demand in European markets. Its ability to withstand long sea voyages, indeed, improved through ocean travel, made it an ideal provision for naval and merchant fleets.

Although less destructive than sugar cultivation, wine production still had considerable environmental repercussions. Terracing steep slopes necessitated the movement of vast quantities of earth and rock, while irrigation systems required engineering feats comparable to Roman aqueducts in the form of levada channels carved into mountainsides that significantly altered the island's hydrological system. Furthermore, the clearing of native vegetation for grape cultivation resulted in a marked reduction of the original forest ecosystem.

By 1600, pre-contact Madeira was scarcely recognizable, with only isolated patches of laurel forests surviving in inaccessible mountain valleys shielded by the difficulty of access. Several indigenous bird species persisted in these sanctuaries. Nonetheless, the landscape had undergone a profound transformation, being reshaped to serve European agricultural interests, specifically the cultivation of crops for export markets.

This same pattern unfolded across the Atlantic world: Brazil's coastal forests were cleared for sugar production using Madeira's pioneering techniques; Caribbean islands were deforested for sugar cultivation; North American forests were cleared for tobacco, cotton, and later for industrial agriculture. The underlying logic was consistent: land in its natural state was considered wasteland, productivity meant cultivating European crops for export, and development demanded environmental alteration, regardless of ecological cost.

The impact on people was far more distressing than any environmental damage. By 1600, the slave trade across the Atlantic was fully operational, with roughly ten thousand Africans being transported each year – numbers that would soar in later decades. Many were sent to Brazil or the Caribbean; however, Madeira played a role both as a stopover for supplies and as a destination where slaves worked on plantations and estates. It is vital to understand that the system's cruelty wasn't an accident but rather its core function. Enslaved individuals could be pushed beyond their limits of endurance and replaced more economically than maintaining free workers, making slave labor extremely profitable. Narratives such as those by Olaudah Equiano detail the Middle Passage horrors, where around 15% perished due to disease, malnutrition, and mistreatment. Survivors like Kofi faced days filled with backbreaking field work under constant threat of violence from overseers, living in overcrowded and unsanitary conditions. Despite these harsh realities, Madeira's colonization facilitated the establishment of plantation systems that relied on slavery, showcasing how tropical crops cultivated by force could yield enormous profits. Zarco and Teixeira, while not directly creating Atlantic slavery – which predated their endeavors – laid the groundwork for future colonizers to replicate the Madeiran model on a larger scale, perpetuating a cycle of exploitation sustained by global demand for plantation products. This widespread knowledge of how the system worked did not lead to change; rather, it persisted and expanded over three centuries after Madeira's colonization. The immense economic benefits for merchants, planters, slave traders, and governments played a significant role in its continuation, while alternatives were deemed less profitable. Despite resistance from enslaved Africans through various means, they lacked the power to stop this system due to the overwhelming force of overseers, colonial militias, naval power, and legal systems that classified them as property rather than individuals. It is essential to note that Zarco and Teixeira did not create Atlantic slavery; however, their successful colonization of Madeira helped establish the plantation model upon which slavery was built. Their sugar economy demonstrated the profitability of tropical crops produced by forced labor, while their administrative structures showed how overseas territories could be governed for maximum wealth extraction. These lessons were later taken up by other colonizers who expanded upon this framework.

THE STORM'S ECHO

1750. The Atlantic Ocean. En route from Lisbon to Brazil.

A Portuguese merchant vessel glided over gentle waves three days offshore from Lisbon, its cargo consisting of manufactured goods intended for Brazil, and hoped to return heavy with sugar, tobacco, and gold. The captain devoted his attention to studying his charts, cross-referencing them against the pilot's estimations of their current location. These mid-eighteenth-century maps clearly outlined Madeira as a mere day's sail southward; they were close enough to discern the island's peaks but not near enough to approach the dangerous near-shore currents.

Madeira was accurately depicted in these charts, with its exact position established with precision over the span of three centuries, thanks to centuries of navigation. No longer an exotic discovery but a well-known waypoint so ingrained in Portuguese maritime knowledge that sailors scarcely gave it a second thought, ships would frequently dock there for supplies such as wine and water.

The charts the captain perused so nonchalantly were built upon a foundation laid down by Zarco and Teixeira's accidental landfall, each subsequent journey adding data such as verifying positions, noting currents and prevailing winds, identifying safe anchorages, and warning of potential hazards. Among the seminal contributions to fixing Madeira on the navigational maps was the work of Pedro Reinel, a pioneering Portuguese cartographer of the late 15th and early 16th centuries. His navigational manuals and charts exemplified the meticulous accumulation of empirical data over decades, which was crucial to accuratcly positioning islands like Madeira. This framework not only facilitated Madeira's exploitation but also cemented its place as a vital waypoint in Portuguese maritime routes.

This knowledge system had been painstakingly constructed over time, albeit at the cost of ships and lives when calculations proved incorrect. This was Madeira's true legacy, not just the island itself but the framework that facilitated its exploitation. Portuguese navigation in the 1700s rested on the understandings developed by Zarco and Teixeira in the 1400s, further refined and improved upon by countless pilots and captains, meticulously documented and preserved by institutions like the Casa da Índia, which oversaw overseas commerce.

The same was true for colonial governance; the captain-donatário system employed on Madeira had been applied throughout the Portuguese empire, adapted to fit local conditions but clearly rooted in Zarco and Teixeira's original model. The plantation economy that dominated Brazil, the slave systems underpinning it, and the social order of colonial society all had origins traceable back to patterns set on the Atlantic islands during the fifteenth century.

By 1750, however, Portugal's empire began its decline. Brazil was growing restless, inching toward eventual independence. African trading posts faced stiff competition from English, French, and Dutch adversaries. Asian territories like Goa found themselves in a vulnerable position amidst more powerful neighbors. The era of Portuguese expansion was drawing to a close, although the empire lingered for two more centuries before dissolving in the 1970s.

Yet Madeira endured as an island that had been stumbled upon by accident, colonized through ad-hoc arrangements, shaped by ecological devastation and slave labor. Despite this checkered past, it remained Portuguese, continuing to produce wine, serve as a crucial stopover for Atlantic shipping, and support a populace descended from the original settlers and their imported slaves. The repercussions of Zarco and Teixeira's decisions continued to be felt, entrenched in both landscape and society, influencing lives that would never know the names of those who had first set foot on its shores.

The tempest that prompted two Portuguese voyagers to seek solace in Madeira during 1418 was a scene of utter turmoil. The elements, unmoved by human intent, abide by the laws of physics and meteorology, spawning circumstances that just happen to converge with a modest vessel crewed by less than a score, embarked on an endeavor that pales in significance when juxtaposed against the grand tapestry of history. Yet from this pandemonium arose patterns that would influence centuries.

This might be the profoundest takeaway from Zarco and Teixeira's odyssey: history is not solely forged through monumental schemes and deliberate tactics; it also unfolds in responses to mishaps, in ad hoc remedies for unforeseen predicaments, and in decisions rendered under duress when all alternatives seem unpalatable. The navigators did not embark with the intent to discover Madeira, nor did they intend to set the precedent for transatlantic colonization or engineer systems that would displace millions across the sea. They were merely contending with a storm and subsequently endeavoring to derive benefit from its bequest.

Fast forward to today, and the echoes of their serendipitous discovery resonate in Madeira's thriving tourism industry, drawing globe-trotters to its metamorphosed landscapes. The island, now characterized by its vine-clad terraces and dramatic coastlines, stands as a testament to historical junctures that reverberate into contemporary times, sculpting both the local economy and cultural identity.

However, their resilience and subsequent choices catalyzed a succession of events neither could have envisaged or dictated. Individuals with divergent agendas amplified and extended the frameworks Zarco and Teixeira had haphazardly laid forth. The Portuguese crown bolstered colonization efforts for reasons aligned with statecraft; merchants capitalized on the lucrative prospects of sugar production; settlers were drawn by available land; and slaves were introduced to meet the labor-intensive demands of sugar cultivation. Each decision, borne out of comprehensible motivations from individuals acting in accord with their interests and values, converged into systems of such magnitude and consequence that they redefined the world's contours.

The navigators themselves slipped swiftly from collective memory. Within a single generation, their discovery became banal; within another century, their names resonated only faintly amongst descendants and historians. They bequeathed no monuments, basked in no enduring fame, nor did they leave an indelible imprint of celebrity. Their legacy was simply that of Madeira's discoverers, a fact chronicled but not dwelled upon, left uncelebrated and unstudied for deeper significance.

Despite this obscurity, their journey through tempestuous seas held profound implications. The island they stumbled upon became a crucible for colonization; the provisional mechanisms they devised served as blueprints emulated across the Atlantic sphere; the governance and developmental choices they made reverberated in continents beyond their sight. Unintentionally, they left a legacy that shapes our present-day world, both deeply and ominously.

The tempest has long since abated. The men who weathered it are now dust of the ages. Yet the patterns they etched, the decisions they made, and the paradigms they pioneered persist in the landscape and society, manifesting in disparities of wealth and power across the Atlantic realm and beyond. The stone markers they would have set as a symbol of Portugal's claim to Madeira are likely weathered or lost. Nevertheless, as chronicled in one early royal decree, "por bem e serviço nosso, vos damos e concedemos... a capitania da ilha da Madeira," the transformations they initiated endure, etched into the very fabric of land and society, shaping the present in ways that often elude our perception, woven into all we've ever known.

Across Madeira's peaks, the Atlantic breeze persists, bearing rain from the sea to irrigate vineyards that have supplanted forests. Vessels still navigate along routes charted centuries prior. The reverberations of that tempest persist, though now faint; they remain audible to those attuned, shaping the world in subtle yet enduring ways, reminding us that happenstance can beget change, that fortuitous encounters can reshape history, and that choices we forge amidst uncontrollable circumstances may have far-reaching implications beyond our imagination. What storms today might be setting similar patterns in motion, calling into question whether we are prepared to navigate the complexities they present?

Gil Eanes

THE EDGE OF THE KNOWN WORLD

May 1434. Somewhere off the coast of West Africa.

Whispers of dread permeated the tales of sailors who braved the tumultuous seas. They spoke of currents that consumed ships without a trace, waters that roared menacingly, and hidden reefs lying in wait to shatter hulls with merciless force. This sense of terror hung heavily in the air as the caravel labored and pitched violently. Each wave hurled emerald spray high into the sky, drenching those on deck. Amidst this chaos, a superstitious whisper emerged, a tale of an ancient curse said to ensnare any vessel daring to breach the Cape's domain, turning the sea into a howling void. Gil Eanes gripped his tiller so fiercely that his knuckles turned ghostly white, unyielding gaze fixed upon where the Atlantic furiously clashed against Cabo Não, the Cape of No. As the ship groaned ominously against a sudden gust, the ominous sound of a hull creak underscored the peril, making even the most seasoned sailors flinch. For fifteen years, it stood as the world's edge, a boundary from which ships limped back to Lagos with their captains haunted by what they'd witnessed and experienced. Prince Henry dispatched expedition after expedition into this perilous realm, all swallowed by failure. The year before, Eanes himself had retreated, steering northward while his crew whispered of the Green Sea of Darkness, an expanse where men were believed to be irreversibly changed, and the ocean roared as if descending into hell's mouth.

Yet this voyage differed from those before. Eanes vowed to honor Prince Henry's call and silence the doubts that once made him turn back. Redemption became his goal. He wanted to prove to his crew, his prince, and himself that fear could be overcome. But within him stirred a deeper need; he longed to resolve an inner verdict that had haunted him: was he truly a coward, as his previous retreat suggested, or could he summon the courage to defy this label and redefine his own identity? This resolve sparked action. It carried him past psychological barriers and became central to the story's theme: personal courage can challenge deep-rooted cultural fears and change history. Eanes knew this journey would define his destiny and could shift collective thinking. As white froth loomed and the wind screamed, Gil Eanes, remembered in noble records as a minor figure, steered boldly into the unknown.

What unfolded next would send ripples far and wide, reshaping the world.

A MAN FROM THE SHADOWS

Who was Gil Eanes? Historians have puzzled over this question for six centuries. Though Portuguese records from the fifteenth and sixteenth centuries mention him fleetingly, we know he hailed from Lagos, a coastal city on the Algarve, where Henry had his base. Likely tied to Henry's household as a squire or minor official, Eanes commanded respect and a ship, yet was not noteworthy enough for chroniclers to document his life in any detail. He entered history as an adult, fully formed and capable, as if rising from obscurity solely to achieve one great deed before fading into anonymity.

The chronicler Gomes Eanes de Zurara, writing about two decades later, offered the most detailed account available. Zurara had access to Prince Henry's archives and may have interviewed those familiar with Eanes. Still, he revealed almost nothing of Eanes's attributes or background. We glean only that Henry "encouraged" him, that Eanes was loyal, and that his 1433 failure left him feeling shamed. Recent scholarship emphasizes the propagandistic nature of Zurara's work. Scholars such as Peter Russell and Bailey W. Diffie argue that Zurara's Chronicle aimed to glorify Prince Henry and Portugal's African campaign, framing their efforts as a Christian mission (Russell, 2000; Diffie & Winius, 1977). Modern historians caution against blindly accepting Zurara's narrative. Instead, they advocate a more balanced approach considering other contemporary accounts, documentary evidence, and archaeological findings. This comparative method, favored by current scholarship, seeks to address distortions or omissions and create a more nuanced portrait of Eanes and his era.

This scarcity of detail is both vexing and fittingly mysterious. Gil Eanes became the prototype for a new kind of hero: the unassuming yet accomplished subordinate whose deeds altered history even as his own story receded into obscurity. He was no grand noble or learned scholar, but a man intimately versed in the sea's fury and its rhythms, a seasoned leader tested by uncertainty. History records his triumph while keeping him shrouded in shadow.

Yet fragments of information can be pieced together. To earn command under Prince Henry, Eanes had to have been forged in the crucible of years at sea. Perhaps he mastered sailing among the Algarve's fishing boats as a youth or traded along routes to Andalusia and Morocco. Confronting the challenge of Bojador was both an obligation and the culmination of skills honed through unyielding perseverance, driven by his own burning desire to confront the unknown. The trial reflected not only his determination to conquer fear but also his yearning for recognition. Eanes had learned to read the Atlantic's moods, decipher its winds and currents, and navigate by sun and star. Such mastery does not emerge overnight; it is sculpted by salt, sweat, and time. By the time Henry entrusted him with Bojador's command, Eanes was likely in his late thirties or early forties, seasoned enough to command respect yet robust enough to face the unknown with unflagging resolve.

He would have carried the old tales as well, with which every sailor was familiar. The Cape of No. The Green Sea of Darkness. Currents that ensnared ships until they vanished forever into the deep. Eanes had heard these legends whispered in smoky taverns and shouted above the tumultuous roar of the sea, shaping his imagination as much as his seafaring skills. Similarly, West African seafarers spun their own myths about the ocean's mysteries. Mandinka mariners spoke of waters inhabited by spirits who guided or misled, while Wolof beliefs recounted powerful currents that could change the fate of a voyage. These shared legends wove a rich tapestry of fear and wonder, affecting sailors from both sides of the Atlantic. When Eanes turned back from Bojador in 1433, it was not out of cowardice but rather a response to fears deeply embedded in every Portuguese sailor's psyche.

THE FIRST ATTEMPT: 1433

The specifics of Gil Eanes's initial expedition are somewhat murky. Zurara only alludes to it in order to contrast it with the success that followed. What we do know is that King Henry equipped a vessel and appointed Eanes as its commander, instructing him to sail south beyond the Canary Islands. We can envision the preparation: loading the ship with essential supplies like water, wine, salt meat, and biscuits; conducting routine maintenance on the rigging and sails; selecting a suitable crew. At the time, Lagos was a bustling port, though not one known for its beauty. The scent of fish and tar hung heavily in the air, intertwined with the unmistakable smell of the sea. Shouts from gulls filled the sky above as merchants loudly haggled over their goods; this cacophony clashed against the rhythmic groans of wooden masts flexing with the waves. Various ships were pulled up onto the shore, their hulls being scrubbed and sealed for another voyage. Sailors crowded into the narrow streets, sharing stories of far-off lands while the noise of dice rolling and coins clinking provided a constant backdrop to their discussions about potential passage on outgoing vessels.

Eanes's ship, likely a barinel, a single-masted vessel of twenty-five to thirty tons or possibly an early caravel, was small in size, fifty to seventy feet long, and carried a crew ranging from twelve to twenty men. Both types were equipped with a single mast and a large square sail; more advanced models featured lateen sails for improved performance against the wind. The ships had no real decks; rather, planks were laid across beams where officers could stand or find refuge. Below deck lay the hold, loaded with supplies. During storms, the crew sought shelter in this cramped space, inhaling air thick with the smells of bilge water and mildew. These vessels offered little in terms of comfort, were designed for coastal exploration, and were built to be cheap and disposable.

The duration of the voyage south varied depending on wind conditions; it could last anywhere from a week to several more. As Eanes approached the approximate location of Bojador, tension among the crew increased. Conversations turned to potential dangers ahead. Men scanned the horizon for signs of shoals or changes in sea color that might indicate the start of the cursed waters.

At some point, an important decision was made. It may have been due to fear within the crew or his own apprehension, or possibly a legitimate concern, such as a shift in wind direction or a perilous reef. Regardless, Eanes chose to turn back. He did not attempt to circumnavigate the cape; instead, he returned to Lagos without fulfilling King Henry's orders.

Zurara records that Prince Henry was displeased but still opted to support Eanes rather than abandon him. The prince spoke with him at length, attempting to bolster his confidence by downplaying concerns about Bojador, suggesting no ship had truly been lost there (as none had ventured with sufficient determination), and promising glory and reward for the first person to succeed. In essence, Henry sought to rebuild Eanes's self-assurance.

This approach proved successful; in 1434, Gil Eanes agreed to embark on another attempt.

THE TURNING POINT: STRATEGY AND RESOLVE

The second voyage differed from the initial one in terms of strategy; whether this idea came from Henry or Eanes remains unclear. The key to rounding Bojador was understood to be avoiding coastal currents rather than battling them. In place of hugging the shore, where dangers and psychological pressure were heightened, Eanes decided to sail far out into the Atlantic. He would travel past the cape's latitude, then turn eastwards to make landfall beyond the dreaded point. This plan, while simple in concept, was radical in execution, requiring navigation across the open seas, out of sight from land, for several days. Such an endeavor unsettled the crew at a time when navigation relied more on intuition than on scientific precision. However, this strategy neutralized threats posed by shoals and currents and lifted the psychological burden of seeing the cape, which sailors deemed ominous because of its white foam. In retrospect, Eanes's offshore approach can be seen as pioneering the 'volta do mar', a navigational method that would later transform maritime exploration. This tactic of venturing into open waters rather than adhering strictly to coastlines prefigured the techniques used by Vasco da Gama nearly a century later when he forged a sea route to India around the African continent. Thus, Eanes's strategy not only enabled immediate success in rounding Bojador but also laid foundational principles for future expeditions into the Indian Ocean, illustrating a crucial link in the evolution of long-term maritime strategy.

Selecting a crew for the 1434 voyage likely involved meticulous choice. Eanes required not only skilled sailors but also men who could remain calm when the coast was out of sight. Yet, the extent of his selection ability is unclear; Lagos offered few volunteers for expeditions beyond Bojador. He probably assembled whomever was available and then relied on his leadership to maintain cohesion.

The voyage commenced from Lagos in spring 1434, though the exact date remains undocumented. Nonetheless, spring was deemed optimal for southward voyages due to the prevailing northerlies, which facilitated travel along the African coast. Eanes navigated south initially, keeping land in sight through waters familiarized over a decade. As he approached Bojador's latitude, he veered westward, distancing from land. The coastline receded into the distance as the ship embarked upon the vastness of the open Atlantic.

This transition must have profoundly unsettled the crew; the ocean stretched endlessly in all directions, desolate and expansive. Sky and sea melded at a distant horizon, with wind and waves being the only constants. Eanes depended on stars to maintain orientation by night, tracking the North Star, while the sun served as his guide by day. Endurance and resisting the urge to return were put to the test.

The days merged into one another as the ship sailed west before veering southward. Eanes awaited the moment he was sure they had surpassed Bojador's invisible boundary, then turned eastwards. Slowly, land emerged from the mist, unfamiliar and stark, devoid of any Lagos memory. This was beyond the Cape of No; it marked the threshold of the unknown.

LANDFALL BEYOND THE VEIL

Zurara's account is brief: Gil Eanes, having rounded the cape, landed on untrodden shores, finding no dwellers but remnants of their sojourn. Yet beneath these terse words, stir the undercurrents of emotion. Imagine Eanes stepping onto uncharted ground, his boots imprinting sands never before trodden by Europeans, real sand proving fear was only rumor. Did they waver with history's weight or stand firm in resolve? The desert wind carried arid scents; the sun burned high above. But the world endured. The sea was tranquil, his crew unchanged. Beyond Bojador lay more lands, barren, austere, yet palpably awaiting discovery.

Traces of human passage footprints, remnants of old hearths, suggested that Berber nomads traversed this desolation. Eanes gathered botanical specimens to demonstrate the possibility of habitation and the potential for future voyages to find sustenance or engage in commerce. Yet, these verdant sprouts symbolized more; they confuted ancient fears and cartography's limits. They were testament to the fact that the torrid zone wasn't an execrated void but a frontier ripe for exploration.

Eanes did not linger post-achievement. His mission was accomplished, rounding the cape and surviving to recount it. To venture further south with one vessel and an apprehensive crew would have been foolhardy. He retraced his course north past Bojador, homeward bound for Lagos. There, he returned not only botanical treasures but a tale that dispelled enchantment. Prince Henry welcomed him jubilantly and rewarded his endeavor. The insidious barrier had fallen; the path south was beckoning.

THE RIPPLE BECOMES A WAVE

Eanes's triumph didn't spur an immediate surge of expeditions. Henry, methodical in strategy, deemed equipping ships too costly for such a rush. Yet, a psychological bulwark had crumbled. In 1435, following Eanes's success, Henry dispatched him south with another captain, Afonso Gonçalves Baldaia. Together, they ventured past Bojador, reaching coastal stretches marked by human footprints and camel tracks, evidence of habitation, though no inhabitants were seen. The coast was not lifeless; it was distant, frequented by nomads whose lifeways differed from European norms.

In 1436, Baldaia journeyed south again, identifying Angra dos Ruivos ,the Bay of Mullet, for its abundant sea life. These ventures became routine. Once-reluctant captains now charted courses beyond Bojador, scrutinizing coastlines, mapping outcrops, hunting for signs of kingdoms or riches. Exploration's tempo accelerated in the 1440s. In 1441, Portuguese seafarers Antão Gonçalves and Nuno Tristão captured Africans along the shore, the first enslaved people brought from Atlantic shores to Portugal. This ominous pivot marked the slave trade's inception but assured the Portuguese crown that African ventures could yield profits where glory and curiosity had lagged.

By 1444, Portuguese mariners had navigated to the mouth of the Senegal River, far beyond the Sahara into West Africa's realm of kingdoms and sizable populations. Trades in gold dust, ivory, and enslaved people commenced earnestly. Henry's dream of outpacing trans-Saharan commerce was becoming a reality. And it all commenced from that pivotal 1434 spring when Gil Eanes propelled his vessel past a waterline holding back Christendom for over a generation.

THE MAN WHO FADED

Post-1435, Gil Eanes fades from historical annals unmentioned by Zurara for further endeavors. Other captains took up the mantle of exploration, Baldaia, Tristão, Dinis Dias, Álvaro Fernandes, whose names would be affixed to capes, bays, and islands as Portugal ventured ever southward. Eanes's deeds opened their pathways, but he receded into oblivion. We do not ascertain his demise or whether he persisted in sailing, settled comfortably in Lagos, or garnered land or title. The records are mute.

This silence embodies a soft tragedy. Eanes did not marshal armadas or chart routes to India; neither did he seize territories nor accumulate wealth. His feat was singular; he breached a mental barrier stronger than any tempest or expanse. He unveiled a broader world than anyone envisaged. Yet, the route he pioneered would lead to a tangled legacy intertwining future pursuits with humanity's darker urges for dominance. History often glorifies finishers over those who first dared commence.

In Lagos, during the late fifteenth and early sixteenth centuries, as it swelled with riches from African commerce, as ships embarked towards evermore distant shores, did Gil Eanes linger in remembrance? Did aged men in taverns narrate his Bojador return? Did the youthful cohort setting sail for India or Brazil even know his name? The chronicles intimate not. According to Zurara's chronicle from the 1450s, Eanes was already a fading figure in living memory, someone who required explanation rather than an actively lauded individual.

THE LONGER ARC: WHAT EANES MADE POSSIBLE

To fully appreciate Eanes's journey, you must trace a line from 1434 forward into the next century. In 1488, Bartolomeu Dias proved that a sea route to the Indian Ocean existed by rounding Africa's southern tip. This paved the way for Vasco da Gama to reach India in 1498, opening direct maritime trade between Europe and Asia while shattering Venetian and Muslim monopolies on spices. In 1500, Pedro Álvares Cabral claimed Brazil for Portugal. By the mid-sixteenth century, Portuguese ships sailed from Lisbon to Nagasaki, establishing a global trading network. This influx of wealth transformed Portugal into a major European power, funded magnificent architecture, and backed voyages of discovery around the globe.

None of this could have happened without Eanes's unlocking of the African coast. For generations, Bojador served as a barrier steeped in fear. His voyage was the key that fit perfectly. It marked the moment when direct experience surpassed old fears, when curiosity and ambition outstripped caution. This wasn't the first or last act of maritime bravery; however, it happened at an opportune time for Europe to gain confidence. Eanes delivered that crucial spark.

It's also important to reflect on the devastating human toll. The Portuguese voyages down the African coast were not purely altruistic explorations. They were commercial enterprises, with one of their most profitable goods being enslaved human beings. This trade, beginning in the 1440s with a few captives taken from

Mauritanian shores, grew into an expansive system over the next four centuries, resulting in millions of Africans crossing the Atlantic. The anguish caused by this system is immeasurable. What price did humanity pay for opening that door? When we honor exploration, we must also confront and acknowledge the exploitation that was intricately linked to it.

However, it's vital to recognize that these enterprises were perceived differently by various parties. While Portuguese chroniclers often portrayed exploration as a civilizational or religious mission, many Africans saw these incursions as violent invasions that disrupted their societies and economies and severed kinship networks. Oral traditions recall the arrival of Portuguese ships with dread, detailing losses from raids and the destabilization of local governance and security. Some African leaders initially sought to negotiate or resist, using diplomatic or military tactics, but were often met with overwhelming violence or deceit. The trauma of enslavement reverberated through generations, influencing communal memory and identity long after initial contact. Some European contemporaries viewed these journeys as an extension of Christian duty, while later critics, including African voices in the historical record, emphasized devastation caused by slavery and colonization.

Gil Eanes himself did not enslave anyone as far as records show. Yet his voyage set off a chain of events that led to a system with immense and destructive consequences. It's not fair to blame him personally; he lived in a society with different moral values and could not have foreseen the scale of what his journey would unleash. Nonetheless, it's essential to acknowledge that the interpretation and impacts of exploration vary greatly depending on perspective, often extending far beyond the explorer's intentions. This invites us to weigh achievements against the diverse and sometimes contested moral costs.

ZURARA'S CHRONICLE AND THE CONSTRUCTION OF MEMORY

Our primary source for information on Gil Eanes's voyages is Gomes Eanes de Zurara's Chronicle of the Discovery and Conquest of Guinea, a text written between 1453 and 1460 by an appointed chronicler named Zurara who documented Portuguese African expeditions with the intention of promoting Prince Henry and justifying Portugal's expansion as a Christian crusade against Islam and paganism. It is essential to approach this account critically because it appears that Zurara exaggerates Portuguese virtue, downplays or ignores any instances of Portuguese cruelty, and intentionally minimizes other individuals' contributions while maximizing Henry's role.

Despite the propagandistic nature of Zurara's work, his chronicle remains invaluable as he had access to records that have since perished. He also conducted interviews with participants in early expeditions or spoke with people who were acquainted with them. His descriptions of the African coast and encounters with indigenous peoples carry significant weight as firsthand testimony, though filtered through his moralizing perspective.

When it comes to Gil Eanes's achievements, Zurara provides detailed accounts that indicate he is drawing from reliable sources. He mentions the offshore strategy employed by Eanes as well as the plants brought back from expeditions. Furthermore, Zurara conveys a sense of relief and triumph upon Eanes's return.

However, Zurara also reveals the limitations of historical memory, particularly in how quickly events can fade into the background. Just twenty years after the voyages, he had to clarify who Gil Eanes was and remind readers that Bojador had once been considered an insurmountable obstacle. As time passed, the drama surrounding 1434 seemed increasingly quaint, with later chroniclers like João de Barros largely ignoring Eanes's accomplishments in favor of grander achievements in India and Brazil.

This shift in focus by later chroniclers exemplifies how rapidly a significant moment can be forgotten as new, more impactful events come to light. Each generation seeks meaning in the past, relegating previously vital information to mere footnotes. While Zurara's chronicle keeps Gil Eanes's name relevant, it also highlights his rapid fade from prominence and near oblivion over time.

THE WORLD THAT WAS OPENING

To stand on the coast of West Africa in 1434 and gaze towards Europe signifies a civilization that regarded itself as central, yet was minor and defensive. Christian Europe in the early fifteenth century grappled with the aftermath of the Black Death, which had decimated nearly a third of its population in the mid-fourteenth century. It was fragmented into rival kingdoms and principalities and threatened by the Ottoman Empire's expansion into the Balkans, culminating in the 1453 conquest of Constantinople. The trade with Asia was managed by Muslim merchants and Italian city-states, leading to high prices for Asian commodities like spices and silks and limited direct access.

Viewing southwards towards Africa revealed a world vastly unknown to Europeans. They had only rumors of kingdoms such as Mali and Songhai, and of cities including Timbuktu, where gold supposedly thrived. They recognized the trans-Saharan trade routes linking the Mediterranean and sub-Saharan regions but remained unaware of the intricacies of African societies, the sophistication of their political structures, and their cultural richness. Africa was perceived as an uncharted territory, valued only for its exotic products, devoid of its own historical significance.

Eanes's voyage gradually dismantled this European ignorance, not always for the better. The Portuguese arrived not merely as neutral explorers but as traders, conquerors, and enslavers, imposing their ambitions upon those encountered. They sought wealth, disregarding the respect due to other societies. Yet, they also transmitted knowledge that challenged long-held myths of European supremacy. Discovering organized kingdoms, skilled artisans, and astute negotiators, they learned that Africans were not mere savages but a diverse array of complex individuals.

Slowly and unevenly, this newfound information permeated Europe, largely ignored by the public but significant to sailors, merchants, cartographers, and scholars. Africa was transforming from a blank space on a map into a land with tangible geography and history. This shift, an unintended legacy of Eanes, compelled Europe to confront its theories with reality, leading to the realization that the torrid zone could be inhabited, that the sea wasn't boiling, and that the world didn't end. Observation replaced old beliefs, paving the way for the Scientific Revolution. The chain of causality is intricate, but it commences with one man circumnavigating a forbidden cape, proving the world was far from imagined.

THE QUESTION OF COURAGE

The audacity of Gil Eanes: a seemingly straightforward inquiry, yet one fraught with nuance. Bravery encapsulates the act of proceeding despite trepidation, an aspect we are privy to, given that Eanes initially retreated. Nevertheless, bravery also signifies a deliberate decision, and whether this choice was entirely Eanes's to make is open to interpretation. Subordinate as he was to Prince Henry, declining a second expedition could have spelled career ruin, financial loss, and social disgrace. Additionally, other influences may have been at play; the lure of potential economic benefits, such as land allocations, pensions, or other incentives, might have weighed on his mind, further complicating his choices. At that critical moment, standing on the deck with the salt air biting at his face and the creak of ropes echoing this inner tension, Eanes's mind whirled with conflicting visions: a future marred by the whispers of cowardice if he turned back, or the silent, vast horizon that promised unknown dangers. Yet, it was within his power to undermine the mission, to retreat once again, and lay blame at the feet of the weather, crew, or vessel. Instead, he chose to persevere, pushing forward even when instinct likely urged retreat. This decision, taken amidst isolation and fear on the open sea, became a pivotal moment in history.

This invites contemplation regarding exploration itself. Commonly, we perceive explorers as heroes propelled by curiosity and valor. Indeed, some were, yet numerous others were driven by necessity, ambition, self-interest, or mere obligation. They may not have been paragons of virtue, but their deeds held significance. Eanes's voyage past Bojador could well have stemmed from a desire to avoid disappointing his prince. This does not diminish his accomplishment; rather, it underscores the realization that history is sculpted not merely by visionaries but also by individuals of ordinary stature yet extraordinary in action who rise to the occasion when required.

THE WORLD AFTER

Gil Eanes's voyage might have been the turning point, but what followed was explosive growth. Post-1434, Portuguese exploration along Africa's coast escalated rapidly. Prince Henry, now convinced of Bojador's fables, moved forward with vigor. He had resources, ships, and proof that conquering the Atlantic coast was plausible. Now it was about results: gold, trade, conversions, and vindication for his costly dream.

Expeditions from the late 1430s to early 1440s pressed southward relentlessly. Alfonso Gonçalves Baldaia, who had accompanied Eanes in 1435, ventured south again in 1436. He reached Pedra da Galé, a spot notorious for its barren landscape that showed no signs of the cities or gold they sought. That evening, as the sun disappeared below the horizon, the crew congregated around a dull cookfire. The meager flames barely pierced through the encroaching darkness, mirroring their dwindling spirits. They whispered among themselves, faces by the fire sullen and fatigued, accepting the grim reality: they had braved life-threatening odds to pass beyond Bojador's psychological barrier only to find an unrelenting desert stretching endlessly.

But Henry was undeterred. In 1441, he dispatched Antão Gonçalves, a young man without prior command experience from his household, to collect seal skins and oil along the African coast. It began as a modest commercial endeavor, barely distinguishable from routine trading voyages. Yet, Gonçalves harbored greater ambitions. Upon reaching the coast, in what would later be known as Rio do Ouro, the River of Gold, though devoid of rivers and precious metals, he organized a raiding party. Under the cover of darkness, he and nine men infiltrated the interior. They ambushed a small nomadic encampment and captured a man and a woman amidst the chaos. The man was wounded in the scuffle. The Portuguese hauled their captives aboard and sailed homeward.

This marked the first recorded instance of Portuguese slavers seizing captives from Africa's Atlantic coast. Although not the inception of slavery itself or Portugal's involvement in it, this event signified a new phase: a systematic, maritime-based traffic in African individuals that would proliferate exponentially over ensuing decades and centuries. Zurara recounted the raid matter-of-factly in his chronicle, offering a practical solution for reconnaissance: the captives could be interrogated for information about the interior and to facilitate contact with African kingdoms. Their humanity seemed inconsequential to him, or was silenced.

Gonçalves's return to Portugal with his captives pleased Prince Henry, who interrogated them. One captive held a significant status among his people and proposed an exchange: arrange for gold dust and the release of additional captives in return. Henry agreed. The following year saw a small quantity of gold dust and ten more enslaved individuals acquired through this exchange, many likely prisoners from inter-tribal warfare.

This transaction set a precedent. Both Portuguese and African parties found incentives to partake in this trade. For the Portuguese, it offered profit and intelligence about the African interior; for certain African political entities, access to European technology and goods, as well as an outlet for war prisoners or enemies. The morality of this arrangement went unchallenged by most participants. Slavery was a norm in the 15th-century world; what was novel was the scale and organization enabled by European maritime technology.

NUNO TRISTÃO AND THE PUSH SOUTH

While Gonçalves was actively conducting raids and negotiating, another captain ventured further south to expand Portuguese understanding. Nuno Tristão, a member of Prince Henry's household and a knight, undertook several voyages during the early 1440s. Each journey extended Portugal's reach along the African coast. In either 1441 or 1442, he arrived at Cabo Branco, the White Cape, a significant headland that marked the end of the lengthy Saharan coastline. At this point, the landscape began to transform. Vegetation began to appear, and the water grew greener, richer with life. The coastline curved eastward, suggesting that Africa might not stretch endlessly south but could have limits or a turning point.

Tristão continued his journey south in 1443 and 1444, reaching the islands at the mouth of the Senegal River before proceeding upriver. The Senegal represented a major milestone. Beyond it lay the region known as Guinea, characterized by dense forests, vast rivers, and sizable populations. It was here that the Portuguese encountered sophisticated, organized kingdoms, professional armies, and extensive trade networks far more advanced than anything found in the desert. These were not merely nomadic herders or small raiding parties but societies with established hierarchies, specialists, and political systems capable of negotiating, resisting, or collaborating as circumstances dictated.

These encounters frequently turned violent. In 1444, an expedition led by Lançarote de Freitas attacked the coast near Senegal, returning to Portugal with over two hundred captives. Zurara vividly describes the scene upon their arrival in Lagos and the subsequent division among investors who funded the expedition. Families were torn apart; children were separated from their mothers. People unfamiliar with Europeans, violently abducted from their homes, stood in the Portuguese sun, weeping and disoriented, as merchants and nobles negotiated over them like property. Zurara, writing for a Portuguese audience, emphasizes Prince Henry's piety, taking a share of the captives not for profit but to ensure they would be baptized and converted to Christianity, portraying this action as an act of mercy. For a captured woman, taken from everything she knew, the sea became a vast, indifferent expanse that mirrored her despair, her thoughts filled with the faces of those lost to her forever. The captives themselves left no written records of their experiences.

Nuno Tristão embarked on his final voyage in 1446. He surpassed all previous Portuguese explorers by sailing further south than anyone before him, passing the mouth of the Gambia River and entering a region characterized by dense tropical forests where the coastline transformed into a complex web of rivers and inlets. At some point, though its exact location remains uncertain, Tristão's expedition was ambushed by local inhabitants using poisoned arrows. He was fatally struck, with most of his crew perishing as well. The few survivors managed to navigate their ships back towards Portugal but lost Tristão during the voyage. His body was consigned to the sea. He was roughly 30 years old at the time of his death, having devoted his adult life to expanding knowledge of Portuguese, a pursuit that ultimately led to his demise.

THE HUMAN COST OF EXPLORATION

Nuno Tristão was not alone in his fate. The 15th-century explorer faced perilous conditions. His ships were small, fragile, and susceptible to the fury of storms. Safe havens along the African coast were scarce. Lethal diseases such as malaria, dysentery, and yellow fever claimed the lives of Europeans with no resistance to these tropical afflictions. Shipwrecks were commonplace. Clashes with coastal inhabitants were a regular occurrence. Out of the numerous men who embarked on Portuguese expeditions following Eanes's groundbreaking journey, many did not return. Some fell in battle, others succumbed to disease, some drowned, and an unfortunate few disappeared without a trace, their vessels lost in tempests or dashed upon unknown shores.

The accounts from this era rarely shed light on these deaths. Zurara briefly notes the losses after each voyage, a grim count of lives forfeited. Later records might reference a ship's demise or a captain's death, but the ordinary seamen, the ones who labored tirelessly at ropes and oars, who perished from thirst or fever, remain unnamed. Their grieving families remained unseen in the pages of history. This is one of exploration's unspoken tragedies: countless lives risked and lost for the ambitions of the select few.

And what of the African victims? Those taken by force, those sold into bondage, the communities fractured by the Portuguese incursion, their suffering is woefully underdocumented. Zurara and other European chroniclers tally up numbers of captives seized and treasure amassed. They occasionally report resistance, usually to highlight the valor of the Portuguese in overcoming it. Yet rare glimpses of African perspectives emerge in oral traditions from the Senegambian region, where descendants recount the terror wrought by Portuguese raiders. For example, Mandinka folklore tells of children being hidden and of villages scattering at the sight of foreign vessels. It recounts how "people were taken like cattle, never to return," and remembers the elders mourning the splintering of families and the ruination of livelihoods. Though these narratives are piecemeal and passed down outside the European annals, they offer testament to the scale of trauma inflicted on African communities. Still, the majority of contemporary records stay silent about such experiences. We are aware that it transpired, and through these oral testimonies, we can surmise the anguish, but the written voices of those directly impacted are mostly absent from the historical record.

This uneven tapestry of history reflects the era's stark power imbalance. The Portuguese held sway over ships, quills, and parchment, their stories destined to be told. West African cultures, rich in oral heritage, were largely absent from these recorded narratives. Hence, history favors the Portuguese viewpoint; we know their actions and intentions, but can only speculate on the African reality. Rectifying this imbalance demands commitment, creativity, and a readiness to listen for what remains unsaid.

THE GENERATION AFTER EANES

As the 1440s gave way to the 1450s, a new cohort of Portuguese mariners emerged, embracing the ongoing endeavor. Men such as Dinis Dias, Álvaro Fernandes, and Diogo Gomes ventured further south than what Tristão had managed previously, arriving at the shores of present-day Guinea-Bissau, Guinea, and Sierra Leone. They meticulously charted the intricate coastline, the river deltas, and the offshore isles. These pioneers established the initial fortified Portuguese trading stations, secure bases where ships could replenish their supplies and engage with African merchants in relative tranquility. Consequently, the maps of Africa began to fill in, erasing the voids by assigning them names and distinctive characteristics.

In 1456, the Venetian seafarer Alvise Cadamosto embarked on an expedition under Portugal's aegis, reaching the unoccupied archipelago known as the Cape Verde Islands. This collection of islands, situated several hundred miles off the West African shore, was swiftly colonized. They were cultivated with sugarcane and grapevines, metamorphosing them into pivotal stopovers for vessels en route to even more southern locales. The template of island colonization, characterized by capturing lands that were either uninhabited or sparsely populated, introducing

European agriculture and livestock, and importing enslaved labor to tend these plantations, would serve as a blueprint for future endeavors across the Atlantic. Regions such as Brazil, the Caribbean, and parts of North America would emulate this model initially pioneered in Madeira and the Cape Verde Islands during the mid-15th century.

Prince Henry passed away in 1460 at the age of 66. He had devoted four decades to orchestrating and guiding Portugal's expansion into the Atlantic sphere. He never ventured beyond Morocco nor beheld the shores explored by his captains. However, his role was instrumental; he facilitated a steady flow of funding and strategic insight amid uncertain near-term profits and skepticism from Portugal's ruling class. Without Henry's unwavering dedication, these expeditions might have fizzled out after a few abortive attempts. With him, they became an enduring venture that revolutionized Portugal's standing on the world stage.

THE PARADOX OF THE MINOR FIGURE

There exists a paradox at the core of Gil Eanes's tale. He achieved something transformative, significantly impacting the world. He overcame both intellectual and physical obstacles that had long hindered Europe. His actions unlocked Africa for exploration, facilitated access to India, and set the stage for global cultural exchange leading to modernity. Despite his pivotal role, Eanes himself fades into obscurity; his life is largely undocumented, his character a mystery. How can someone be so crucial yet remain largely unseen? Does obscurity safeguard or squander legacy?

The answer partly lies in the complexity of historical causality. Major events are often the result of multiple factors coming together: individual actions, technological advancements, economic motivations, political structures, social attitudes, and fortune. Eanes's voyage was indispensable for what followed, but it alone did not guarantee success. It created a possibility, but others had to seize that opportunity. It demonstrated the feasibility of something, yet others had to invest the labor to make it standard practice. In this sense, Eanes acted as an enabler or catalyst rather than the sole driving force. His contribution was genuine yet specific: he shattered the psychological barrier. The colonization, trade, and empire that followed were the work of thousands more.

Another aspect is how history is chronicled and remembered. History tends to highlight climactic moments and notable individuals. We remember Columbus due to his voyage to the Americas, overlooking earlier medieval Norsemen who reached North America but did not persist. We recall Vasco da Gama for reaching India, while ignoring those who charted the African coast that facilitated his journey. History remembers emperors, conquerors, revolutionaries, those whose immediate, visible, and dramatic actions altered the course of events. The unsung heroes, the technicians, problem solvers, and support staff who enabled these grand achievements often go unnoticed. Eanes falls into this category of overlooked enablers. His achievement was subtle; it changed what people believed was achievable. Modifying beliefs is foundational work but lacks glamour and does not create the engaging narratives that history preserves.

Yet it might be appropriate that Eanes is mostly forgotten. He did not pursue fame or eternal recognition. He was a pragmatic individual who faced tough circumstances and proved what could be accomplished. His accomplishment altered the world, yet this was nearly secondary to his personal ambitions. He aimed to keep his ship afloat, ensure his crew's safety, maintain a steady course, and return home. These are not the aspirations of visionaries but those of an artisan. Perhaps that is Eanes's true legacy: history is influenced by both dreamers and those who quietly execute what needs to be done.

THE HISTORIOGRAPHY OF EXPLORATION

How have historians portrayed Gil Eanes and the Portuguese voyages of exploration? How have they evolved over time? The answer to this question reflects broader shifts in historical methods and cultural values. In the fifteenth and sixteenth centuries, when Zurara and others documented these expeditions, the prevailing narrative was one celebrating Christian triumph. The Portuguese were seen as bearers of Christianity's light into pagan lands, expanding Christendom while doing God's work. However, they often justified or downplayed the violence, enslavement, and exploitation that occurred during these voyages as necessary steps toward a divine goal.

Fast forward to the nineteenth and early twentieth centuries, the era of European imperialism, and historians now portrayed Portuguese explorers as pioneers who boldly ventured into the unknown. They were seen as trailblazers extending knowledge, connecting the world, and laying the groundwork for global trade with European dominance at its core. This narrative glossed over the negative consequences of exploration, dismissing them as unfortunate but necessary byproducts on the path to progress.

In recent times, especially in the late twentieth and twenty-first centuries, scholars have come under fire for their complicity in the colonial enterprise. Historians now scrutinize the Age of Exploration's inherent violence and exploitation more critically. They endeavor to unearth African and indigenous viewpoints, amplifying voices that were historically marginalized or erased. These historians interrogate narratives of progress and civilization, revealing them as veiled justifications for conquest. While this shift has greatly advanced our comprehension, it also poses new questions: How can we fairly represent figures like Gil Eanes? How do we honor their accomplishments while acknowledging the systems they inadvertently facilitated?

There's no straightforward solution to this dilemma. We recognize that Eanes himself did not engage in enslavement (as far as records indicate), yet his voyage laid the groundwork for the slave trade. We can commend his skill and bravery while simultaneously castigating how others wielded these traits. By presenting his story as a human one, a tale of facing fear, making tough decisions, and succeeding, we also place it within the broader context of exploitation and empire. This approach requires embracing complexity and resisting the urge to reduce history into simplistic tales of heroes and villains. People and historical events are far more intricate than such binary portrayals allow.

THE CARAVEL AND THE CONQUEST OF SPACE

If Gil Eanes was the human link in exploration, the caravel was the technological one. This vessel carried Portuguese explorers along the African coast and eventually around the world, a blend of centuries-old maritime knowledge refined through trial and error, optimized for specific tasks. Grasping the essence of the caravel helps us understand how exploration became feasible in the first place.

The Mediterranean galley, which served as the backbone of medieval European maritime trade and warfare, was ill-suited for Atlantic exploration. Galleys were long, narrow, oar-powered ships created for the relatively calm waters of the Mediterranean. They couldn't handle the large waves of the Atlantic. They required large crews of rowers, necessitating the transport of large amounts of food and water, which limited their time at sea. While they excelled in battle due to speed and maneuverability, they were not efficient for long-distance voyaging.

Atlantic shipping customs, developed by fishermen and coastal traders along the Portuguese, Spanish, and French Atlantic shores, prioritized wind-driven sailing ships with smaller crews that could carry cargo efficiently. However, traditional Atlantic sailing ships, equipped with a single square sail, struggled to sail against the wind. They could only navigate effectively when the wind was slightly behind them. This posed a significant limitation on exploration, as they couldn't return from their expeditions.

The solution was the lateen sail, a triangular canvas hung from a long yard on a short mast. The lateen rig, borrowed from Arabic and Indian Ocean maritime traditions, allowed ships to navigate closer to the wind, generally within 60 degrees of the wind direction, compared to about 80 degrees for square sails. This meant that lateen-rigged vessels could tack back and forth, zigzagging into the wind, even when the conditions were unfavorable. Coupled with a light, shallow-drafted hull that could operate close to shorelines and a relatively small crew, reducing supply needs, the caravel became the perfect exploration vessel. These features not only improved navigation but also aligned with royal economic strategies. The efficient design enabled longer voyages at lower cost, meeting the crown's demand for profitable exploration while minimizing financial strain, thereby marrying technological innovation with economic imperatives crucial to sustained maritime endeavors.

The earliest caravels likely emerged in the 1420s or 1430s, roughly around the time Eanes attempted to round Bojador. They rapidly evolved with innovations introduced by shipbuilders in Lagos and other Portuguese ports, focusing on hull design, sail configuration, and rigging. By the 1440s, the caravel had become a fully developed technology, and Portuguese explorers began using it to venture deeper into uncharted territories. Eventually, in the late fifteenth century, the Portuguese developed the caravel redonda, which combined lateen rigging for windward navigation with square sails for downwind speed. Nonetheless, the fundamental formula, lightweight, shallow draft, lateen-rigged, cost-effective, remained constant.

Without a caravel or something comparable, the Portuguese voyages wouldn't have been possible. The ship was the unobtrusive facilitator, turning ambitions into action on the ocean's surface. Though we remember captains and royalty, let us also honor the shipwrights, sailmakers, blacksmiths, and rope-makers, those whose hands crafted the vessels of discovery. Individuals shape history, but their instruments define what is achievable.

THE MEANING OF A MOMENT

Let us revisit that moment in May 1434: Gil Eanes stands upon his caravel's deck, Africa's coast stretching to the east, the white froth of Bojador receding in the distance. In this instant, he has accomplished what no one before him had dared to attempt. By surpassing an unseen threshold, the deep-rooted fear and conviction of disaster that shaped generations, he not only transforms his own fate but also initiates a process that redefines cultural perceptions and collective knowledge. Although the physical world remains visibly unchanged, Eanes's act embodies the central theme of this narrative: personal courage can overcome entrenched cultural fears, catalyze historical change, and expand the boundaries of what is considered possible.

This is the impact of firsthand discovery. Eanes didn't engage in debates with sailors who thought the sea boiled beyond Bojador, nor did he write tracts to dispute their beliefs. He simply navigated past the cape and returned. His survival was his own validation. Fears long regarded as insurmountable were dismantled by reality alone. Knowledge progresses not merely through discourse but through the audacity of those who venture into the unknown.

It also underscores that many obstructions exist solely within our minds. Bojador posed real hazards, with its treacherous shoals and currents; yet it was no insuperable obstacle. The sailors who chose retreat were not cowards; they acted based on their knowledge and apprehensions. What they lacked was the empirical evidence Eanes brought: the route could be traversed, risks were less than imagined, and success was feasible. Once this was demonstrated, the barrier dissolved. Over time, Bojador evolved from a formidable obstacle to just another point on maps, no longer a boundary. What constraints today do we accept as merely challenging rather than insurmountable? Which ancient fears deter us from scrutinizing the truth? Eanes's journey is more than historical trivia; it serves as a reminder of the boundaries we impose upon ourselves and how often these are mere illusions.

THE HINGE TURNS

In the grand tapestry of time, 1434 could easily go unnoticed. No monumental battles, no empires crumbling, no earth-shattering revolutions marked its passage. Yet, it was a pivotal year – a moment when a door creaked open and the world began an irreversible shift. A minor Portuguese nobleman embarked on a voyage past a forbidden cape, confronting fear only to reveal it as a mere illusion. The subsequent years saw Portuguese expeditions pushing boundaries: circumnavigating Africa, traversing the Indian Ocean, then bridging the Atlantic and beyond. This opened up new corridors of connection – trade goods, human migration, exchange of ideas, and diseases carved out fresh paths. Cultures collided, fused, and transformed one another. The framework of today's interconnected world, complete with its bonds and inequalities, began to take shape.

Gil Eanes had no inkling of the monumental impact his journey would have. His singular aim was to navigate past a cape, driven by the desire to honor his prince and avoid disgrace from failure. He could not fathom the magnitude of success that awaited him. Yet, the repercussions he set in motion were not scripted; they unfolded organically, influenced by countless individuals, most of whom remained oblivious to their role in shaping history.

After his triumph, Eanes slipped back into obscurity. It remains unknown what thoughts occupied his mind about his achievement or if he grasped its profound implications. His fate and the circumstances of his departure from life are equally shrouded in mystery. However, his absence from historical records may be seen as a fitting coda to his story. Eanes sought neither fame nor glory; he was merely fulfilling his duty by overcoming a formidable challenge. The fact that he executed this feat with competence, valor, and success is what mattered most. The world changed through his actions, whether he stayed to bask in the spotlight or not.

In commemorating the Age of Exploration, we often celebrate luminaries such as Columbus, da Gama, and Magellan. Yet we should also honor figures like Gil Eanes – behind-the-scenes heroes whose diligent efforts laid the groundwork for bold leaps forward. They remind us that history's course is not solely steered by visionary leaders but also by individuals with specialized skills who undertake daunting tasks and forge ahead. Though their names may fade into obscurity, their legacies endure. The thresholds they unlocked remain accessible today. The paths they charted continue to guide our way. The world they helped mold, for better or worse, is the reality we inhabit today.

Diogo Cão

April 1486. The Coast of West Africa. 6°15' South.

The solitary cross emerged alone. A ten-foot pillar made from Portuguese limestone stood tall, adorned with King João II's royal arms, standing alone in the red African soil - a unique monument that had never been seen before. The inscription pointed towards the Atlantic, making a declaration that was unfamiliar to those nearby. Salt spray had already started eroding the stone. Within one generation, the jungle would completely engulf it. Diogo Cão disappeared without witnessing whether the cross could endure. Somewhere between this isolated landmark and the Kingdom of Kongo, caught between his aspirations and the vast unknown, he vanished. No ship returned to Lisbon with news of his fate. No sailor recounted his final moments. He disappeared from history as quietly as morning mist dissipating over the African shore, leaving only these stone crosses to whisper about how far Portuguese exploration had dared to venture. History nearly erased him. Two decades later, Vasco da Gama reached India and claimed the glory that was built on Cão's foundation. Within a century, children in Portugal recited da Gama's name while Cão's faded like the weathered engravings on his crosses. However, in the 1480s, Cão had indeed pushed the boundaries of the known world further than anyone since the Phoenicians. His expeditions surpassed those of earlier explorers like Gil Eanes, who could only sail beyond Cape Bojador and provided valuable insights that led to further exploration along the African coast. Cão was a pioneer in several areas: he established diplomatic ties with a major African kingdom and proved that Africa's coastline did not extend infinitely. By crossing the equator, he opened up avenues for exploring what lay beyond, suggesting that land must curve and reveal paths leading to the riches of the East. This accomplishment was significant during an era when exploration faced geographical and logistical challenges, making him a crucial figure in the age of discovery. Thanks to Cão's accomplishments, the cross represented more than just territory. It symbolized the moment when outdated maps shattered, revealing a world that was both larger and more attainable than anyone had ever imagined.

Here lies the tale of the man who erected those crosses and disappeared into legend. To understand Diogo Cão's journey, we must look back to his origins.

THE MAKING OF A NAVIGATOR

The Lisbon that shaped Diogo Cão was vastly different from the imperial capital it would eventually become. Around 1440, when he drew his first breath, Lisbon still bore scars from plagues, wars, and decades of dynastic turmoil. The waterfront buzzed with fishy odors and tar fumes. In shipyards where slaves hauled heavy timber from Alentejo, the clatter of hammering on caravels echoed along the docks. The air was filled with overseers' barks as slaves toiled relentlessly. Meanwhile, merchants squabbled over salt, cork, wine, and, increasingly, human beings shackled from raids along the Mauritanian coast. Amidst a cacophony of languages - Portuguese, Castilian, Genoese, Arabic, and many African tongues - rivalries flared and deals were cut.

Surprisingly little is documented about Cão's birth or his early years; no baptismal records or family chronicles have surfaced. This silence speaks volumes: it indicates that Cão did not hail from noble stock whose progeny would have been dutifully recorded by scribes and extolled in verse. Rather, evidence points to him being part of a lesser noble lineage with exceptional maritime skills. These men discovered opportunities at sea that remained elusive on land.

The name "Cão," meaning "dog" in Portuguese, has led some historians down the garden path, leading them to speculate that it derives from a coat of arms or noble ancestry. The truth appears much plainer: it was likely a persistent nickname, perhaps earned for an ancestor's loyalty or stubbornness. Whatever its origins, this moniker carried little prestige during Cão's formative years; he had to earn distinction himself, most often by serving the Crown and, crucially, by joining Portugal's African exploration endeavors.

As Diogo Cão matured, Portuguese exploration was already a structured operation overseen by royal monopolies and licenses. Nobility and merchants invested in expeditions with hopes of profiting from slaves and gold. A multitude supported these initiatives: pilots charted coastlines, traders established outposts, interpreters mastered African dialects, and seafarers learned to navigate the winds and currents. For a young man like Cão, gifted at sailing but lacking a fortune, venturing in these undertakings was the gateway to upward mobility.

The first verifiable mention of Diogo Cão surfaces around 1480, suggesting he was nearing forty and seasoned enough for command. Some records imply prior voyages along the African coast, though specifics are sketchy. The crux is that by 1480, within the Portuguese court, probably King João II recognized a rare blend of navigation expertise, diplomatic acumen, endurance, and loyalty in Cão. These qualities cemented his role as the prime candidate to venture beyond charted territories.

João II ascended to the throne in 1481; known for his sharp, strategic mind, he envisioned transforming Portugal into a formidable trading empire. In contrast with his father, Afonso V's misadventures in Morocco and Castile, João II concentrated intently on Africa. Like his great-uncle Henry the Navigator, he understood that dominion over the sea route to India equated to control of global commerce. The challenge was two-fold: first, they had to locate the southernmost tip of Africa; no Portuguese captain had achieved this feat.

The farthest Europeans had reached until then was Cape Catherine, situated in present-day Gabon, just south of the equator. Beyond that point lay uncharted territory. Some ancient geographers posited that the southern seas were too scorching to traverse, claiming the sun would literally boil the sea and doom sailors to death. Others believed Africa extended all the way to Antarctica, making a direct route to India implausible. A medieval geographer once questioned, 'Could the seas near the equator be so fiery that no life could endure such heat?' This skepticism underscored the academic tension between known geography and uncharted oceans. Yet, sailors like Diogo Cão were determined to contest these fears and prove them unfounded. Some dreamt of perilous currents, marine monsters, or distant Christian kingdoms that might serve as allies. The reality was that no one truly comprehended because no voyager had ventured far enough to ascertain the truth.

To unravel this enigma, João II resolved to dispatch someone further than any European had dared. He selected Diogo Cão for this monumental task.

THE FIRST VOYAGE

June 1482. The Tagus River, Lisbon.

The vessels lay at anchor by the Torre de Belém, their hulls sealed tight after caulking, sails tucked away from the morning dew. Two caravels, maybe three, records aren't clear. They were small ships, no more than sixty feet long, each carrying about two dozen men for crew. The essentials of a lengthy journey filled their holds: water and wine barrels, salted meats and fish, biscuits with worm holes despite care, dried beans and onions, vinegar against scurvy, though its effect was not understood. Unique to these voyages were stone pillars, padrões, each ten feet tall, carved from nearby Lisbon limestone.

Estes padrões serviam como marcas de posse. One face bore the royal coat of arms; the other carried an inscription with King João II's name and the discovery date. The third showcased a cross; the fourth awaited inscription for future expeditions' claims. They were more than mere marks on land, legal instruments asserting sovereignty through time and weather, telling rivals: Portugal was here first.

Diogo Cão received explicit orders: sail south past Cape Catherine, exploring and mapping as far as possible. Seek Prester John's lands, a legendary Christian king in the African interior whose support could shift power from Islam. Establish relations with significant kingdoms, gather intelligence on gold sources, trade routes, and populations; wherever new coastal limits were reached, plant a padrão.

Departing in late June or early July 1482, the initial stages of this voyage would've been familiar to seasoned hands: past Cape St. Vincent, where the Portuguese coast ends, catching northeast trades which Henry Navigator's pilots learned to harness, then cutting diagonally across the open sea to the Cape Verde Islands, the last Portuguese outpost before mainland Africa.

Here, crews replenished supplies and perhaps traded with locals for food. Then back to shore, moving south through Guinea-Bissau, a coastline fragmented by river mouths into numerous channels, waters shifting from blue to brown as sediment loads increased, mangroves crowding muddy shores. The climate was harsh, oppressive heat, dense humidity causing leather rot and swift food spoilage; men unacclimatized would fall ill with fevers, some dying at sea amidst priest-performed rites barely audible over the wind.

Passing Sierra Leone, Pepper Coast, Ivory Coast (names to come), and Gold Coast, where copper bracelets and cloth were traded for gold dust. This was known territory charted, fraught with malarial dangers, demanding swift in-and-out trading to avoid deadly fevers. Beyond Cape Three Points into the Bight of Benin, navigation became more challenging as the Guinea Current's eastward flow made headway difficult. Cão's pilots managed by tacking and anchoring when necessary to keep gains despite contrary winds.

By the time they reached Cape Catherine, previously the Portuguese limit, they had sailed for months since Lisbon, crews depleted by death and illness, provisions low, especially in fresh water. Ships required upkeep: re-caulking seams where tropical heat dried pitch, repairing rigging from ocean strain, and patching leaks from warped planks. The choice loomed clear: retreat with modest gains or advance into the full unknown.

Diogo Cão chose to press onward.

The coastline beyond Cape Catherine gradually unveiled itself, bit by bit. Our latitude shifted, but the land's nature remained constant: thick forests extending to the water's edge, scattered beaches with white or brown sands, and river mouths both small and large. Cão's crew were vigilant for any sign of human life - wisps of smoke from fires, canoes leaving shore. Interacting with locals was crucial; they held secrets of the interior lands, gold sources, and kingdoms worth forging alliances with. Yet, interactions posed risks too. Earlier Portuguese expeditions met hostility, and men were killed or taken captive. Their protocol was cautious: approach armed but not overtly threatening, send a few ashore while keeping the bulk of the crew aboard, engage any interpreters to initiate basic talks.

Crossing the equator likely occurred in August 1482, though Cão may not have pinpointed the exact date. Navigating then mainly relied on dead reckoning, estimating position based on speed, time, and direction, with latitude checks using quadrants or astrolabes for the sun's high noon measures. Calculating longitude accurately eluded them, leaving east-west positions unclear. What Cão did confirm was that he surpassed all previous European expeditions in the southern reach, a land seemingly boundless.

Around 6° South, everything altered dramatically. The water turned murky from sediment. Debris floated by - branches, vegetation, possibly remnants of structures. The taste shifted from salty to slightly fresh. Seabirds filled the skies en masse. Cão stumbled upon the mouth of an enormous river, its flow so vast it altered ocean waters miles out.

This was the Kongo River, unnamed to Cão then. What mattered was its significance - such a vast body suggested broad territories and populous lands upstream. There, kingdoms and resources awaited, justifying this voyage entirely.

The Portuguese navigated carefully into the river's mouth, not a single channel but a sprawling estuary miles wide. The current fought back with immense strength; the Kongo, the world's second-largest by discharge, poured over forty thousand cubic meters per second. Fighting upstream was futile for ocean-built ships. Cão likely anchored near the south shore, dispatching boats to scout and connect.

Initiating contact with the Kingdom of Kongo radically shifted the course of Portuguese exploration. Unlike most African groups encountered who resided in small-scale societies centered on villages or loose confederations, the Kongo stood as a centralized kingdom with complex governance. It boasted Mbanza Kongo, its capital, days inland from the coast, governed by the ManiKongo through provincial governors. Its economy thrived on agriculture, ironwork, and textile crafts. Diplomatic networks, military organization, monotheistic yet non-Christian religious beliefs, all suggested potential for Portuguese overtures.

Cão couldn't absorb all this at once. He grasped that locals welcomed interaction, unlike coastal groups, which preferred structured dialogue. Through interpreters and simple gestures, he gathered that a powerful king ruled from inland and might engage in Portugal's pursuit of relations.

Here, Cão gambled, either with brilliant diplomacy or reckless overconfidence. He seized several Kongolese hostages - sources debate if voluntary or forced - promising their return aboard his next voyage. The message: Portugal was returning, not raiding; beginning a relationship. These hostages would journey to Lisbon, learn Portuguese, witness the kingdom's might, and return as interpreters for future dealings.

Before departing from the Kongo River, Cão erected one of his stone padrões on the southern shore near modern Soyo in Angola. The inscription chronicled 1482 under João II's reign, claiming possession for the Portuguese crown. As the stone rose against the sky, a mix of fear, curiosity, and indifference rippled through the local populace. To some, it was a mystical signal from a world beyond their shores. This marker stood unyielding, a testament to expanding horizons and the beginning of new cultural and territorial dynamics.

Cão didn't return straight to Lisbon; he ventured further south. The coast curved almost due south for many miles, presenting an appealing prospect in the quest for Africa's southern tip. Landscapes evolved: forests to savanna, then arid scrubland, hinting at desert onset. Climate transitioned from hot and humid to hot and dry, and population density waned.

Approximately 13°26' South, Cão planted another padrão near Cape Santa Maria in contemporary Angola, marking his southernmost exploration milestone. He extended Portuguese reach by over a thousand miles, discovered the Kongo River, initiated contact with a major African kingdom, proved coast continuity far south of the equator, and debunked ancient geographers' horror prophecies.

The homeward voyage would be quicker with winds and currents favoring a northbound course, but it would be fraught with dangers. By the time Cão's ships likely docked in Lisbon, circa early 1484, they'd endured eighteen months at sea or more. Crews were decimated by disease and accidents; ships were barely seaworthy.

Yet, they hauled a more precious cargo than gold: evidence of navigable southern waters and kingdoms worth sustained efforts to engage with.

King João II greeted this news with measured satisfaction, confirming his strategic calculations. He knighted Cão into the Order of Christ, advancing exploration's banner further still.

THE KINGDOM OF KONGO

Before Diogo Cão's return to Africa, an unusual diplomatic situation unfolded in Lisbon. The Kongolese hostages he had brought back, members of noble families possibly related to provincial governors or court officials, became local celebrities in the Portuguese capital. They were being instructed in Portuguese and Christian doctrine, and shown the magnificence of King João II's court, with the intention that they would return to their homeland as living proof of Portugal's power and Christian civilization.

This strategy illustrated King João II's broader ambitions; he sought not just plunder and trade but to establish Portugal at the center of a global alliance network, with African kingdoms as junior partners in both commerce and their shared crusade against Islam. The Kongolese kingdom, known for its centralized governance and adaptable religious views, seemed ideal for this vision. If ManiKongo were to convert to Christianity and Kongo became an ally, it would secure the Kongo River region as a base for further exploration and trade. It would also serve as an exemplar to other African rulers of what collaboration with Portugal might bring.

The Kongolese were not passive recipients of Portuguese interest; they were proactive players in this interaction. They sent young men to Lisbon to learn about Portuguese technology, including firearms, shipbuilding, and advanced ironworking techniques. These individuals observed the military organization and courtly customs of their hosts. They established connections that could prove invaluable upon returning home. Their king had explicitly dispatched them or at least consented to their departure, to gather this intelligence. Kongo was not simply accepting Portuguese advances; it was strategizing on how such an alliance might further its own interests.

By the end of 1484 or early in 1485, preparations for Cão's second voyage were complete. The mission had multiple objectives: first, returning Kongolese hostages to demonstrate good faith and establish formal diplomatic relations with

ManiKongo; second, continuing exploration southward to find either the southern tip of Africa or information about Prester John's kingdom; third, establishing a more permanent Portuguese presence on the Kongo River, potentially including missionary work; fourth, gathering detailed intelligence regarding gold sources, population distribution, potential allies, and threats.

The expedition departed Lisbon, likely in late 1484. Unfortunately, records are again scarce. We know Cão led it and that he carried gifts for ManiKongo, including cloth, copper, brass items, wine, and flour, meant to impress and foster trade relations. It is confirmed that he had the authorization to negotiate on behalf of the Portuguese crown. However, specifics remain unknown: the exact number of ships under his command, who else held authority in the expedition, or any private instructions given by King João II.

Early 1485. The Kongo River Estuary.

The return of the hostages became an elaborate diplomatic display intended to make a significant impact. Rather than simply sailing into the river mouth and leaving the young Kongolese men there, Cão took another approach. He dispatched messages ahead, likely through interpreters and local leaders, announcing his arrival and requesting an audience with the ManiKongo. After waiting for these messages to reach the interior and receive authorization, he recognized that African kingdoms had their own diplomatic protocols that needed to be respected to achieve Portuguese objectives.

Upon receiving permission, a delegation set out towards Mbanza Kongo. Cão himself may have led it, though some sources suggest he stayed by the ships while his subordinates undertook the journey. The path they took ran parallel to the Kongo River at some distance from its banks, passing through landscapes that alternated between forest and savanna. This route was well-maintained, showcasing the administrative capabilities of the Kongo kingdom. Villages were situated regularly along their path, with cultivated fields of manioc, plantains, and palm trees in sight. The delegates witnessed evidence of advanced activities such as iron-working, textile production, and pottery-making, indicating a complex civilization that had developed over centuries.

Mbanza Kongo, when the Portuguese finally arrived there, exceeded their expectations in every way. It was not a European-style city with stone buildings and defensive walls, but rather an expansive collection of compounds and structures made primarily of wood, thatch, and woven materials, accommodating around 50,000 people. The royal enclosure stood at its center, an elaborate complex where the ManiKongo held court. During their reception ceremony, drummers and musicians entertained them while demonstrating Kongolese authority to impress the visitors and, at the same time, to evaluate the Portuguese intentions.

The ManiKongo during this period was likely Nzinga a Nkuwu, but the chronology of rulers is somewhat uncertain. He received the Portuguese delegation with cautious interest, engaging in complex negotiations through interpreters. The Portuguese offered them goods and technology, proposed a potential military alliance against Kongo's adversaries, and presented their offer to send Christian missionaries who claimed that their religion was the true faith. In return, they sought trade access, particularly to sources of copper and possibly gold in the interior, as well as permission to establish a permanent presence at the river mouth and gather information about territories to the south and east, potentially belonging to Prester John's realm.

The ManiKongo did not make an immediate decision. African rulers had learned through bitter experience that agreements with Europeans could have unexpected consequences. He required consultations with his advisors to assess the risks and benefits and calculate how the Portuguese alliance might affect Kongo's relationships with neighboring kingdoms. However, he did not reject their offer outright. Instead, he accepted their gifts, agreed in principle to future contact, and even sent his own ambassadors to accompany Cão back to the coast and possibly as far as Lisbon.

For Diogo Cão, this diplomatic triumph was vital. He had achieved more than merely discovering a river mouth; he had established the beginning of a sustained relationship with a significant African kingdom. The potential was enormous: Kongo could become a base for further exploration, a source of trade goods, and potentially a Christian kingdom that would validate the entire Portuguese enterprise in Africa. King João II would be pleased.

"The ManiKongo paused, then said, 'If your king seeks friendship, let him prove it through honest trade and respect for Kongo's people.' Cão responded carefully, 'Our intentions are to build alliances and to learn with respect, as well as to share the fruits of mutual understanding.' With this exchange, a tentative accord was reached, and the framework for future engagements was delicately set."

However, Cão's exploratory journey was not yet complete. With the diplomatic mission accomplished, he turned his attention southwards, ready to discover more uncharted territories.

THE VANISHING

The coast along the Kongo River is one of the most desolate areas in Africa. The cold water from Antarctica, brought by the Benguela Current, flows northward here, creating fog banks, supporting rich marine life, but making the nearby land extremely arid. This is the Namib Desert, which has not recorded rainfall in some places, making it one of the oldest and driest deserts on Earth. The shoreline showcases a harsh beauty, with white-sand beaches, dunes that drop directly into the ocean, rock formations carved by wind into strange shapes, and shipwreck bones bleaching under the sun.

In 1485 or early 1486, Diogo Cão ventured into this desolate area. He followed his established pattern from his first voyage: he stayed close to the shore, observed carefully, charted meticulously, and planted padrões in significant spots. However, the coast here offered fewer opportunities compared to the more northern regions. There were almost no inhabitants to contact, scarce freshwater sources, poor harbors susceptible to Atlantic swells that could destroy ships against the rocks, and nothing was revealed about its interior except for being a desert.

At Cape Cross, which is now in Namibia near 21°47' South, Cão planted another padrão. The limestone cross stood out brilliantly against the brown desert landscape, with an inscription announcing Portuguese possession of a coast where it seemed nothing was worth owning. Although some accounts say 1485 and others claim 1486 as the date inscribed, this intriguing discrepancy invites us into a puzzle of historical records. Did Cão make two separate visits, or did the chronology become blurred soon after these events? Such questions linger, adding a touch of mystery to our understanding of his voyages.

Centuries later, European explorers found the cross at Cape Cross still standing and eventually moved it to a museum in Germany. It remains one of the few physical artifacts related to Diogo Cão's voyages, connecting us to when Portuguese ships entered one of the most inhospitable coastlines on Earth.

However, the distance Cão ventured further is unclear. Some sources suggest he reached Cape Santa Maria at approximately 13° South, while others claim he went even further south, possibly up to Walvis Bay at 22° South or nearly up to Cape Cross at around 22° South. A few writers speculated that he might have extended his journey as far as the Cape of Good Hope itself, achieving a breakthrough before Bartolomeu Dias did a few years later.

The facts we know are limited. Cão planted padrões in several places along the Angolan and Namibian coasts. He returned to the Kongo River region because later Portuguese expeditions found evidence of his presence there. He sent or brought information about his discoveries back to Lisbon, as King João II acted upon that intelligence for subsequent expeditions. However, what we do not know is when, where, or how Diogo Cão died.

The most plausible scenario is that Cão advanced southward until limited by the provisions, the conditions of the ships, and the endurance of his crew. When he reached Cape Cross or slightly beyond, he realized he had achieved the limits of this expedition. Although the coast continued trending southward with no sign of Africa's southern terminus, further exploration risked losing ships and men without guarantees of finding anything different.

Cão turned back from there. However, during the return voyage along that hostile coast or perhaps in the Kongo River region, something went wrong. It could have been due to disease, accidents, or violence. Regardless, Diogo Cão never returned to Lisbon. Some of his ships returned later with news of his discoveries and, potentially, information about his death. A solitary, battered logbook, perhaps set adrift by the waves, might have spoken volumes of his final days, but it never reached the shores of history's annals. Nor did the echoing silence of an empty captain's cabin, forsaken and haunting in its void, find its way back to Lisbon. The records do not mention how the Portuguese court reacted to this loss. João II had more pressing matters to attend to: consolidating power against restive nobles, planning further African exploration, and managing complex diplomatic and commercial networks.

By 1487, preparations were underway for an expedition aimed at surpassing Cão's achievements. Bartolomeu Dias, another experienced African pilot, received command of a mission specifically tasked with finding the southern cape of Africa. He departed Lisbon in August 1487 with three ships. By early 1488, he had sailed far beyond Cão's furthest point and rounded the Cape of Good Hope, although stormy conditions meant that he actually sailed past it without seeing it; only on the return voyage did he realize that the coast finally turned northward and eastward. Dias returned to Lisbon in December 1488 with definitive proof: the sea route to India was confirmed, Africa had limits, and the southern ocean could be navigated. His triumph overshadowed Cão's pioneering work almost immediately.

Nonetheless, without Cão's voyages, Dias's success would have been impossible. Cão demonstrated that European ships could survive and navigate the South Atlantic. He established a pattern of exploration: conducting systematic coastal surveys, carefully noting landmarks and hazards, establishing relations with local populations, and planting markers to signify possession. Cão gathered essential intelligence about winds, currents, the layout of the African coast, population distribution, and kingdoms. He proved that the fearsome latitudes south of the equator were not deadly for European navigation. Every lesson Dias learned was partly derived from studying what Cão had accomplished.

Moreover, Cão's diplomatic efforts in Kongo created a lasting Portuguese presence in Central Africa. The relationship he initiated continued to deepen. In 1491, Portuguese missionaries arrived in Kongo and began serious efforts at conversion. Within a few years, ManiKongo Nzinga a Nkuwu was baptized as João I, adopting the name of the Portuguese king. His son, Afonso I, became one of the most fervent African Christian converts, corresponding regularly with Portuguese kings and popes, building churches, and sending Kongolese students to study in Lisbon. The Kingdom of Kongo remained a staunch ally of Portugal for over a century, serving as a base for slaving operations, missionary work, and trade that profoundly shaped Central African history, mostly negatively; this was not apparent in the 1480s.

THE WEIGHT OF STONE

1504. The Coast of Angola.

Eighteen years after Diogo Cão erected his stone cross at Cape Cross, Portuguese ships continued to explore the African coast and push further southward and eastward in search of reliable routes and trade opportunities. An expedition led by an unnamed captain stopped at the mouth of the Kongo River for supplies and trading. During their stay, they discovered one of Cão's fallen crosses, which had toppled, possibly due to natural forces or intentionally knocked down by locals as a symbol of opposition to Portuguese claims to their land.

Although weathered and broken, the cross still bore legible inscriptions, including the royal arms of João II and Christian symbolism. The sailors documented this find before continuing their voyage, and the copied inscription was preserved in Portuguese archives, providing crucial evidence connecting Cão to specific times and locations. Historical information from this period is often fragmented, with sources including copied inscriptions, brief mentions in later chronicles, occasional references in diplomatic correspondence, and the physical remains of padrões found centuries later.

Despite the lack of personal accounts or letters from Diogo Cão, the padrões themselves offer valuable insights into his expeditions. These carved stone crosses were remarkable objects that required significant effort to produce, transport, and erect thousands of miles away. They served as messages for future generations, establishing Portuguese claims, aiding navigation for subsequent expeditions, and reminding crews of their broader enterprise.

The existence of these markers in the twentieth century and the local legends surrounding them among indigenous populations highlight their importance to historians seeking to piece together Cão's voyages. The geographical position, inscriptions, weathering patterns, carving quality, and other details provided valuable data points for comparing them to other padrões from the same era.

However, further discoveries of these markers also revealed the darker side of Portuguese exploration in Central Africa. The trade relationships established along the Kongo River coast quickly evolved into massive slave trading operations, causing devastating consequences for local populations. This progression has been the subject of significant historiographical debate. Scholars such as Joseph C. Miller and John Thornton have emphasized that, although Diogo Cão likely did not anticipate the scale of exploitation that would ensue, his initial diplomatic engagements nonetheless set in motion structural changes that facilitated the expansion of the Atlantic slave trade. Other historians debate the degree to which individual explorers, as opposed to broader economic or political forces, should be held responsible for these outcomes. Thus, the question persists within the scholarly literature: to what extent did Cão's seemingly benign diplomatic overtures inadvertently contribute to the emergence of a tragic human legacy in Central Africa?

Exploration during the Age of Discovery was never innocent; it was closely tied to motives of power and domination from its beginning. Explorers like Cão, who faced immense challenges venturing into the unknown, served as instruments of broader imperial systems with profoundly destructive consequences for the societies they encountered. Reconciling Cão's achievements with the horrific outcomes that followed requires acknowledging both his courage and navigational skills and the exploitative dynamics he inadvertently supported.

The stone crosses also represented a claim of European possession over lands with their own religions, political systems, and ways of life. These proclamations of Christian dominion ignored existing civilizations and rulers in these regions. Thus, exploration during this period entailed complex moral calculations that differed from those of later eras, leaving deep scars on affected societies even to the present day.

THE SILENCE OF SOURCES

The gaps in historical documentation regarding Diogo Cão's expeditions offer insights into how information was controlled and preserved in fifteenth-century Portugal. The available records were not designed for future historians; they served state policy, commercial interests, and royal administration. King João II, known for his secrecy regarding the African enterprise, grasped that geographical knowledge equaled power. He tightly regulated access to navigational data, imposing severe penalties on those who shared it with foreigners.

The Casa da Índia vigilantly safeguarded its archives. This policy led to much of the early voyage information not being recorded for posterity. Pilots maintained personal notes, but they were practical guides, not formal records, and were often lost or destroyed. Some survive, but many do not. Official reports submitted to the king were stored in royal archives inaccessible to scholars. Many such documents perished in the 1755 Lisbon earthquake and fire, erasing invaluable records of Portugal's early exploration efforts.

Firstly, brief mentions in official chronicles by João de Barros, Rui de Pina, and Garcia de Resende, works written decades later, with political agendas that prioritize Portuguese achievements over precise details about individual explorers. These accounts contradict each other on key points and offer little on Cão himself. For instance, Barros simply notes 'Cão did the King's will', leaving his motivations a mystery.

Secondly, the inscriptions on the padrões and the copied ones made available geographical data but provided no insight into Cão's thoughts or experiences. Lastly, indirect references in diplomatic correspondence and administrative documents offer a fragmented view of his activities, leaving significant gaps.

Unlike Christopher Columbus, who meticulously documented his voyages and fought for recognition, Cão's voice is absent from the historical record. His descriptions of discoveries, reflections on their significance, and insights into the people he met remain unknown. The silence surrounding him contrasts sharply with Columbus's self-promotion and detailed accounts of his travels.

This stark difference in how their legacies have been preserved raises questions about why Cão's contributions have been overshadowed by those of less significant explorers. The answer remains elusive, caught within the limitations of surviving records and the capricious nature of history's remembrance.

1492. Palos de la Frontera, España.

Six years following Diogo Cão's likely passing, Christopher Columbus embarked on a journey from Spain with three ships, charting a course westward in pursuit of an alternative sea route to Asia. In his possession were replicas of Portuguese maps, insights gleaned from their exploration endeavors, and navigation methods refined by the Portuguese. Upon reaching the Caribbean archipelago, he erroneously perceived it as the outer fringes of Asia, thereby paving the way for Europe's colonization of the Americas. This event initiated a series of conquests and exploitations that eclipsed even the transatlantic slave trade in both magnitude and repercussions.

Columbus's expedition was facilitated partly by the Portuguese accomplishments, which demonstrated the viability of prolonged ocean voyages; the capacity of vessels to bear sufficient provisions for extended periods at sea; and the efficacy of precise navigation techniques in ensuring the safe return of crews across vast, uncharted waters. The methodologies employed by Cão, such as meticulous coastline mapping, the placement of possession markers, and the establishment of diplomatic ties with indigenous populations, became models emulated by Spanish explorers in the Americas.

Despite these shared methodologies, Columbus's name endures while that of Cão fades from memory. Why? In part because Columbus's discovery appeared more extraordinary: an entire new world as opposed to another stretch along the African shore. In part due to Spain's significant investment in disseminating Columbus's achievements, which served to justify their subsequent conquests in the Americas and their rivalry with Portugal for global supremacy. Additionally, Columbus's capacity to return and defend his findings stands in stark contrast to Cão, whose fate remained entwined with the African coast, leaving him unable to assert his claims.

Beyond these factors lies another element at play, the way we construct historical narratives. We gravitate towards stories with clear beginnings and endings, in which heroes achieve their ambitions. Columbus ventured westward and found land; Da Gama journeyed eastward to reach India; Magellan accomplished a circumnavigation of the globe, each narrative offering a definitive resolution. In contrast, Cão's tale lacks such closure: he embarked southwards, but then there is only silence, vanishing without a trace. He did make contact with new lands, yet his journey lacked an ultimate destination. His story feels uncompleted because it remains so.

History leans towards totality. It cherishes the explorer who navigated past the Cape of Good Hope, while relegating those who nearly achieved such feats to obscurity. It exalts the navigator who reached India over those who made progress but fell short. History covets triumph, not near-success or preparatory efforts for victory. This predisposition influences our remembrance of history, elevating particular individuals whilst allowing others, even when their contributions were crucial, to fade from collective memory.

THE VIEW FROM MBANZA KONGO

If Diogo Cão's account is scarce in Portuguese records, its elusiveness becomes even more pronounced from an African standpoint. During this period, the Kingdom of Kongo relied on oral traditions rather than written documentation. Consequently, the Kongolese view of Cão's arrival, the negotiations he conducted, and the overall impression he made were all transmitted orally and transformed through retelling. By the time Portuguese missionaries and chroniclers documented Kongolese history in the sixteenth century, the inaugural arrival of the first Portuguese ships had taken on mythic proportions, becoming an integral part of origin stories surrounding Kongo's conversion to Christianity. These narratives were filtered through decades of subsequent relations between Portugal and Kongo.

It can be surmised that from a Kongolese viewpoint, the advent of the Portuguese must have been profoundly enigmatic. These men suddenly emerged from an unfamiliar domain; the ocean was not perceived as a thoroughfare but rather as a boundary, signifying the end of the known world to the Kongolese. They appeared in vessels deemed impossible, structures that moved upon water propelled by winds that seemed to possess magical qualities. Additionally, they possessed copious metal objects, indicative of immense wealth, yet their interest seemed to be fixated on relatively ordinary trade commodities: ivory, copper, and foodstuffs. Their languages were unintelligible to the Kongolese, and they insisted on communicating through interpreters who struggled to convey concepts that lacked equivalence in either language.

The Kongolese rulers had to gauge whether these newcomers posed a threat. Were they potential allies? Could they be manipulated or controlled to their advantage? The decision to engage with the Portuguese, rather than repel or ignore them, suggests an astute political calculus. Kongo was indeed a powerful kingdom, but it faced rivals and adversaries. Access to Portuguese technology, particularly firearms, could tip regional power balances in Portugal's favor. Engaging in trade with these unusual foreigners might accrue wealth and enhance prestige. Embracing their faith may have facilitated diplomatic relations and opened doors to a broader world.

Nonetheless, the Kongolese could not have foreseen the full repercussions of their engagement with the Portuguese. They had no idea that Portuguese missionaries would challenge the very foundations of Kongolese religious life. It was beyond their imagination that Portuguese merchants would initiate the buying and selling of individuals on such an extensive scale, eventually decimating the local population. The Kongolese could not predict that aligning with Portugal would entangle them in wars with neighboring kingdoms, that Portuguese military backing would come with conditions, or that Christian conversion would spark internal conflicts between traditionalists and converts, thereby weakening the kingdom's unity.

From the Kongolese perspective, Diogo Cão was probably not a particularly significant figure. He was merely the first of numerous Portuguese captains who would arrive at the river mouth during subsequent decades. What mattered was not Cão himself but the relationship that his arrival initiated. The Kongolese recollected patterns rather than individual identities. They might recall an old saying: 'When the ocean beckons, even the mighty river changes course.' This captures the sense of adaptation and new possibilities his arrival brought with them. Later Kongolese accounts of early encounters with Portugal often conflated different expeditions, combined multiple Portuguese captains into composite characters, and struggled to establish accurate dating. This was not due to any unreliability in Kongolese oral traditions; they remarkably preserved crucial information about their own history with precision, but from their viewpoint, these distinctions between individual Portuguese explorers held little significance.

What proved meaningful was the subsequent transformation. Within two generations of Cão's arrival, Mbanza Kongo had been renamed São Salvador and adorned with churches. The ManiKongo was drafting letters in Portuguese to European rulers and popes. Kongolese nobility were studying in Lisbon and Rome. However, Kongo was simultaneously losing its population, as thousands upon thousands of individuals were sold into slavery each year. Portuguese missionaries denounced traditional religious practices. Portuguese merchants demanded concessions in trade. And Portuguese military strategists influenced Kongolese foreign policy.

By the seventeenth century, the Kingdom of Kongo had diminished to a shadow of its former power, its strength sapped by internal conflicts, partly driven by competition for Portuguese patronage, depleted by the slave trade, and its economy distorted by dependence on European goods. By the nineteenth century, it had completely disintegrated, with its territory fractured among rival factions and its independence all but extinguished. And by the twentieth century, it was merely part of Portuguese Angola, its history nearly forgotten, its once-great capital reduced to a provincial town.

None of this can be directly attributed to Diogo Cão. He did not establish the slave trade; Portuguese ships had been transporting African slaves since the 1440s. He did not create the detrimental patterns of European-African interactions; these patterns were rooted in Mediterranean commercial practices spanning centuries. He did not determine Portuguese policy toward Kongo; those decisions were made by kings, merchant companies, and colonizers who arrived after him.

Yet his explorations were the fulcrum upon which the fate of the Kingdom of Kongo swiveled. Prior to Cão, Kongo existed in relative seclusion, primarily engaging with neighboring African peoples and kingdoms. Following Cão, Kongo was drawn into the Atlantic world system, encompassing all its associated wealth, violence, and cultural metamorphosis. One trade concession granted by Portugal altered the agricultural landscape of Kongo, prompting farmers to focus on cash crops demanded fervently by the European market. This shift fostered dependence on Portuguese goods and precipitated local food shortages, illustrating but one of many cascading effects of the Atlantic trade. The stone cross that Cão planted at the river mouth served both as a marker and an entrance, symbolizing a connection that would prove ruinous and transformative.

THE NAVIGATION

To fully appreciate Diogo Cão's accomplishments, one must first understand the context of navigation in the fifteenth century. There was no use of GPS systems or radio beacons to determine longitude, and accurate chronometers were unavailable at the time. Instruments such as the magnetic compass (known since medieval times but prone to variation), the quadrant for measuring sun altitude, the astrolabe for calculating latitude, the lead line for depth sounding and sample collection, and dead reckoning, estimating position based on speed, time, and direction, were commonly used.

Dead reckoning was the primary method of navigation, and while it required constant adjustments, it also allowed pilots to make the most of their surroundings. Estimating one's speed accurately was crucial; any slight miscalculation in knots could lead to significant positional errors over hours at sea. The accurate measurement of time using half-hour sand glasses was essential but posed its own challenges, as missing a turn would nullify all calculations.

Maintaining direction with the compass and accounting for magnetic variation were also critical tasks. Simple yet demanding math played a role, and even small estimation errors could result in being off course by miles after several hours. The lack of understanding about currents added another layer of complexity to navigation. These unpredictable elements made sailing conditions varied and dangerous.

Determining longitude proved to be the most significant challenge since accurate clocks for comparing local time with reference points did not exist until the eighteenth century. Thus, east-west positions were mainly estimated based on experienced guesswork. The Portuguese navigators relied heavily on luck while having an approximate idea of their latitude. Their charts depicted coastlines vaguely and distances inaccurately, leading to a gradual learning process through each voyage.

Cão operated at the edge of this knowledge, charting accurately up to Gabon's Cape Catherine but venturing into the unknown beyond it. His journey demanded constant observation, measurements, and recordings of river mouths, distinctive headlands, hazards, and survival against unforgiving elements, including hostile populations and harsh weather conditions. Each compass reading felt like rolling dice with the Atlantic, as navigational uncertainty was ever-present, turning each course correction into an act of courage and a gamble.

The achievement of reaching Cape Cross, which was approximately 6,500 nautical miles from Lisbon by his route, cannot be understated. The voyage involved navigating the Atlantic's favorable winds, navigating the dangerous West African coast, crossing the equator, exploring the Kongo River, and then moving south through desolate, waterless territory.

Cão had to make informed decisions with limited information at each stage, such as whether to continue or return, explore river mouths, and trust local pilots, among other critical choices. Moreover, the risk of error was deadly; ships that went aground could break apart in the waves, while those caught too far from shore by storms lacked shelter, and those lingering in tropical waters accumulated marine growth leading to unseaworthiness.

Cão's ability to strike a fine balance between caution and boldness led him to extend Portuguese knowledge further than any previous navigator. The charts and sailing directions he compiled served as the foundation for future voyages. Bartolomeu Dias' 1487 journey south and Vasco da Gama's 1497 voyage to India built upon Cão's discoveries.

This incremental accumulation of knowledge by explorers like Cão, who have largely been forgotten, is how the Age of Exploration unfolded. It was not characterized by individual brilliance but rather by institutional persistence, systematic data collection, and repeated expeditions funded by organizations such as the Portuguese Crown, each of which slightly pushed boundaries. The dramatic achievements were made possible by decades of unheralded coastal surveying efforts by navigators whose names are scarcely remembered.

THE ARTIFACTS

The padrão Diogo Cão established at Cape Cross endured over four centuries in the harsh desert climate. It served as a beacon for sparse European ships navigating this forsaken coast, as well as for slavers, whalers, and traders exploring its potential riches. Seafarers occasionally carved their names alongside his monumental mark, adding personal touches to such an imposing tribute.

In 1893, a German naval mission relocated this padrão from the site to Berlin's Deutsches Historisches Museum. This act exemplified colonialism in miniature: European nations removing African relics for display in their museums under the pretense of curating and interpreting history. The space where it once stood remained barren except for its foundation stones.

The newly independent Namibia, freed from decades of German and then South African rule, petitioned for its return in the 1980s. Their request was denied; thus, this piece of African heritage remains in Berlin, a testament to a past era held captive within European walls, its absence now as telling as its historical presence.

A replica was erected at Cape Cross by the South African government in 1980, standing as a tribute but differing from the original in craftsmanship and proportions, constructed with local rather than Portuguese stone. Although not exact, it continues to signify Cão's reach and Europe's initial exploration of this remote shore. Adjacent to it lies a nature reserve protecting the cape's seal colony, drawing tourists for photos of this symbolic cross, though its historical context is vague.

Other padrões planted by Cão met various fates: one near the Kongo River vanished entirely amid colonial warfare and political upheaval; another at Cape Santa Maria in Angola was discovered weathered but identifiable by a Portuguese team early in the last century. It resides now in Luanda's museum, surviving tumultuous years following Angolan independence. The supposed placement far within the Kongo at Yelala lacks confirmation; some attribute it to Cão, while others suspect misattribution or myth-making.

These physical remnants, the few preserved and many lost, are our strongest links to Diogo Cão, symbolizing the ambition and logistics of late medieval exploration. They entailed significant effort: quarrymen extracting stone near Lisbon, craftspeople carving royal symbols and inscriptions into blocks, which were then transported by ship across perilous journeys to distant shores and erected with primitive tools and human strength. Yet, the story of these padrões would be incomplete without considering the hidden workforce behind their creation and transportation. Whose hands labored to quarry, carve, and carry these stones? By highlighting the often-unseen laborers, quarrymen, ship crew, and local inhabitants who aided in their erection, we can deepen the exploration of imperial logistics and their moral implications. Indeed, these stones did not merely symbolize claims of possession by the Portuguese crown; they also embody the silent toil of those whose efforts underpinned such claims.

Such crosses held substantial costs in time, labor, and resources; their erection was not taken lightly. They marked a serious investment by Portugal's monarchy in exploration, serving as potent symbols of its presence and claims to newly acquired lands. Though useful to some extent for navigation, their primary purpose was to assert rights through psychological and legal means, proclaiming territory as their own without the locals' actual consent.

Yet, these crosses are preserved in museums or replicated in images, testaments to an era when European expansion reshaped the world. They represent both remarkable feats and arrogant claims, human capabilities for great deeds and dire consequences. While not exempt from critique, they cannot be disregarded either; like Cão himself, they embody complexities from a tumultuous period of history.

THE NAMIB

1486. Cape Cross, Namibia. 21°47' South.

The seal colony extended for miles along the coastline, with thousands upon thousands of creatures packed together so tightly that the sand beneath them became invisible. A cacophony of sounds filled the air: barks, roars, and the piercing cries of pups, mixed with the deeper growls of bulls vying for territory.

The stench was unbearable, a pungent, organic odor resulting from countless animal bodies and excrement baking under the relentless sun at Cape Cross. This remote location, reached by Diogo Cão, represented a place as far removed from human civilization as any on Earth, where the vast ocean met the unforgiving desert and life clung to existence in this narrow strip between two inhospitable environments.

The crew who assisted in planting the limestone cross here must have pondered its purpose. Why mark such a desolate spot? Why assert ownership over a coast devoid of anything but seals and sand? Why persist in the search for a southern cape that might not even exist, venturing ever further from home into increasingly hostile geography?

The answer lay hidden from their sight, the larger patterns of coastline and ocean slowly revealing themselves. Africa's shape was emerging degree by degree, and evidence was accumulating that somewhere, somehow, land must give way to sea, opening the path to India. They were observing individual trees while their captain, king, and the entire apparatus of Portuguese exploration sought to perceive the forest. Each padrão served as a data point, a measured fact, a small piece of an enormous puzzle that would require decades to assemble completely. What then was the crew truly marking?

Diogo Cão likely passed away without knowing whether his endeavors would prove significant. He pushed the boundaries of the known several hundred miles further south, yet did not discover the cape he so desperately sought. He established diplomatic ties with Kongo, unaware of how beneficial or detrimental this relationship might ultimately be. He gathered valuable information about winds and currents, the layout of Africa's coast, and the distribution of peoples and kingdoms; however, he had no way of knowing whether future expeditions would build on his discoveries or fade into obscurity.

This uncertainty perhaps represents the most human aspect of Cão's story. We view him from the perspective of centuries, aware that Vasco da Gama would soon round the cape, that da Gama would reach India, and that Portuguese ships would eventually traverse all oceans. We can place Cão's achievements in context, understanding their role within the broader pattern of European expansion. Yet, Cão himself lacked such foresight; he could only sail south, meticulously charting his course, planting his crosses, and hoping that what he was accomplishing served a greater purpose.

The cross at Cape Cross endured for centuries, outlasting not only Diogo Cão but also the Portuguese Empire in Africa and the certainties of its age. It weathered blistering sun, torrential winds, and relentless salt spray. It bore witness to passing ships from various nations, Portuguese, Dutch, English, German, and eventually Namibian, and observed colonialism's arrival and eventual departure. It remained standing while the world transformed around it; the meanings associated with exploration, discovery, and conquest underwent radical shifts; and triumphalist narratives of European expansion gave way to more complex understandings that acknowledged both achievements and atrocities.

Eventually, the cross was relocated to a German museum, leaving behind an empty space where it once stood. Perhaps there is some appropriateness in this. The Age of Exploration that Cão helped initiate was always characterized by absence as much as presence. The absence of knowledge was replaced with maps, the absence of European power was supplanted by colonies, and the absence of European goods was replaced by trade networks that ultimately drained more than they provided. The crosses marked absences: the absence of explorers who planted them and sailed away; the absence of African voices that might have contested the claims inscribed upon stone; and the absence of any certainty about the consequences stemming from these first encounters.

Diogo Cão vanished into one of those absences, disappearing from history as completely as he disappeared from Africa's coast. No tomb marks his grave. No contemporary portrait captures his visage. No detailed account describes his character, motivations, or thoughts. He exists for us solely as a name connected to accomplishments, a shadow cast by stone crosses planted at the edge of the known world.

Yet in that disappearance, an enigmatic power resides. Because we know so little about Cão as an individual, he becomes emblematic of all those anonymous sailors, pilots, and seamen who carried out the actual work of exploration. The renowned names Columbus, da Gama, and Magellan were supported by countless men whose names remain unknown, whose stories were never documented, who perished from disease, shipwrecks, or violence, and who were subsequently forgotten. Cão nearly became one of those lost souls in the annals of history. He came close to vanishing completely from the record of human events. It is primarily due to the stone crosses he planted that Cão is remembered today, a testament to the enduring power of physical objects when documents and memories falter.

Those crosses, standing or fallen, or relocated to museums, persist. They signify not only geographical points but also historical junctures marking moments when the world began contracting, when barriers between continents started crumbling, and when isolated human cultures' independent development yielded to a turbulent, productive, destructive amalgamation that characterizes our contemporary world. All outcomes of this transformation, both advancements and atrocities, genuine exchanges and brutal exploitation, can be traced back through chains of causation to moments like the one where Diogo Cão planted his crosses along Africa's coast.

The Atlantic wind sculpting dunes at Cape Cross continues to blow. The seals still swarm beaches. The desert stretches inland, indifferent to human assertions and counterclaims. And beneath those sands or beneath South Atlantic waters lie Diogo Cão's scattered and forgotten bones, marking a grave no one can locate. Yet the repercussions of his voyages still ripple outward, spanning five centuries, shaping the world we inhabit today, reminding us that even vanished individuals leave traces that persist long after memory fades.

The stone cross at Earth's edge remains, replicated in photographs and preserved in museums, a testament to human ambition, our insatiable thirst for knowledge driving us beyond secure boundaries, and the courage to venture into the unknown. However, it also serves as a cautionary reminder: that exploration bears consequences we cannot predict; that opening new worlds often leads to the destruction of old ones; that the crosses we plant signify not only discovery but also dispossession, not only achievements but also the inception of catastrophes. We would do well to consider modern space exploration, where our pursuit of comprehending and inhabiting novel frontiers must be balanced by introspection on potential impacts, a reminder that past lessons remain pertinent. What we unearth today in our endeavors might reshape worlds, echoing cautionary parables etched upon stone crosses from centuries past.

Diogo Cão's legacy: profound, disquieting, indispensable. He contributed to shaping the modern world. Whether we ought to thank him or condemn him for this accomplishment, history provides no straightforward answer. Perhaps all we can do is acknowledge what transpired, neither extolling it unreservedly nor dismissing it entirely, recognizing instead the dreadful complexity of human exploration, the entwining of courage and cruelty, achievement and exploitation, knowledge and destruction as interdependent forces.

The cross persists. The man has vanished. Their shared creation endures.

Pêro de Covilhã

May 7, 1487 – Santarém, Portugal

The gentleman lingering within the antechamber had traversed six realms, and the remnants of foreign soil clung to his footwear. The room was saturated with the fragrance of beeswax tapers, their wavering flames casting a mesmerizing dance upon the ancient walls. A delicate reverberation from spurs striking the tiled floor persisted, an indication of couriers' presence in that space. Pêro de Covilhã remained perfectly still, hands entwined behind his back, gazing intently at the robust oak entrance between him and King João II. On the other side of this portal, the sovereign meticulously scrutinized maps that morphed into legendary narratives. The ocean had been surmounted; now Bartholomeu Dias was on the brink of circumnavigating Africa's extremity. Still, the passage to India, where spices surpassed gold in radiance, was jealously protected by Muslim merchants guarding their dominion fiercely. At this watershed juncture in civilization, Covilhã's forthcoming endeavor would be instrumental in redefining Portugal's worldwide influence: his undertaking was to dismantle this obstruction through a clandestine journey beyond Portuguese territories, blending seamlessly with the merchants and potentates of the East, and reclaiming intelligence pivotal to remold future exploration and trade.

Covilhã remained ignorant of the certainty that he would ceaselessly wander. This unknowing set the stage for the moment about to unfold.

The door opened grandly, and a scribe summoned him. At this precise instance, an uncelebrated middle-aged man without the lineage of nobility or inclusion in tales of admirals or conquerors stepped into the annals of history. This mission was destined to span two decades and transmute the destiny of three continents. History would nearly blot out his existence; however, the world he orchestrated would forever hold a place in memory.

THE MAKING OF AN INVISIBLE MAN

Pero de Covilha was born circa 1460 in the mountain town of Covilha, nestled within Portugal's Serra da Estrela. The winters in this location were prolonged, and wolves roamed freely across the terrain. His family lacked wealth, titles, or any significant connections to influential families in Lisbon. Yet they relied on their necessity-driven ingenuity, which poverty hones. The social and economic conditions of late-medieval Portugal, particularly in rural mountain towns like Covilha, often shaped individuals through hardship and resourcefulness. Living at the periphery of Portugal's political centers, Pero experienced firsthand the limitations and social barriers imposed by birth. As a youth, Pero worked for the Duke of Beja, who would later become King Joao II, performing tasks such as delivering messages and cleaning the stables. These formative experiences not only exposed him to the expectations and workings of court life but also offered him unique insights into power dynamics and social mobility. Despite these humble duties, he was observant and learned from each experience, developing skills that would later enable him to navigate complex social and political environments beyond his home.

Pero had an extraordinary gift, a rare linguistic talent. He mastered Castilian Spanish at court, absorbed Arabic from Moorish servants and traders, and acquired Hebrew through conversations with Jewish merchants whose networks extended from Lisbon to Cairo and Constantinople. His talent was so captivating that one afternoon, as he navigated the bustling market in Lisbon, a heated exchange broke out between a Moorish merchant and a local official. The tension was palpable, the air thick with misunderstanding. Pero, standing nearby, smoothly interjected in flawless Arabic, satisfying the merchant's concerns and offering a bridge of communication. The official, surprised yet relieved, watched as Pero's voice shifted seamlessly to Hebrew to reassure a Jewish money lender standing anxiously by. Each new language unlocked another door for Pero, and he amassed them with relentless determination. In the rigid social order of late-medieval Iberia, where one's birth often predetermined one's life's trajectory, his linguistic capabilities afforded him an unusual degree of mobility. While others were bound to their assigned roles by birth, Pero could navigate these societal boundaries due to his cross-cultural communication skills, which proved transformative and radical.

The young King Joao II recognized Pero's exceptional abilities. Known as "O Príncipe Perfeito" (the Perfect Prince) for his forward-thinking vision, he envisioned Portugal's future not in territorial conflicts with Castile but in the exploration of unknown lands beyond current maps. Joao understood that information held more power than military might; a well-placed spy could outperform an army of a thousand soldiers. He established a network of covert agents whose identities remained undisclosed, missions were kept confidential, and their achievements were kept silent.

Pero de Covilha became one of these anonymous operatives. In the late 1480s, his diplomatic endeavors evolved into acts of espionage. Discreetly, he ventured to Spain to observe the military intentions of Ferdinand and Isabella. He entered North Africa covertly, sketching ports and fortresses under the guise of a merchant. His command of Arabic grew so proficient that he could convincingly pass as a native speaker. To avoid detection as a Christian, he learned local customs, including attire, diet, and religious practices; awareness of being discovered could lead to enslavement or death. This ability to blend into foreign environments both safeguarded his life and propelled him onto a path of solitude and exile.

By 1487, King Joao faced his most daunting challenge. The Portuguese had been gradually advancing along Africa's western coast for decades, establishing trade outposts, subjugating local leaders, and seeking a sea route to India to disrupt the Venetian-Mamluk monopoly on spices. Yet no European had successfully reached India by sea; it remained uncertain if such a journey was feasible. Arab and Indian merchants clearly navigated between East Africa and India, their dhows capitalizing on the monsoon winds across the Indian Ocean to carry cargoes of pepper, cinnamon, cloves, and nutmeg. However, they guarded these routes jealously, deceiving European questioners, diverting information, and obscuring facts. It was only natural for them not to relinquish such a valuable advantage; the spice trade was worth incalculable riches.

King Joao required genuine intelligence on these Muslim trade routes. He sought someone capable of traveling through these paths, reaching India, and determining whether the Indian Ocean was navigable or impassable. He needed precise knowledge about winds, currents, ports, rulers, and uncharted territories. This individual had to accomplish this without being recognized as a Portuguese agent; alerting the Ottomans, Mamluks, or other groups opposed to European presence was not an option.

The king selected two individuals for this challenging mission. One was Afonso de Paiva, whose historical record is minimal except for his name and fate. The other was Pero de Covilha, who awaited instructions in Santarem to embark on a journey to the farthest reaches of known civilization.

INTO THE LABYRINTH

In a quest that seemed both straightforward and fraught with peril, Covilhã and Paiva embarked on their journey in May 1487, setting sail from Lisbon. Their path led them to Barcelona and then Naples as they sought to blend into the bustling merchant community, meticulously gathering intelligence while keeping a low profile. Upon their arrival in Alexandria in late 1487, the stark contrast between the city's former glory and its current state became apparent. The once-majestic lighthouse had vanished, and the renowned library was long gone, yet the port bustled with ships from all corners of the Mediterranean.

The markets offered a dazzling array of exotic goods: silks from China, Chinese porcelains, indigo from India, frankincense from Arabia, and slaves from Africa. Most coveted of all were spices, invaluable not only for their flavor but also for preserving food and their purported healing properties. During their time in Alexandria, Covilhã and Paiva absorbed as much information as possible about the spice trade. This trade was tightly controlled by the Mamluk Sultanate: Indian merchants sold spices at ports like Jeddah, after which the goods were transported by Arab traders to Cairo and Alexandria. There, Venetian merchants acquired them for European markets, with prices rising significantly at each stage.

By early 1488, the two Portuguese explorers went their separate ways in Cairo. Paiva set off towards Ethiopia in search of Prester John, a legendary Christian king whose existence had long fascinated Europeans. Meanwhile, Covilhã journeyed eastward, joining a caravan destined for Aden by the Red Sea's mouth. It was at this point that their paths forever diverged; they would never reunite. Paiva's ultimate fate remains an enigma of history, with some accounts suggesting he perished in Ethiopia, while others hint at him reaching his target but failing to return.

Now on his own, Covilhã found himself more adaptable and less conspicuous as a single individual. To avoid drawing attention, he blended seamlessly into the crowd and boarded a ship from Aden bound for India. The voyage was arduous; it took weeks, carried along by the southwest monsoon that favored trade between the Middle East and the Malabar Coast. During the sweltering heat, cramped conditions, and the stench of unwashed bodies and spoiled cargo, Covilhã's thoughts often returned to his homeland and the uncertainty of his mission.

Gradually, the outline of land emerged on the horizon - the Western Ghats rose sharply from the sea, adorned with coconut palms and villages. Calicut, the world's preeminent spice port, finally came into view, its harbor bustling with ships from far and wide.

THE PEPPER KINGDOM

In the year of our Lord 1488, Calicut was a marvel akin to the grandeur of ancient Alexandria. Under the Zamorin, who ruled over the Malabar Coast, lay a vast trading empire stretching across the Indian Ocean. Merchants came from all corners: Arabia, Persia, East Africa, Southeast Asia, and even China. The markets were awash with pepper, quantities so vast that they astounded European travelers accustomed to only a few pounds of such a precious commodity. In Calicut, tons of pepper were sold without fuss as if it were merely another crop. As Covilhã meticulously documented prices, shipping costs, quantities, and trading seasons, the pressure of time became tangible. The imminent shift in the monsoon winds loomed overhead, threatening to close the brief window that allowed for favorable navigation across the Indian Ocean. This impending deadline instilled a sense of urgency in Covilhã's observations, compelling him to act swiftly, aware that any delay could jeopardize the success of his mission and the intelligence vital to Portugal's ambitions.

The monsoon dictated all comings and goings. From April to September, the southwest monsoon allowed navigation from the Red Sea and Persian Gulf to India's shores. The northeast monsoon from October to March forced ships back towards the west. This ebb and flow of winds set the rhythm for trade, voyage timing, and the accumulation of goods in warehouses awaiting favorable winds. Covilhã immediately grasped the implications for Portuguese ambitions. By circumnavigating Africa, they could disrupt this established system by employing the same seasonal cycles that had long granted Arab and Indian merchants control over maritime commerce. Buying spices directly from Indian suppliers would dismantle existing trade monopolies, reduce European dependence on intermediary powers, and redefine economic exchanges. The market was flooded with affordable pepper, threatening the financial stranglehold of Venetian and Arab traders, shifting wealth and influence to Portugal. This knowledge was crucial for Vasco da Gama's historic voyage in 1498, which revolutionized global trade networks and contributed to the decline of traditional powers in favor of emerging maritime empires.

But first, a survivor had to bear witness to these truths and bring this knowledge home to Portugal. Covilhã spent months in Calicut before moving on to other Malabar ports - Cochin, Cannanore, Quilon. He observed shipbuilding techniques using coconut-fiber rope instead of iron nails, seemingly fragile but surprisingly seaworthy to European eyes. Covilhã learned about India's caste system, religious divisions between Hindus and Muslims, and political rivalries between the Zamorin and neighboring rulers. All this gathered intelligence was vital for Portuguese strategists planning their incursion into these waters.

From India, Covilhã journeyed to the Persian Gulf, visiting Hormuz, an almost impossibly wealthy island that controlled access to Persia and Central Asia. Its merchants were adorned in silk and jewels, and its markets were stocked with goods from across Asia Minor. The Portuguese would seize it by force in 1507, only recognizing its true value after Covilhã's reports. Yet in 1488, they remained unknown here, still grappling with rounding the Cape of Good Hope and decades away from the violent conquest that would establish their dominion over the Indian Ocean.

Covilhã returned to East Africa and arrived at Sofala, now located in Mozambique. Arab traders had frequented these shores for centuries, exchanging fabric and trinkets for gold and ivory with landlocked kingdoms. Covilhã documented harbors, trade goods, and political landscapes. While European maps depicted much of Africa as a mysterious coastline, he was assembling an exhaustive picture of the entire trading system: from African interior gold mines to Malabar spice warehouses, from the Persian Gulf to the Red Sea.

In Sofala, Covilhã found himself in a tense negotiation scene that vividly illustrated the complex trade dynamics. Amid a bustling marketplace, he encountered an Arab trader unwilling to part with a shipment of ivory unless he paid an exorbitant price. The air was thick with anticipation as voices rose and fell in a symphony of bargaining. As Covilhã haggled, he drew on his extensive knowledge of local customs and languages, seamlessly blending into the surroundings. The scene not only highlighted his skills but also embodied the intricate web of commerce he sought to understand. More significantly, this episode underscored that his ability to navigate such exchanges was not simply a personal triumph but a testament to the indispensable role of cultural fluency and adaptability in gathering intelligence within protected, competitive trading networks. No European had carried such comprehensive knowledge within memory, forced to commit everything to mind for fear of discovery and exposure of his true intentions. In this moment, the negotiation became both a microcosm of the global struggle for economic influence and an illustration of how Covilhã's methods foreshadowed the increasing reliance on critical, firsthand intelligence to inform European expansion.

By 1490, after three years of exploration, Covilhã returned to Cairo to meet Paiva as agreed, but was left in suspense; Paiva never arrived. Covilhã anxiously awaited news, pondering whether Paiva had fallen ill, been apprehended, or lost along the journey. Eventually, he encountered two Portuguese Jews, Rabbi Abraham of Beja and Joseph the Shoemaker, who had been dispatched by King João with fresh instructions. They informed him that Paiva had passed away, though details were scarce due to the communication challenges.

New orders awaited Covilhã: continue the pursuit for Prester John by traveling to Ethiopia and establishing contact with its Christian kingdom. Only after completing this mission could he return home.

At this juncture, Covilhã likely realized his life had irrevocably changed. The first assignment, fraught with peril yet promising a return home, had now morphed into a significantly longer commitment. Ethiopia lay distant and surrounded by Muslim territories, presenting formidable barriers to entry and even greater challenges for departure. Nevertheless, Covilhã pledged service to the crown, armed with vital intelligence he was duty-bound to share. Instead of seeking escape, he prepared detailed letters to be conveyed back to Portugal by the Jewish messengers.

These letters were potentially the most significant intelligence of the century. Covilhã meticulously described the Indian spice trade, mapped key ports, elucidated monsoon patterns, and confirmed African traffic to India. He demonstrated that the Indian Ocean was navigable and connected it to the Atlantic near Africa's southern tip, proving access to India by sea from Europe was within reach. These missives, reaching King João circa 1491, though now lost to us, undeniably shaped Portuguese strategy for years. Vasco da Gama's successful voyage to India in 1498 relied on Covilhã's comprehensive reports.

After dispatching his findings, Covilhã embarked on his Ethiopian mission. He navigated up the Nile and eastward across the desert, arriving in the Ethiopian highlands around 1493. There, Pêro de Covilhã vanished from historical records for thirty years.

THE GILDED CAGE

Ethiopia during the 1490s stood as an extraordinary cultural and political entity: a Christian realm surrounded by Muslim nations, preserving its ancient form of Orthodox Christianity in seclusion from Rome and Constantinople. The lineage of Ethiopian emperors was traced back to King Solomon and the Queen of Sheba, who ruled over mountains with rugged terrain and fierce defenders. Limited contact with the outside world occurred through occasional pilgrims visiting Jerusalem and travelers who dared to venture through Muslim-held lands separating Ethiopia from the Mediterranean.

Upon Covilhã's arrival, Emperor Eskender, or possibly another ruler at that time, eagerly received the Portuguese traveler. Here was a European Christian adept in Arabic, well-traveled within the Muslim world, bearing news of Portugal and its rulers. Honest in his presentation, Covilhã divulged his mission, spoke of Portugal's quest for a sea route to India, and proposed potential alliances and trade.

The emperor welcomed him, showered him with honors and titles, and unequivocally communicated that Covilhã was not permitted to depart. This situation did not equate to conventional imprisonment; he neither faced confinement in dungeons nor physical restraint. Instead, he was granted lands, income, and eventually married an Ethiopian woman who bore his children. His living conditions were comfortable, possibly even luxurious for the era's standards.

Yet, despite this comfortable existence, Covilhã remained a prisoner, detained by a government that valued his knowledge too much to risk his sharing it elsewhere or, if he were set free, its potential use by rival powers like the Ottomans or Mamluks. This situation mirrored traditional Ethiopian practices of indefinitely retaining foreign visitors to prevent them from disclosing vulnerabilities about the kingdom.

Time elapsed, with King João II passing in 1495 and Manuel I succeeding. The Portuguese had already rounded the Cape of Good Hope by Bartholomeu Dias in 1488, followed by Vasco da Gama's arrival in India in 1498, effectively opening the sea route that Covilhã's intelligence had facilitated. Portugal established their Estado da Índia (Asian empire), conquering places like Goa, Malacca, and Hormuz, destinations that Covilhã had personally described.

Portugal won pivotal naval battles, defeating Arab and Mamluk fleets and dominating global spice trade routes as King João had envisioned. Da Gama was lauded with titles, large pensions, and fame spanning continents, while Covilhã remained in Ethiopia, recognized only through minor possessions awarded by the local emperor. This stark contrast highlighted the uneven rewards of exploration; da Gama's name became synonymous with the Age of Discovery, overshadowing Covilhã, who faded into obscurity despite his crucial contributions to Portugal's ambitions.

Covilhã was ignorant of these events, living a peculiar life in the Ethiopian highlands, belonging neither fully there nor able to return to Portugal. He fathered children fluent in Amharic and Ge'ez, only familiar with their father's homeland through tales. They were entirely African while Covilhã remained perpetually foreign.

In 1520, a Portuguese embassy finally reached Ethiopia. King Manuel I sought an alliance against Ottoman expansion in the Red Sea, and the delegation was led by Rodrigo de Lima, accompanied by soldiers, priests, and, notably, the priest-chronicler Francisco Álvares. His detailed account of Ethiopia, later published, became a crucial European source on the kingdom.

The Portuguese likely encountered Covilhã in 1521 or 1522; now in his sixties, weathered by the highland sun, with a family that had never known Portugal and speaking languages learned long ago. He retained a worn Portuguese rosary, a small memento of home and faith, anchoring him to his past yet isolating him from his present.

Álvares documented their encounter, moved by both fascination and pity. Covilhã was a man who had departed Portugal in his prime, only to find himself stranded at the edge of civilization. He had fulfilled his mission, yet instead of returning to glory, he lived in comfortable exile. Covilhã shared his story with Álvares, recounting his travels through the Muslim world, journeys to India, and his trials in Ethiopia. He longed to hear news of Portugal's achievements and the king and wondered if his efforts had been honored. Álvares assured him that his service had made a significant impact, echoing the hope Covilhã held as he first stepped through the door in Santarém. The anticipation he felt at the start of his journey found a bittersweet resonance; his contributions had indeed paved the way for Portugal's successes, though he could only witness them from afar. Nonetheless, knowing he had played a vital role was a source of quiet pride amidst his enduring solitude.

Despite the presence of fellow Portuguese, Covilhã's situation remained unchanged. Emperor Dawit II declined to release him; the political climate demanded Ethiopian cooperation, and pressing for his freedom risked jeopardizing broader diplomatic objectives. Furthermore, as an older man with an Ethiopian family, Covilhã likely realized that returning to Portugal would mean leaving behind the life and children he had built there.

Covilhã last appeared in historical records in 1526, still in Ethiopia but now considered a senior statesman held in comfortable captivity. After this, silence enveloped him; the date of his death remains unknown, and any possible grave is untraceable. He simply vanished as he was trained to do, dedicating his life to becoming unremarkable, invisible, forgotten.

THE ECHOES

Yet, Covilhã's journey extended its effects far beyond his era. The information he gathered transformed Portugal from a minor European kingdom into a formidable global power. His insights into navigation encouraged the Portuguese to fund expeditions that culminated in Vasco da Gama's groundbreaking voyage. Covilhã charted vital Indian ports, enabling da Gama to arrive early and establish trade relationships in dominance. The detailed intelligence on Indian commerce and politics provided by Covilhã gave the Portuguese an unmatched advantage when they arrived. They knew exactly where to go and who to negotiate with or intimidate, leveraging that knowledge to their strategic benefit.

This dramatic shift in global trade routes was not just about economics; it devastated older networks, such as Venice's, significantly impacting the Mamluk Sultanate's revenue and weakening its position, making it more susceptible to Ottoman conquest. Portuguese ships overwhelmed markets with pepper, displacing Arab merchants who had long dominated the Indian Ocean trade due to their superior military might and trading posts.

Covilhã's journey also demonstrated a more subtle yet profound lesson: European espionage could succeed where conventional warfare could not. By disguising himself and navigating Muslim territories undetected, he showed that information gathering through stealth could be just as effective as force. This insight was instrumental for subsequent European powers as they built sophisticated intelligence networks to further their global ambitions.

The methods Covilhã pioneered, language skills, adopting local customs, using commercial covers, operating alone in hostile lands, became the blueprint for future spies and explorers like Richard Burton or Charles Doughty, who explored Muslim territories in later centuries. They followed paths that he had first forged, though without his recognition.

Yet, Covilhã's contributions were largely overlooked in his lifetime. Unlike Vasco da Gama, whose achievements were celebrated lavishly, Covilhã died in obscurity, known only to a few Portuguese officials who had read his reports. His role was shrouded in secrecy by the very nature of his mission, leaving him without recognition or honors.

For centuries, Covilhã's story remained relatively unknown even in Portugal, overshadowed by the more dramatic tales of navigators and conquistadors. It wasn't until late into the 19th century that Portuguese historians began to appreciate the significance of his intelligence work, recognizing it as pivotal to opening the route to India.

Even today, Covilhã remains largely obscure, with few places outside his hometown in Portugal knowing his name. His story occasionally appears in academic histories but is often swiftly overshadowed by accounts of more famous figures. Unlike many adventurers who inspire novels and films, Covilhã's story lacks dramatic battles or triumphant returns; it's a slow-motion tragedy that doesn't capture the public imagination.

Perhaps this obscurity is fitting for someone whose profession was to remain unnoticed. Covilhã succeeded because he could be forgotten, mastering the art of leaving no trace. History treated him as he had treated himself during his travels, acknowledging him briefly before moving on, leaving him in the margins while others basked in the spotlight.

May 7, 1487 – Santarém, Portugal

The doorway swung open, admitting Pêro de Covilhã into its embrace. At that precise moment, an act of obedience carried him from his known existence into the vast expanse, the unfamiliarity, and solitude he could never have anticipated. Maps would emerge under his guidance, unveiling new worlds, while the empires he helped shape spanned continents. His acquired knowledge revolutionized global commerce and laid the foundation for Portugal's ascent as a world power. Despite the transformative impact of his mission on exploration and trade, Covilhã himself became but a footnote, a faint echo, an obscure figure relegated to history's margins, highlighting how significant contributions might go unrecognized even as they redefine nations' destinies.

He could not foresee, while standing on that threshold, witnessing Arabian deserts, Indian monsoons, or Ethiopian highlands, yet never beholding the mountains of Portugal again. Unaware then that his descendants would speak languages he had learned as an outsider, and that his grandkids would be African in essence except for their distant recollection of a Portuguese grandfather. The success he achieved felt like failure; fulfilling his duty meant enduring perpetual exile, offering impeccable service at the cost of everything he held dear.

Yet perhaps there was a glimmer of understanding within him. Perhaps Covilhã grasped that the world was shifting, that ancient certainties were crumbling, marking the end of isolated kingdoms and uncharted seas. He may have recognized the necessity for pioneers, those willing to venture into the unknown without maps or guides, becoming a link between an old world and the one that was coming into being. Maybe he knew the price involved and paid it voluntarily.

Or maybe he was just a servant who followed orders, executing tasks without pondering their outcomes. He stepped through the door in Santarém and continued onward for forty years until his journey concluded with a burial in the highlands whose location has gone unrecorded.

Regardless, he walked forward. The world transformed. Covilhã faded into obscurity. Though spices still travel along new paths, guided by unfamiliar hands, the allure that once beckoned men to farthest horizons has dwindled to mere profit margins in a world where pepper is cheaply acquired. The system Covilhã helped dismantle is now so distant we cannot visualize the world it once shaped. And his presence has almost vanished, remembered primarily by scholars and a few compatriots who recognize that profound journeys seldom truly end, acknowledging the service of those who disappear into shadows, understanding history often favors obscurity over prominence.

He deserved better than to be forgotten. He deserved to return home. But history grants no one their just deserts. It merely documents what transpired or omits it, moving on to the next challenge, the subsequent voyage, the next overlooked individual stepping through a doorway into the uncharted. Empires remember ships, not the footfalls of those who ventured forth.

Bartolomeu Dias

February 3, 1488. Somewhere off the southern tip of Africa.

The vessel quaked tumultuously. Bartolomeu Dias clutched the railing, his knuckles ashen against the slick wood that threatened to relinquish its hold with every rain-drenched tremor. Sailors cried out amidst the tempest, their voices swallowed by the relentless wind. An hour prior, the main sail had torn asunder, fabric tearing like parchment ripped in fury. Now, the São Cristóvão was adrift, propelled by waves that towered ominously, each one a specter of dread.

Visibility waned, obscured by the departure of the supply ship early on, swallowed by the previous night's chaos. The São Pantaleão, under Pêro's command, vanished behind an impenetrable curtain of rain. For thirteen relentless days, they were battered by the storm, adrift from land for as long as they had ventured beyond the last marked cross and stone that delineated Portugal's known world.

From the stern, the navigator scrutinized the heavens for a hint of direction amidst the churning clouds. The sun was obscured, hidden so completely its whereabouts could not be discerned. Estimating their position had become a gamble. Dias surmised they might have been carried south and then east, or perhaps west, direction itself now a conundrum. In this turmoil, even the compass spun aimlessly, useless without the sun or stars to guide.

A wave crashed over the bow, water thundering across the deck, swirling around waists and tugging at the crew's legs with an appetite for their souls. Dias watched a sailor lose his grip, sliding towards the precipice, only to be saved by two companions' desperate intervention. Their exhaustion was palpable; sleep snatched in fits between the relentless waves, never enough to mend their weary bodies or minds.

The captain closed his eyes, envisioning Lisbon: the Tagus River glimmering at dawn, King João II's voice firm and resolute, issuing a decree. "Find the southern tip of Africa," he had commanded, "Round it and open the passage to India." These words, once so clear and direct, now seemed an insurmountable task. Portuguese explorers had ventured down Africa's unending coast for 60 years, each voyage pushing boundaries and claiming new havens. Yet, Africa stretched endlessly, veiling its secrets with enigma, humbling those who dared to advance.

Dias opened his eyes. Dawn was a concept uncertain in the face of persistent darkness. Yet, shadows retreated and gray crept in. Despite the ongoing tumult, an imperceptible shift had emerged. He felt it within his bones before rationality caught up: the storm's grip was waning.

By noon, patches of blue dared to manifest. By evening, the wind diminished to a persistent though less savage gale. Noon gifted them with blue skies peeking through, while by evening, the breeze steadied in forcefulness. That night, stars reemerged as silent witnesses after the tempest's fury. Under their watchful gaze, Dias took his bearings and realized an astonishing truth: they had been carried far south, then eastward beyond the extent Africa was supposed to reach on any chart. They unknowingly traversed through the storm at a southern point, thus opening the way to India. The dream that fueled Portugal's navigators for half a century might be realized; the world could very well differ from anyone's imagination.

Yet, land needed to be found and survival ensured. Most critically, they must understand what transpired during their thirteen days of struggle amidst the seas' darkness.

This tale unfolds the realization of this epoch. It narrates how one man and three small ships reshaped the world's map not through conquest or sudden enlightenment but through endurance and skill. Central to this narrative is an inquiry: How did resilience redefine the map? They exemplified the rare mettle required to cross every known boundary, venturing into waters that, on European charts, were nothing but voids. This exploration of endurance's power promises to heighten anticipation with each chapter, revealing how the relentless pursuit of an impossible goal reshapes destinies and geopolitics alike. The forthcoming chapters will trace this transformation, delving into historical context, the voyage itself, its repercussions for both European and African realms, and the lasting legacy of Dias's achievement.

THE MAKING OF A NAVIGATOR

Bartolomeu Dias is an understated historical figure. Unlike Columbus, who became renowned for his discoveries, Dias left behind minimal personal accounts. In contrast to Vasco da Gama, whose fame was largely due to Dias's accomplishments, Dias's success did not gain widespread recognition. Unlike Magellan, who lent his name to a strait, the nature of Dias's legacy remains ambiguous. Although overshadowed by others, he is known for mentions in others' accounts and for significant changes that occurred after his voyage. However, piecing together his identity poses challenges, as it relies heavily on sparse references and official documents.

Born circa 1450, likely in the Algarve, southern Portugal, where the land meets the Atlantic and the scent of salt permeates everything. The Algarve was renowned for producing sailors, much like other regions produced farmers or merchants. Growing up, boys witnessed ships arriving and departing from the port, heard tales of Guinea and the islands, and learned navigation skills as they learned to walk. Bartolomeu's lineage had strong ties to both the sea and the court. His father, Dinis Dias, sailed to Africa in the 1440s, reaching Cape Verde during early Portuguese exploration. His brother Pêro would become a navigator, while his relative Diogo Dias accompanied Vasco da Gama. The seafaring spirit ran deep in their blood.

Bartolomeu's career began on merchant vessels. He made voyages to Flanders or England, gaining knowledge of how ships responded to wind and currents. He mastered weather predictions and maintained crew well-being during long expeditions. Following this, he served in the royal fleet, transitioning from a merchant sailor to a king's navigator. By the 1470s, Dias received royal assignments. In 1474, he was appointed as sailing master of São Jorge da Mina fortress, the Portuguese trading post on Africa's Gold Coast. This assignment demonstrated his competence and adaptability, showcasing a blend of diplomatic acumen and navigational expertise.

Life within the fortress at Mina was demanding. Maintained through trade, force, and tactful interactions with local rulers, it was a challenging environment where sailors learned to handle more than just the sea; they also had to navigate local politics. The realization dawned upon them that Africa was vast and intricate, indifferent to European ambitions. They gained insights into resources such as gold and slaves, which would fuel Portugal's growth, alongside learning about deadly diseases like fevers that claimed more European lives than conflicts ever did, making the coast of Africa perilous.

Dias not only endured these challenges but distinguished himself. By 1486, he had been honored with the title Knight of the Royal Court, a distinction symbolizing royal favor and trust. When King João II began planning the most ambitious expedition down the African coast yet, Dias's name was among those considered for command. The king meticulously prepared for this journey. He consulted charts and accounts from past voyages, commissioned ships tailored for long-distance exploration, and handpicked crew members with a careful balance of experience and youth, caution and daring.

By the time Dias reached adulthood, Portuguese ships had crossed the equator and ventured further south. In 1482, Diogo Cão reached the Kongo River, suggesting the existence of vast kingdoms within the interior. Two years later, Cão planted his markers at Cape Cross in present-day Namibia, deepening the mystery of Africa's extent. Its southern boundary eluded them, curving slightly east but seemingly endless. How far did it stretch? Did it link to a large southern landmass, or did it end in the Indian Ocean?

This question obsessed King João II. Access to India meant securing lucrative spice trade routes and coveted commodities like pepper, cinnamon, cloves, and nutmeg, which were worth more than gold by weight. These spices were as valuable as a month's rent in Lisbon, rivaling the market value of precious metals. Portugal envisioned itself as a crusading nation extending the Reconquista across the sea. Each marker planted was a Christian triumph, every African contact a potential ally against Islam. Rumors of Prester John, a legendary Christian king in Africa with an empire to ally with against Muslim forces, further motivated navigators, despite being unfounded.

In essence, Portugal's navigators were not merely exploring; they were waging a holy war through their maritime maps and accounts.

Bartolomeu Dias ventured into a multifaceted scenario in 1487, driven by a range of forces: economic necessities, religious fervor, scientific inquiry, national pride, and personal aspirations. King João II convened him and outlined the mission. Dias was entrusted with leading an expedition aboard three ships, with the ambitious goal of venturing farther along Africa's coastline than any voyager had previously. The objective encompassed discovering the continent's southern extremity and, if feasible, circumnavigating it to access the Indian Ocean. The king provided meticulous instructions, drawing upon decades of navigational expertise. Dias was tasked with hugging the shoreline closely, conducting soundings, observing significant landmarks, and documenting the positions of river mouths and harbors. Engagements with African communities were mandated to gather intelligence about southern kingdoms and any Christian rulers they might encounter. Markers were to be planted at strategic points to assert territorial claims and chronicle advancements.

Above all, Dias was required to meticulously record navigational data wherever possible. The Portuguese had honed advanced navigation techniques that incorporated multiple methodologies. Dead-reckoning computed positions through careful measurements of speed, time, and direction. Celestial navigation determined latitude by measuring the sun's or the North Star's altitude above the horizon. Navigators maintained detailed journals, noting weather patterns, currents, coastal features, and any other relevant data for future expeditions. This methodical approach distinguished Portuguese exploration from earlier ventures; it was not just a journey southward but an act of mapping a viable route and creating enduring knowledge.

The ships prepared in Lisbon in 1487 were tailored to their mission, reflecting decades of Portuguese shipbuilding innovations. The lead vessels, São Cristóvão and São Pantaleão, were caravels, compact yet agile ships adept at navigating near the wind due to their lateen sails, which allowed for tacking against the wind, a critical skill when coast-hugging. They measured around seventy feet in length, displaced roughly fifty tons, and each carried approximately thirty crew members. The third craft was a supply ship, wider and slower, carrying surplus provisions and designed to be abandoned once its cargo was transferred.

Caravels revolutionized ship design with their lateen sails, triangular canvas sheets hoisted on long yards that enabled them to sail against the wind, crucial when navigating close to shore. Their shallow draft allowed navigation into estuaries and near coastlines without grounding. Despite these advantages, life onboard was austere; space below deck was limited, accommodating cargo and supplies with barely enough room for rest, let alone comfort. Sailors found wherever they could sleep among provisions in the hold or on deck under makeshift shelters. Privacy was nonexistent amidst a constant environment of unwashed bodies, bilge water, spoiled food, and human waste.

Provisions for long voyages consisted primarily of ship biscuit, salted meats and fish, dried beans, wine, and water. The biscuit, an extremely hard bread that could last months if kept dry, provided sustenance and bulk. Salted provisions provided protein but required soaking before consumption, resulting in tough, unpalatable textures. Fresh food preservation was impossible; within weeks at sea, all greenery or freshness rotted. Sailors developed scurvy due to vitamin C deficiency, leading to swollen gums, loosened teeth, reopened wounds, and dysentery from contaminated water, along with mysterious fevers. Mortality rates on extended voyages could reach twenty or thirty percent.

Dias was acutely aware of these perils; so were all seasoned navigators. The crux was not whether sailors would fall ill or perish but how many would survive the journey. Speed to the objective and swift return before the disease took hold were paramount. However, haste and careful exploration were conflicting demands. King João required comprehensive records, necessitating stops for measurements, shore visits, and marker placement at significant locations. Each halt cost time, and each additional week at sea escalated the danger.

The three ships set sail from Lisbon on a summer day in 1487, the precise date lost to oblivion. It fell between July and August. Dias commanded the São Cristóvão, his brother Pêro helmed the São Pantaleão, and João Infante led the supply ship. They carried stone crosses blessed by the king's chaplain for marking critical points. Two interpreters accompanied them, Africans who had been in Portugal and trained as intermediaries.

These facts are preserved in later records and royal documents. Yet, we remain ignorant of Dias's emotions as Lisbon receded from view, his apprehension or confidence, let alone the sentiments of his sailors as they departed for history's most distant Portuguese voyage. Did they sing hymns, pray, or sit silently as Portugal disappeared behind them? What is known is their southward trajectory following a familiar route shaped by decades of repeated journeys. Through Cape Saint Vincent and past the Canary Islands, they embarked on the Atlantic's Volta do Mar, a circular wind-and-current pattern that facilitates navigation. Portuguese pilots adeptly utilized these patterns instead of combating the ocean's forces, swinging wide into the Atlantic before returning towards Africa, harnessing northeast trades and the Canaries Current, then southeast trades closer to land. Though it added distance, this method saved time and effort.

INTO UNKNOWN WATERS

The initial passage of time was defined by regular travel procedures. The crew navigated efficiently and arrived at the Cape Verde Islands without incident, replenishing their supplies and resources. These islands, under Portuguese control, were inhabited by colonists who managed to cultivate some crops despite challenging weather conditions that frequently led to droughts. Dias would likely consult with fellow sailors familiar with these waters to update his maps and carry out necessary repairs on the rigging affected by Atlantic winds. Additionally, he might have granted his crew a brief rest before continuing their journey south.

Beyond Cape Verde, they entered less familiar territories but ones still mapped by Portuguese explorers. The African coast was known to them, with interactions made with local villages and documentation of coastal features. Trading occurred at these stops, exchanging Portuguese goods such as textiles, metal tools, and wine for products like gold dust and ivory from Africa. These exchanges adhered to established protocols; however, Portuguese sailors did not venture far inland due to the risks posed by numerous kingdoms that guarded their territories against uninvited outsiders. The interactions along the coast were carefully managed, proving profitable yet tense.

Dias would likely have sought information about lands further south through local interpreters. Responses would be limited by language barriers and the coastal dwellers' lack of knowledge regarding distant regions. Africa was comprised of various kingdoms, ethnic groups, and trading networks, so while someone on the Gold Coast might know their neighbors, information about the Kongo region or beyond was likely secondhand.

By December, they had surpassed Diogo Cão's farthest marker at Cape Cross, entering uncharted territory where every mile brought new possibilities. Every headland could potentially be their turning point toward the south. They felt intense excitement mixed with apprehension as they navigated unknown waters, aware of the risk of unforeseen hazards, including reefs and unpredictable currents that could ground ships.

The coast trended southwest, still far from their destination. As they moved further south, colder temperatures surprised them; conventional wisdom held that southern latitudes should be hot, but cold water from the Benguela Current, a cold stream originating in Antarctica, led to cold air. This brought both support for vast marine life and conditions unsuitable for human habitation, which are accustomed to Portugal's Mediterranean climate.

The landscape was harsh, with the Namib Desert meeting the ocean in towering sand dunes. There were no natural harbors or river mouths suitable for shelter and freshwater sources. The surf was dangerously powerful, making landing perilous or impossible. The ships continued southward, facing nothing but dunes and crashing waves.

On December 8th, Dias marked a spot at Angra dos Ilhéus (Bay of the Islands) with a stone and bronze cross, symbolizing Portugal's claim over this remote coastline. They proceeded onward as the supply ship became impractical due to its excess weight and crew needs. Near present-day Lüderitz in Namibia, they left nine men ashore with provisions while others continued south. The agreement was for Dias to return within a certain timeframe; if he didn't, those left behind were to make their way back north, seeking friendly territories until reaching Portuguese lands.

Leaving men behind on long voyages reduced the number of mouths to feed and established presences capable of gathering information. However, these men faced isolation, vulnerability, psychological strain, and uncertain rescue prospects.

The two remaining ships pushed south with mounting urgency as supplies dwindled daily. By January 1488, five or six months into their journey, the crew battled scurvy and low freshwater reserves amidst harsh conditions that took a toll on their health and morale.

Late January or early February brought a storm from the southwest, battering them for thirteen days with unpredictable winds, monstrous waves, and the constant threat of being overwhelmed by water. They had to run before the wind, barely maintaining steerage with minimal canvas sail as they were driven southward. Separated during this ordeal, each captain fought alone against the elements, sacrificing cargo to lighten their vessels and praying desperately for survival.

After thirteen days of turmoil without sunlit navigation, the storm gradually subsided, leaving them in calmer waters but devoid of land sightings. Dias calculated their position, realizing they had drifted far south and were potentially beyond the southern tip of Africa. Turning northeast to find any coastline, they sailed for several days without success until on February 3rd, a lookout spotted land, marking the end of their ordeal thirteen days after the storm ceased.

THE TURNING POINT

The coastline that appeared before them was like nothing they had seen. It did not follow the usual southwest trend but instead ran east to west. This clear sign meant they had indeed passed the southern end of Africa, and now they were on a coast facing the Indian Ocean. The impossible had become reality, as the barrier blocking European access to the East for centuries had fallen before three small ships and a storm.

The bay where they landed was sheltered and had fresh water from a nearby river. Dias named it Bahia dos Vaqueiros or Bay of the Herdsmen because he saw Africans herding cattle in the hills above the shore. These were the Khoikhoi people who lived along this coast as pastoralists. The Portuguese attempted to make contact, but communication failed because the interpreters they had brought spoke West African languages that were of no use here. Therefore, communication devolved into gestures. Initially, the Khoikhoi were not hostile but wary of these strange men who suddenly appeared from the sea.

The encounter quickly deteriorated as one of Dias's men exploring inland was struck and wounded by a thrown stone. In response, Dias ordered a crossbow to be fired, killing one of the Khoikhoi. The others fled. Although it seemed like a small incident with only one dead man, worth barely noting in the chronicles, this event established a pattern that would repeat endlessly in years to come: European arrival meant violence, contact led to conflict, and discovery often resulted in death.

From the Khoikhoi's perspective, this land might have been viewed not just as territory but as a living tapestry interwoven with sacred sites and seasonal rhythms. Their pasturelands were vital to their herds and way of life. The sudden appearance of strangers with unfamiliar voices and gestures, both curious and cautious, created a moment of shared wonder against the backdrop of the sun casting elongated shadows on the hills. As the Khoikhoi observed these Europeans inspecting their surroundings and gesturing in ways that hinted at a desire to communicate, there might have been a fleeting spark of shared curiosity, a brief moment where two worlds considered bridging the unknown dividing them. However, this fragile peace shattered abruptly when a stone thrown in defense or confusion interrupted the scene, wounding one of Dias's men. Shock swept through both groups, turning the air tense and swiftly replacing any emerging connection with conflict. In retaliation, Dias ordered a crossbow to be fired, resulting in the death of a Khoikhoi individual. The others fled, foretelling a pattern that would repeat in the years to follow: European arrival often incited violence, their contact led to conflict, and discovery tragically resulted in death.

Dias did not linger at the bay; he had confirmed the crucial fact that the coast extended eastward. Now he needed to determine how far it continued, whether it curved back south or maintained this new direction. The ships sailed on for several hundred miles along the shoreline, stopping occasionally to take soundings and readings. At a small island near the mouth of what is now known as the Great Fish River, Dias planted another marker. This would prove to be the farthest east he reached, roughly around latitude 33 degrees south, similar to Buenos Aires or Sydney.

Beyond this point, his crew's morale began to deteriorate. They were exhausted and sick, and the ships were battered and dangerously low on supplies. Although they had accomplished their primary mission of finding and rounding the southern tip of Africa, continuing east toward India as King João suggested seemed like suicide.

Dias called a council meeting aboard one of the ships on February 12, 1488, which was standard practice for Portuguese expeditions. While the captain held ultimate authority, major decisions were made collectively with input from officers and experienced sailors. Dias proposed sailing further east for several more days to be absolutely certain that the coast maintained its direction. However, the officers refused, demanding to be returned home. Dias could have overruled them by invoking his authority as captain, but was outnumbered and probably agreed in his heart. The crew had been at sea for months, and returning would take even longer, risking further loss of life.

A compromise was reached: they would sail east for three more days, after which, if the coast still favored an eastern direction, they would turn back. The three days passed with the coastline maintaining its easterly trend. And on February 15, at a small island where Dias had planted his last marker, they turned the ships west and began their journey home.

The return voyage retraced their route, but now filled with knowledge, replacing any lingering mystery. They saw Africa's southern tip and named it Cabo das Tormentas or Cape of Storms, acknowledging the storm that drove them past this point. Later, King João renamed it Cabo da Boa Esperança, the Cape of Good Hope, emphasizing its role as a gateway to the East rather than an insurmountable barrier.

The headland jutted into the ocean like a fist with cliffs rising from waves that never stopped moving. The water here was dangerous, representing the meeting point of two oceans where currents collided and weather systems brewed storms. Dias carefully studied it, noting its coordinates, hazards surrounding it, and the best approach for ships attempting to round it in either direction. This information would be crucial for future expeditions.

They later stopped to retrieve the nine men left on a supply ship, with only three surviving out of the original nine. Six had died, likely from attacks by local people, disease, or despair. The details are lost, but these three survivors were emaciated and traumatized, barely functional. They were taken aboard, and the journey north continued.

The journey home was challenging. Scurvy had spread rapidly across the ships, causing men's gums to swell until eating became unbearable. Their teeth fell out, and old wounds refused to heal properly. Their gums wept dark blood, a tangible symbol of their suffering. Sailors of this era were baffled by the ailment; they couldn't comprehend that it stemmed from a vitamin C deficiency and could be prevented with fresh fruits or vegetables. They only knew that long voyages caused such suffering, and once we reached land and ate fresh food, many recovered.

Dias navigated northward at the pace our ships permitted. We made stops at the Gold Coast for trading and resupplying, then again at the Cape Verde Islands. At each stop, we learned that other Portuguese vessels were actively engaged in trade and exploration in these waters. Dias gained insight into what had transpired globally during his absence: political shifts, novel expeditions, minor conflicts, and burgeoning commerce.

In December 1488, exactly sixteen months after our departure, the two caravels entered the Tagus River and sailed up to Lisbon. The city appeared largely unchanged from how we remembered it, yet those of us returning were visibly aged and emaciated by our experiences. Out of approximately sixty men who had accompanied Dias, around forty returned; twenty others succumbed at sea or were abandoned along the way. These figures, though sorrowful, represented a better fate than some expeditions suffered, which lost half or more of their crews.

King João II welcomed Dias to court, and while details are scarce, their meeting must have been remarkable. Dias presented his findings, charts, and navigational records, detailing the coast beyond Cape Cross, the great tempest we endured, the realization that the shore turned eastward, the markers we planted, and the distances we traversed. He offered irrefutable proof that Africa could be circumnavigated, affirming the feasibility of a route to India.

João's response was measured; he expressed contentment. He granted Dias a modest pension in recognition of his service. However, rather than immediately commissioning another expedition, reasons for delay remain unclear, perhaps consultation with other navigators and astronomers, financial constraints, or an understanding that rounding the cape was merely the first step toward reaching India, requiring yet another ambitious endeavor he wasn't ready to sanction.

Dias received his title, his pension, and his place in the annals of history. Yet, the glory expected didn't manifest. Within a few years, King João II would commission another navigator, Vasco da Gama, to follow Dias's route and complete the journey to India. Da Gama's expedition succeeded, reaching Calicut in 1498 and returning with spices that confirmed the commercial viability of the sea route. Da Gama was lauded, celebrated, and enriched. Meanwhile, Dias remained overshadowed, his achievement recognized but somehow incomplete; he found the gateway but did not traverse it himself.

THE MEANING OF DISCOVERY

To comprehend Dias's achievements, one must first grasp what the world looked like before his journey commenced. The understanding of geography in late fifteenth-century Europe was an amalgamation of classical knowledge, medieval myths, contemporary observations, and inherent inaccuracies. This foundation rested on Ptolemy's Geography, a compendium composed in the second century AD that was rediscovered by Western scholars in the fifteenth century. Ptolemy meticulously charted known lands, provided coordinates for major cities, and described continents; yet, his work was profoundly mistaken on critical points.

The decks held ancient charts, and below them, an intense debate unfolded among Dias's crew. The Ptolemaic scrolls were unrolled, revealing precise illustrations of a landlocked Indian Ocean, in contrast to the saturated, dripping-wet compass that defied these depictions with each movement of the São Cristóvão. "Though stars may be veiled, and ancient texts revered," argued the pilot, "this needle does not deceive; it points to a world distinct from Ptolemy's." Such dialogue encapsulated the growing tension between established ancient theories and the practical challenges navigators faced as they charted unknown waters with their intellects and instruments.

However, Portuguese exploration systematically refuted classical geography. Each voyage introduced new information contradicting earlier sources. The southern ocean did not boil; equatorial Africa teemed with life rather than the scorched wasteland depicted by certain authors; the African coast did not extend endlessly southward as Ptolemy's map suggested. This map was in error, and how and why became significant inquiries.

Dias's expedition addressed a pivotal question: Africa was not linked to a southern continent but instead formed a vast yet finite peninsula with an open ocean beyond its southern tip, proving the route to India navigable for Portugal's shipbuilding capabilities. The theoretical transitioned into practicality; the dream materialized as achievable.

This revelation quickly dawned in certain circles. Portuguese navigators realized that everything had shifted. Astronomers and mathematicians perceived a world larger and differently arranged than ancient records claimed. Merchants discerned commercial opportunities, while King João recognized Portugal's immense strategic advantage. Directly reaching India by Portuguese ships could shatter the Venetian and Muslim monopoly on spices, transforming Portugal's economy.

Yet, these implications took time to percolate. Information dissemination was sluggish during the fifteenth century. The news of Dias's voyage trickled into other European courts in mangled forms. Spain knew, undoubtedly, as did Venice; German and Italian geographers were aware too. However, ordinary people, even educated ones, remained oblivious to the world map being redrawn, clinging to Ptolemy's landlocked Indian Ocean, unknown southern continents, and impassable seas.

Portugal maintained secrecy over its discoveries; navigational intelligence was a valuable asset not to be shared with rivals. The Portuguese crown regulated the spread of charts and reports from expeditions. Foreign cartographers seeking accurate maps of the African coast had to either pilfer the data or conduct their own exploration. As such, Lisbon alone knew the true shape of the world while Paris, London, and Rome remained uncertain.

Dias's post-expedition career involved continued service in Portugal's maritime sector, including appointments and committee roles concerning navigation and shipbuilding. In 1497, during Vasco da Gama's preparation for his voyage to India, Dias participated in planning and supervised the construction of da Gama's vessels, offering insights into the route, hazards, weather patterns, and optimal approaches to the cape.

The relationship between Dias and da Gama is obscure: whether they were friends, competitors, or colleagues; whether Dias harbored resentment toward another navigator garnering acclaim for his discovery, or if he embraced a division of labor in which each contributed to an overarching enterprise. Absent personal correspondence, diaries, or testimony, we can only speculate within the bounds of plausibility.

Da Gama's voyage exceeded expectations; he reached India, traded spices, and, upon returning in 1499 with profits many times the cost, validated the commercial viability of the route. Portugal commenced organizing regular fleets to sail for India, establishing trading posts, negotiating with local authorities, and building an empire in the East. The Age of Discovery entered a new phase marked by exploitation rather than mere exploration.

Dias participated in this altered era. In 1500, he was captain on one of several ships in Pedro Álvares Cabral's large fleet, whose destination was India via the route Dias had charted previously. Thirteen vessels departed Lisbon in March with soldiers, merchants, missionaries, and trade goods. The plan included leveraging Atlantic currents and winds mapped by Dias and prior explorers.

However, Cabral steered much further westward than anticipated, landing in South America. Whether this was an intentional concealment of known land or a genuine accident is debated among historians; what is confirmed is that Cabral claimed Brazil for Portugal, sent news back to Lisbon, and then proceeded toward India.

The fleet rounded the Cape of Good Hope in late May, familiar waters to Dias, who knew them better than anyone. Yet disaster struck in the form of a storm of horrifying speed, named after the cape's original name, leading to the loss of four ships. Among these was Dias's vessel.

Surviving captains documented fragmented details of storms overnight with monumental waves and wind so fierce it tore rigging. Ships vanished into darkness and spray; at dawn, four vessels were gone without trace or survivors. The ocean swallowed them whole, leaving no remnants.

Bartolomeu Dias was around fifty years old when he perished within sight of the cape he initially circumnavigated, taken by the same waters that brought him triumph. This symmetry between man and sea is almost literary fate, completing a circle; yet it offers no solace to the drowned or compensation for lives abruptly severed.

Dias's death arrived with Cabral's surviving fleet in Lisbon; records noted his passing officially, but an outpouring of grief characteristic of famous explorers' passings did not occur. He was respected, indeed; his achievement recognized. Yet he did not elicit mourning akin to cultural reverence for heroes. Dias was a public servant who fulfilled his duties exceptionally, nothing more.

The irony is profound: the man who broke Africa's barrier, proved the existence of the India route, and provided geographical knowledge that would reshape European dominance and spawn the modern global economy, passed unremarked upon and largely forgotten within a generation. His name might appear in the annals; historians acknowledged his voyage; yet popular imagination fixated on other figures, such as Columbus, who sailed westward to the New World; Da Gama, with riches from India; and Magellan, whose fleet circumnavigated the globe. These were the enduring legends.

Dias became a footnote.

THE ARCHITECTURE OF EMPIRE

Yet the path Dias blazed laid the groundwork for Portugal's dominion over lands afar. For well over a century, Portuguese vessels traversed around the Cape of Good Hope in droves, ferrying eastern trade goods and western spices. The

Carreira da Índia, the India Run, emerged as one of the pivotal trade arteries of history. Yearly, fleets set sail from Lisbon in springtime, journeyed to India, and returned laden with riches beyond imagination. The pattern Dias cemented between 1487-88, veering wide into the Atlantic, catching propitious breezes, approaching the cape from the west, then rounding it into the Indian Ocean, became standard operating procedure.

Navigating this route never promised ease. Its notoriety for peril endured. Storms frequently wrecked ships. Eventually, rival European nations such as the Dutch, English, and French mastered the route and challenged Portuguese hegemony, sparking naval skirmishes and shifting the balance of power. Nonetheless, Portugal maintained a virtual monopoly for decades, and the wealth amassed transformed its kingdom.

Lisbon bloomed as a premier European city, fattened by commerce with the East. The Portuguese established enclaves across the Indian Ocean, including Mozambique, Goa, Malacca, and Macau. Rather than conquering broad swathes of land like Spain did in the New World, they fashioned a maritime empire, a web of fortified trading bases connected by sea lanes, shielded through naval might, sustained by a mix of trade and coercion. This network extended far beyond Portugal's shores, intertwining with existing commerce routes and reshaping them. The Cape route, unveiled by Dias, not only tied Lisbon to Asia's wealth but also threaded it into markets within the Ottoman Empire, among Gujarati merchants in India, and even the spice-laden isles of Java. Portuguese ships did more than carry commodities; they facilitated cultural exchanges and wielded influence over local economies, disrupting age-old trade patterns and altering regional power landscapes. Thus, Lisbon's gains were mirrored by transformations worldwide, underscoring the global repercussions of Portugal's imperial pursuits.

The human toll was colossal. Thousands of Portuguese mariners perished on the India Run. Disease, tempests, shipwrecks, and warfare accounted for their demise at a rate deemed acceptable in the sixteenth century but horrifying by today's standards. Populations in Africa and Asia paid yet another price for Portuguese arrival, which brought warfare, slavery, economic upheaval, and cultural erosion. The spice commerce profited Lisbon but impoverished producers. Moreover, the Christian mission that accompanied Portuguese expansion ransacked temples, persecuted local faiths, and imposed European values through force.

None of this was Dias's aim. He sought only a route, not an empire. But in history, intentions often count for little against consequences. Dias unearthed a path; others followed it armed with weapons, chains, and crosses, turning his geographical breakthrough into an instrument of conquest.

The moral intricacy persists. Dias's voyage showcased unparalleled navigation skills, bravery, and fortitude. It broadened human understanding, confuted ancient geographical certainties, and demonstrated the power of diligent observation and methodical strategy, achieving feats once deemed impossible. These are achievements worthy of admiration. However, alongside these triumphs lay immediate costs. In the wake of Dias's discovery began centuries of colonial exploitation, initiated contacts that ravaged societies, and instituted patterns of race-based hierarchy and economic extraction that linger to this day. Each mile of uncharted waters conquered heralded new geopolitical realities, a contradiction that called for contemplation of the double-edged sword of discovery.

Can we laud the navigation while deploring its aftermath? Can we extol the skill yet acknowledge the harm inflicted? History offers no simple solutions. Dias himself might have been appalled by what his discovery enabled or seen it as divinely willed, justifying any torment in the spread of Christianity. We cannot know for certain. The man is silent, his thoughts lost to us, leaving only the facts of what he did and its sequels.

THE TECHNOLOGY OF DISCOVERY

Behind Dias's accomplishment lay decades of Portugal's investment in the enhancement of maritime technology and navigation knowledge. The ships he commanded were remarkable feats of engineering, continuously honed over time. A caravel was conceived, melding elements from various cultures, the lateen sail originating from Mediterranean and Arab shipping, the hull design from Atlantic fishing boats, and rigging systems inspired by northern European vessels. Portuguese shipbuilders adeptly synthesized these influences into a new type of vessel, one specifically tailored for long-distance exploration.

However, ships alone were not enough; navigation also demanded specific tools. Fifteenth-century Portugal led Europe in developing and refining instruments crucial to this art. Instruments like the quadrant and astrolabe, used to measure the angles of celestial bodies above the horizon, became standard aboard Portuguese ships. Astronomers meticulously compiled tables that displayed the sun's declination at different latitudes and times of year, enabling pilots to calculate their exact positions with increasing precision. Although the compass had been known in Europe since the thirteenth century, Portuguese sailors refined it for reliability.

Portuguese pilots mastered the art of dead reckoning, calculating positions based on speed, time, and direction. To measure speed, they utilized a chip log, a wooden board attached to a rope with knots at regular intervals, by throwing it overboard and counting how many knots passed through their hands in a set amount of time; this practice is whence the term "knots" for nautical speed originated. Timers like sandglasses were used to measure time accurately, and logbooks meticulously recorded every alteration in course, estimates of speed, and landmarks observed.

These combined technologies enabled daring voyages akin to those of Dias. Without precise navigation techniques, ships would have aimlessly roamed the oceans, unable to return to their discoveries or relay their locations. Without robust, agile vessels, they would likely have met their doom in severe storms. Without systematic record-keeping, knowledge risked being lost with each expedition. Portugal excelled not through a single groundbreaking innovation but by integrating numerous advancements into a cohesive system that could be taught and refined across generations.

The crown actively supported this integrated approach through deliberate policies. King João II established a council comprising mathematical advisors, astronomers, and navigators to scrutinize expedition reports and enhance Portugal's geographical understanding. He commissioned more accurate maps and sailing directions. Successful navigators were rewarded with pensions and accolades, fostering excellence. The king maintained secrecy regarding their findings but ensured efficient information exchange among Portuguese pilots and captains.

This institutional framework set Portugal apart from other European powers such as Spain, England, and France, which all possessed skilled sailors and robust ships. Yet none of them could match the systematic method to navigation and discovery adopted by Portugal in the fifteenth century. The result was Portugal's dominance of Atlantic and Indian Ocean exploration for 50 years, roughly between the 1440s and 1490s. It wasn't until other nations emulated Portuguese techniques that this monopoly ended.

Dias embodied both this system and its practitioner. His expertise as a navigator stemmed from training in Portuguese techniques; his success was a testament to the intelligent and brave application of these skills. Dias's contribution to the collective pool of knowledge included detailed reports, charts he meticulously drew, and insights that later navigators would build upon. In essence, he stood on the shoulders of earlier explorers, and future generations would stand on his.

THE AFRICAN PERSPECTIVE

The African experience is often overlooked in accounts of Dias's voyage. What did the people at Bahia dos Vaqueiros think when Portuguese ships arrived? How did various Africans perceive these strangers who came, traded briefly, asked questions, and then left? No chronicles from Africa document these encounters, nor do we have records from those Dias met. Their perspective is absent in history, visible only through Portuguese accounts that portrayed them as obstacles, curiosities, or commercial opportunities.

Yet, something of their experience can be inferred. The arrival of Portuguese ships wasn't mysterious to coastal Africans by the 1480s. For decades, Portuguese vessels had visited Africa's Atlantic coast, establishing, albeit unequal, trade relationships. Africans were aware that these foreigners sought gold, ivory, and slaves; they knew the Portuguese brought metal goods, cloth, and weapons. They understood dealings with them could be profitable but also risky.

The African kingdoms and communities along the coasts were sophisticated political entities with their own histories, conflicts, and ambitions. The Kongo kingdom, contacted by Portuguese explorers in the 1480s, was a large, centralized state with intricate administrative structures. Wealthy and powerful West African kingdoms dominating the gold trade were far from primitive societies awaiting European civilization; they were functioning civilizations with advanced technologies, arts, political systems, and worldviews.

The arrival of the Portuguese disrupted African societies multiple times. The demand for slaves, initially moderate, grew over time into a catastrophic forced migration that devastated regions. The introduction of European goods reshaped local economies, rendering communities reliant on trade beyond their control. The spread of Christianity, pursued by Portuguese missionaries, assaulted indigenous religions and cultural practices. Furthermore, the mere presence of Europeans, backed by naval power and weapons, redefined power dynamics that local leaders had to navigate with care.

From an African standpoint, Dias's voyage was just another incursion by foreigners promising both opportunity and disaster. The Portuguese were not the first outsiders Africans encountered; Arab and Indian Ocean traders had been active on Africa's eastern coast for centuries. However, the Portuguese introduced different technologies, motivations, and ultimately different scales of violence. They heralded a wave that would eventually engulf much of Africa under colonial rule.

None of this was inevitable in 1488. The Portuguese sought trade at that time, not conquest. They lacked the military strength to conquer large African kingdoms even if they desired to. The relationship was more balanced than it would become. However, Dias's discovery made Africa's integration into European power structures more feasible. The route he charted permitted larger exchanges of ships, goods, and individuals. It fused Africa more tightly with European economies and aspirations.

The tragedy is our inability to hear African voices from this era speaking firsthand. We possess Portuguese accounts that mention Africans, describe them, and document transactions with them. But we lack African perspectives on the Portuguese. We cannot discern the thoughts of the herder killed at Bahia dos Vaqueiros when he threw a stone at these strangers or what his community felt upon learning of his death. All we can do is acknowledge this silence and recognize that every discovery represented an invasion for someone else.

THE SILENCE OF FAME

Bartolomeu Dias should be renowned for his accomplishments. By any standard, he achieved greatness comparable to the foremost explorations of his time. He resolved a challenge that had perplexed Portuguese navigators for an entire decade. He forged a route that would amass vast riches. His skills, bravery, and leadership were exemplary. However, history relegated him to the sidelines, known to scholars but not widely recognized by the general public.

Why? Primarily because he did not complete his journey. Da Gama reached India, while Dias only demonstrated its feasibility. Although this distinction might seem insignificant, it was of immense importance for recognition and rewards. The completion of a voyage often garners more accolades than mere exploration. To illustrate this disparity, consider that Dias received a modest pension, a mere one-tenth of what da Gama was awarded. This stark contrast in rewards deepens the emotional impact of Dias's relative obscurity, underscoring how fame and fortune favored those who could claim ultimate success.

Secondly, Dias lacked a chronicler to document his exploits. Columbus had numerous biographers who propagated his fame, embellished his achievements, and created an epic narrative around his voyages. Similarly, Da Gama's expedition was accompanied by chroniclers who composed vivid accounts. In contrast, Dias had only dry official reports filed in royal archives, which were primarily read by navigators and administrators rather than the public.

Furthermore, the Portuguese crown carefully controlled information regarding discoveries. Spain proudly proclaimed Columbus's achievement, legitimizing Spanish claims to new territories. Meanwhile, Portugal kept its findings relatively quiet, revealing information only when necessary. Consequently, Spanish explorations became international news while Portuguese achievements remained largely secretive, known only to specialists rather than widely celebrated.

Additionally, Dias passed away at an unfortunate time. Had he lived for another twenty years, accumulated further accomplishments, and become a prominent figure in Portuguese exploration, his fame might have grown significantly. However, he died prematurely due to an accident unrelated to his greatest feat, and as such, was overshadowed by subsequent events.

Some of his contemporaries were indeed troubled by this injustice. The Portuguese chronicler João de Barros wrote in the sixteenth century, extolling Dias' achievement while lamenting his obscurity. Nevertheless, Barros penned history for specialists rather than a popular narrative aimed at mass audiences. While he ensured Dias remained within the historical record, he did not elevate him to fame.

Contemporary historians have since attempted to rectify this imbalance. Dias is included in serious accounts of the Age of Discovery, often with proper emphasis on his significance. However, he has never captured the public imagination like Columbus or Magellan. There are no Hollywood films about Dias, and children do not learn his name in school. He remains obscure, a hero to specialists rather than a household name.

Perhaps this outcome is appropriate. It could be fitting that the individual who accomplished the task remains in obscurity while others reap the glory. Perhaps Dias' anonymity reveals something genuine about how history functions, how fame stems from narrative prowess rather than actual achievement, and how the world remembers symbolism over substance.

Or perhaps it is simply a tragedy, another example of how randomness, circumstance, or fortune shape reputation. It is conceivable that with a different chronicler, a longer lifespan, or alternate circumstances, Dias' name could have been etched alongside history's most renowned explorers. Instead, he remains confined to footnotes and specialist monographs.

THE VOYAGE IN CONTEXT

To truly grasp Dias's voyage, it's important to step back and examine the broader pattern of European expansion during the fifteenth and sixteenth centuries. It was a time when Western Europe, once seen as less influential than great civilizations in Asia and the Islamic world, began its ascent to global dominance. This rise was propelled by advancements in maritime technology, capitalism, and systematic violence, carried out through exploration and colonization.

This pattern first took shape in the Mediterranean. Italian city-states such as Venice and Genoa established extensive trade networks across the region, accumulating wealth through commerce. They also pioneered banking systems, marine insurance, joint-stock companies, and other capitalist innovations. Although the Black Death in the fourteenth century disrupted these systems, it ultimately strengthened them by reducing populations and increasing per capita wealth among survivors.

Portugal, located on the Atlantic edge of Europe, was not ideally positioned for Mediterranean trade but did have a significant advantage in Atlantic exploration. Its strategic position allowed Portuguese ships to easily sail south along Africa's coast or venture west into open waters. This geographical benefit, combined with the technological expertise developed in the Mediterranean, such as the compass, portolan chart, and caravel, enabled Portuguese merchants and navigators to adapt these tools for Atlantic conditions.

The initial impulses driving exploration were diverse. Religion played a role, as Portugal viewed itself as continuing the Reconquista against Islam. Economic interests also played a part, with opportunities in African gold and, eventually, Asian spices promising enormous profits. The pursuit of knowledge was another motivation, aligned with Renaissance humanism's emphasis on discovery and the expansion of human understanding. Additionally, national pride was at stake, with each kingdom vying for prestige and power. These incentives reinforced one another, fueling a powerful drive to explore that endured despite individual failures and setbacks.

By the 1480s, Portugal had established an efficient system for systematic exploration. Ships departed regularly following established routes to known destinations, gradually pushing further out each voyage. Information was accumulated in royal archives, charts improved, and navigational techniques were refined. This process was methodical, cumulative, and highly effective.

Dias's voyage was a product of this well-organized system, yet it also represented a leap beyond it. Prior expeditions had expanded boundaries incrementally, cautiously moving down the coast in stages. In contrast, Dias's storm-driven passage around the cape was far from incremental. It was a violent, terrifying, and almost accidental breakthrough that, in just thirteen days, covered more ground than planned expeditions might have traversed in months. Despite this, Dias was prepared to capitalize on this breakthrough precisely because he possessed the skills and knowledge developed by the Portuguese system. Accurate navigation enabled him to determine his position after the storm, while detailed charts of the coast already explored allowed him to know where to find land when turning north.

In the midst of the tempest, a crucial decision showcased the interplay between method and luck with striking clarity. Amidst massive waves threatening to capsize vessels and relentless winds tearing at the sails, Dias made the daring choice to order the jettisoning of cargo. He understood that lighting the ship might offer their best chance to ride the monstrous swells. This decision was not taken lightly; it was a gamble grounded in experience, navigating the fine line between strategy and desperation. With barrels of valuable cargo lost to the sea, the São Cristóvão, now lighter, managed to maneuver with slightly more stability through the storm. It was a perfect moment of pragmatic decision-making amidst chaos, where careful calculations met the whims of nature head-on. This choice underscored the paradox of the age of achieving success through precise navigation and taking unprecedented risks.

LEGACY AND MEMORY

What remains of Bartolomeu Dias? A few scattered mentions in historical writings. Several commemorative sites in Portugal and South Africa bear testament to his journey. The cape was named after a king's patronage, though originally dubbed differently by Dias himself. Ancient routes once navigated by ships for centuries are now abandoned, made redundant by air travel and contemporary maritime routes. And lingering questions about the man behind such an accomplishment.

Did he comprehend the magnitude of what he had accomplished? It is certain that in terms of navigation, he realized he had solved the challenge posed by Africa's coastline, proving it could indeed be circumnavigated. But did he grasp the broader ramifications? Did he envision Portuguese dominion over the Indian Ocean or foresee the wealth that would flow through this new route? Were his thoughts concerned with the impact on African and Asian populations who would encounter their powers?

We cannot know. The historical records fail to provide insight into Dias's personal life. There are no letters, no diary entries, no confessions to shed light on his innermost thoughts. We only have his name inscribed in official documents, his testimonies in reports, and the basic facts of his career. His personality remains elusive, visible only through his actions interpreted by speculation.

Perhaps that is fitting for a navigator. The profession values tangible outcomes above all else - precision over eloquence. A successful captain ensures their vessel reaches its destination, takes precise measurements, documents vital information, and returns safely home. Personal expression holds no significance; what matters are the chart, the logbook, and the knowledge shared with subsequent navigators.

Dias imparted critical knowledge: the route existed, the cape could be circumnavigated, and the Indian Ocean was accessible to ships that followed in his footsteps. Subsequent sailors built upon this foundation, refining techniques, learning from their experiences, and achieving what they had set out to do. In this sense, Dias's legacy is collective rather than individual - he contributed to a pool of knowledge that facilitated European expansion for better or worse.

The "worse" cannot be dismissed or downplayed. The route opened by Dias became a conduit for colonialism, slavery, exploitation, and the destruction of cultures. Millions of Africans were enslaved and transported across oceans partly because Dias charted these waters. Asian societies faced disruptions, attacks, and eventual colonization through routes pioneered by Portuguese navigators. The stark wealth disparities between nations in the global North and South can be traced to patterns established during the Age of Exploration. Dias bears responsibility for this not through deliberate intent but as a consequence of his actions.

Yet he also merits credit for broadening human knowledge, demonstrating that rigorous observation could challenge established authorities and revealing that the world was larger and more diverse than Europeans had previously imagined. The scientific method owes much to explorers like Dias, who consistently proved reality differed from classical texts. The modern achievements in navigation, cartography, and our understanding of Earth's geography are built upon the foundations laid by early explorers.

The moral reckoning surrounding his legacy is complex and perhaps unresolvable. We can acknowledge his accomplishments while condemning the outcomes. We can admire his skill whilst recognizing its harmful consequences. We can study history without endorsing the actions of historical figures. Dias was neither a hero nor a villain but a person shaped by the constraints of his time, achieving something remarkable, enabling dreadful events, and passing away without fully understanding the implications of what he had set in motion.

February 3, 1488

Return to the bay, where land emerged after thirteen days of storm, where Bartolomeu Dias stood on São Cristóvão's deck, gazing at a coastline that seemed impossible, realizing everything had transformed.

Feelings of relief surged - they survived, ships intact, brother's ship reappeared, fresh water, and sheltered anchorage. But soon curiosity took over. What is this place? Who inhabits it? Where exactly are we? Navigator instincts kicked in, taking readings, establishing position, mapping the coast, transforming chaos into knowledge.

Wonder followed, crossing Africa's southern point, entering uncharted waters, theory to reality, dreams coming true. This ordinary bay proved the world wasn't as mapped, opening a path for further discoveries.

Yet, uncertainty lingered - what next? They achieved their goal but were far from home, weakened and in unknown seas. Push on towards India or return, consolidating their discovery? Risks of success or settling for accomplishment?

Dias chose pragmatism, moving forward to confirm the coast's direction, then turning back, a balanced approach between ambition and caution, valuing knowledge over risking death.

Standing on February 3, 1488, Dias couldn't predict his legacy fading, da Gama's triumph, or his own fate. He only knew he crossed beyond known worlds and found it navigable, with that moment's achievement enough.

The sun continued its path, ships' crews attended to their tasks, the Khoikhoi observed from the hills, and ocean waves kept rising and falling, indifferent to human affairs. Dias recorded observations, ready for further exploration.

This moment encapsulated the Age of Discovery - courage, violence, knowledge, exploitation, achievements, tragedy. It marked an era's end and beginning, not with fanfare but a weary navigator making calculations in a safe haven.

Dias matters because he wasn't perfect or heroic, but a skilled professional who altered history by surviving chaos for knowledge, opening doors others raced through uncontrollably. His achievement was real, its consequences huge, his obscurity reminding us of countless individuals shaping our world, regardless of fame.

São Cristóvão anchored in the bay, water gently touching its hull, Dias standing on deck, gazing at an ordinary shoreline, knowing something profound had happened but couldn't foresee the empires, suffering, wealth, knowledge, and complex legacies that followed - a moment when European ships rounded Africa, changing human history's course.

Vasco da Gama

September 29, 1502, Off the Coast of Malabar

The capitulation seemed almost uneventful, a silence preceding the violence yet to come.

From his vantage point on the deck of the São Gabriel, Vasco da Gama observed his men ascending onto the merchant vessel they had stalked for days. The Miri, a member of one of Calicut's wealthiest merchant families, did not resist. Merchant ships did not engage in battles with warships. They carried wealth, and thus, a ransom was expected, a common practice.

But da Gama harbored no intentions of abiding by the customary protocols. The Portuguese sailors surged onto the Miri's deck, their weapons at hand, issuing commands. Approximately 380 individuals were aboard: men, women, and children, pilgrims, merchants, travelers. Many had just returned from Mecca, a journey deemed sacred in Islam. The ship was laden with treasures, silks, carpets, spices, and gold coins, estimated to be around 12,000 in total value, along with merchandise valued at an additional 10,000.

Da Gama directed the transfer of these riches onto his own ships. Panicked messages flew between da Gama and the Miri's owner, who offered any amount as ransom. He pleaded desperately for mercy. He sought not only his own life but also those of his passengers, especially the women and children. His entreaties were fraught with terror. He appealed to God and begged for a semblance of humanity still present in the world. Yet the Portuguese crew remained unmoved. They proceeded methodically: barrels, crates, and sacks crossed between the ships against the backdrop of an autumn sun and the serene Indian Ocean. Fishermen onshore, ten miles away, observed these anchored vessels: a massive Portuguese carrack and a merchant ship from Arabia in a state of eerie calm.

Once the final crate was transferred, da Gama issued his decisive command, colder than steel. His men herded the passengers below deck, sealing them in darkness. The hatches slammed shut above, their impact echoing through the enclosed space. The Portuguese smeared pitch around the edges and lit their torches; flames devoured the dry wood with ravenous hunger, and choking smoke seeped out from every seam. The screams pierced the air, agonizing and relentless. Inside the ship, those who realized the unfolding horror grew more frantic by the second. They battered the hatches with bloodied fingernails, offering wealth and jewels for their lives. Some frenziedly pounded the deck planks until their bones broke, only to be brutally beaten back by Portuguese soldiers, driving them deeper into rising heat and smoke. Women clutched their children desperately, their knuckles white with fear. Men shouted fragmented prayers to Allah amidst the intensifying fog of smoke and flames.

The Miri burned for four days before it finally succumbed to the sea. Its destruction heralded an era of colonial dominion that would extend across the Indian Ocean and beyond for centuries. Vasco da Gama, aged forty-two, stood witness to the consumption of everything he had subjugated. As Admiral of the Seas of Arabia, Persia, and India, and a knight of the Order of Santiago, his countenance mirrored an unyielding resolve. He had not come for friendship or negotiation; his motives were to assert Portuguese authority and secure wealth at any cost. This was no expedition of discovery. It was a conquest, pure and deliberate.

Thus began this saga five years before the calamities off Malabar, with a fleet of four ships and an audacious vision.

THE PORT OF SINES

To truly grasp the figure who flung open the gates of the East to European dominion and soaked that breach in blood, one must first fathom the dual essence of his pursuits, a blend of valor with terror. This duality is vividly portrayed through the life and exploits of Vasco da Gama, beginning with an understanding of his origins.

From the standpoint of strangers, Sines was merely a dot on the map, barely significant. It sat on a rocky outcrop along Portugal's southwestern shoreline, facing the Atlantic. In the mid-1460s, when da Gama was born, this town housed no more than 300 souls. The landscape boasted whitewashed houses with terracotta roofs, and narrow streets filled with the smells of fish and salt. Above the harbor perched a small fortress. The nearby church of Nossa Senhora das Salas may have witnessed his baptism rites. Beyond this hamlet lay the Alentejo, a land of rolling hills dotted with cork oaks, wheat fields, and sparse villages. It was among Portugal's least prosperous areas: blistering in summer, tempestuous in winter.

Yet, Sines held three invaluable assets: a harbor, a fortress, and the boundless lure of the open sea. The anchorage, though modest in size, provided a safe haven from the wrath of the Atlantic storms due to its strategic position sheltered by the cape. Every dawn saw fishing boats set out to meet the day's catch of sardines, sea bream, and all else bestowed by the oceanic embrace. The larger ships, traversing between Lisbon and Cádiz or venturing along Africa's coastline for trade and discovery, made rare but significant pit stops. Dominating the town stood the fortress, replete with Roman foundations and fortified over time, serving as a stronghold under the civil governor's authority. The governor in that era was Estêvão da Gama Vasco's sire. Estêvão, though not among Lisbon's wealthiest nobility, held land, bore a title, and served in the household of Ferdinand, Duke of Viseu, one of Portugal's most influential figures. To govern Sines was an honor yet burdened with the duty to protect this modest yet strategically valuable stretch of coastline. In union with Isabel Sodré, whose lineage boasted connections, Estêvão wed her family bearing English blood. Isabel's grandfather, Frederick Sudley, arrived in Portugal with an English military expedition in 1381 and chose to remain. Through service to the crown and links to the burgeoning Order of Christ, which played an increasingly pivotal role in exploration, the Sodrés ascended. Isabel's kin, brothers Vicente and Brás Sodré, seafarers and warriors destined to accompany Vasco into India, were evidence of their rising stock. Vasco, born roughly around 1460, perhaps 1469 (records are tantalizingly imprecise), was the third son with little in inheritance aside from his own resolve. The firstborn, Paulo, claimed legacy; João, the second son, was destined for the clergy or military knighthood. Vasco's path forward was shrouded in uncertainty. Growing up, he developed an intimate acquaintance with the sea. Day after day, from the fortress walls, he observed the vast Atlantic. Its mood ranged from tranquil mirror reflecting the sky to tumultuous waves crashing against jagged rocks, sending spray into the air. He mastered forecasting storms by reading nature's cues: dense air, birds' flight patterns, and ebb tides betrayed by stranded seaweed on shore. Equally ingrained was a spirit of quest and exploration. By the time Vasco was born, Portuguese mariners had charted their vessels south along the coast of Africa for over four decades. Even after his time, Prince Henry the Navigator's legacy loomed large as he sponsored ventures to outpace each cape and push beyond. With Vasco's upbringing, these expeditions did not wane; under Kings Afonso V and João II, fresh waves of captains set sail. Diogo Cão navigated to the mouth of the Congo River, followed by Bartolomeu

Dias rounding the Cape of Good Hope in 1488, proving the land mass had a southernmost point, effectively uniting the Atlantic with India's waters. These endeavors were not merely for glory; they were commercial enterprises, driven by ambition and calculation. Portugal's lust was to dominate the spice trade, pepper, cinnamon, cloves, and nutmeg, a boon to Arab and Venetian traders. These spices originated from India and the Spice Islands in the East, traversing overland through Muslim realms or via sea with Arab fleets through the Red Sea and the Persian Gulf. Breaking this monopoly by discovering a direct maritime path to India promised unimaginable wealth and dominion. Education was steeped with these aspirations. Vasco learned reading and writing, an uncommon trait for his social standing, alongside mathematics and astronomy, both critical to navigation. He trained in martial arts and horsemanship akin to knighthood. It's plausible he spent time aboard Sines's boats mastering seamanship from the hands-on craftsmen of this fishing hamlet. By adulthood, he stood tall, sinewy with a warrior's mien and calculating gaze, neither charismatic nor eloquent; prideful, imperious, quick to anger, and unbending. In 1492, in his early thirties, Vasco seized upon an opportunity when King João II appointed him on his first significant mission. French privateers were commandeering Portuguese merchant vessels, prompting the king to demand retribution through da Gama's hands. He carried out his orders with ruthless efficiency, capturing several ships and showcasing his command and intimidation over adversaries. This display of prowess cemented da Gama in the king's consciousness. When João II embarked on planning the crowning voyage to finally reach India by sea, Estêvão da Gama was considered as its leader. However, fate saw Estêvão depart before setting sail. The mantle then passed to other hands. And with João II's death in 1495, Manuel I, the young ruler, inherited not only a realm but an unyielding fixation. Manuel I sought India's glory and riches by being the first Christian monarch to forge direct trade links with the Orient. To achieve this, he required a captain who was competent, ruthless, and expendable, someone unbowed by demands ahead. In 1497, Manuel found his man in Vasco da Gama.

THE ARMADA OF DISCOVERY

On July 8th of the year 1497, a quartet of vessels embarked from Restelo beach, just outside Lisbon.

The São Gabriel, flagship of da Gama, was a carrack weighing 178 tons; its length was 27 meters, width 8.5 meters, and draft 2.5 meters. The sails encompassed three hundred seventy-two square meters. Da Gama's older brother, Paulo, led the São Rafael, which had similar dimensions. Smaller yet faster was the Berrio commanded by Nicolau Coelho. The fourth vessel, a supply ship, was laden with provisions such as food, water, weapons, and trade goods. The scale and specifications of these ships reflected the Portuguese ambition of a strategic maritime empire, with vessels crafted to withstand long voyages and carry considerable cargo, vital for expanding trade routes and asserting dominance in foreign waters.

The fleet was manned by 170 individuals: sailors, soldiers, craftsmen, religious figures, interpreters fluent in Arabic and Bantu tongues, and convicts who were to be the first to disembark when landing in perilous locations. Additionally, they carried padrões, stone pillars adorned with Portugal's royal arms, intended to be erected as markers of discovery.

King Manuel had issued explicit orders to da Gama: reach India, forge trade relations, import spices, and gain knowledge of that region, plant the padrões to affirm Portuguese sovereignty, and, when possible, convert pagans to Christianity. Implicit within these directives was a further message: to do whatever was necessary.

By conventional reasoning, this voyage should have been deemed impossible. The sea route from Lisbon to India spanned approximately fifteen thousand miles. No European expedition had previously dared to undertake anything remotely akin to it. Though the Portuguese knew segments of the journey as far as their ships had reached the Cape of Good Hope, the remainder was unknown waters, foreign winds, and uncharted shores. The perils were tremendous: storms, disease outbreaks, famine, hostile encounters with indigenous peoples, and the simple risk of drifting off course into eternal oblivion amidst the ocean's vastness.

Yet da Gama possessed advantages. He had accrued fifty years' worth of knowledge from Portuguese exploration. Expert pilots such as Pero de Alenquer and Pedro Escobar were at his side, veterans who had sailed alongside Bartolomeu Dias. The fleet was equipped with the latest navigational instruments, astrolabes for latitude measurements, compass bearings, and maps. Above all, they benefited from Dias's counsel during the initial leg of their journey.

The armada halted on July 26th at the Cape Verde Islands. There, they lingered a week, replenishing their water and supplies. On August 3rd, they set sail upon perhaps the most daring voyage ever conceived. In contrast to tradition, which called for hugging Africa's coastline, da Gama led his ships into the open Atlantic, boldly embarking on what is known as the volta do mar, the turn of the sea, a risky maneuver perfected by Portuguese navigators. Far from land, they chased fleeting breezes and currents, venturing into boundless blue waters week after week, guided solely by numbers, celestial bodies, and water's wisdom.

For a staggering ninety-six days without respite, they sailed through an infinite expanse of ocean and sky devoid of any other landmark.

The São Gabriel and its companions became diminutive islands of wood floating amidst immeasurable waters. The passing days were marked by routines: watches, sail adjustments, navigation checks, and supplications. Their sustenance remained unchanged; they subsisted on hard biscuits, salted meats, dried fish, and water from barrels which soon spoiled. With each sunset, as they lit lanterns to maintain their formation in the darkness, a palpable sense of both fear and hope mingled with the growing shadows. The sailors often wondered what awaited them at their voyage's end, their thoughts a secret blend of prayers and whispered courage. The constellations appeared altered as they voyaged southward, stars offering a distant promise akin to their own aspirations clinging to the vastness of the sea.

On November 4th, land emerged from the horizon. This was southern Africa's coast, christened Saint Helena Bay by the crew. They had traversed over ten thousand kilometers, six thousand miles, without beholding another shore, setting a record for open-sea travel that surpassed all previous feats of Portuguese seamanship.

However, their true test had only just commenced. Encircling the Cape of Good Hope required three weeks of relentless struggle against contrary winds and currents. The vessels pitched violently in tumultuous seas. On November 22nd, they triumphantly passed the cape, entering waters never before navigated by Europeans. They were now charting uncharted territories.

The coastline teemed with inhabitants; at Mossel Bay, the Khoikhoi traded cattle for trinkets. At the Great Fish River on December 16th, they surpassed where Dias had once stopped. Beyond that point, every mile was a new discovery for them. Da Gama named this stretch Natal, meaning birth, for it was upon this coast that they found themselves on Christmas Day.

By March of 1498, they arrived at Mozambique Island. Here, the world fundamentally shifted beneath their feet.

THE WORLD OF THE INDIAN OCEAN

Mozambique did not lie barren in its solitude. It was, instead, a bustling hub in a vast network that connected East Africa with India and China.

Upon anchoring their ships in Mozambique's harbor, the Portuguese beheld an organized cityscape: stone structures, teeming docks, merchant vessels, and impeccably dressed individuals conducting business. This was not akin to the landscape they had encountered along the southern coast of Africa. In this location, civilization unfolded in its familiar form, characterized by trade activities, architectural marvels, a functioning government apparatus, and an established culture.

However, it was a Muslim civilization that thrived there.

The Sultan of Mozambique, Arab merchants, Swahili traders, and dhow captains who navigated between Africa and India were all integral parts of the vast Islamic world that dominated the Indian Ocean. For centuries, this region served as a cosmopolitan trading hub where diverse merchants, Muslims, Hindus, Christians, and Buddhists traded together. Arabic was the language that facilitated trade, while the monsoon winds guided ships on journeys refined over generations.

For the Portuguese, it was like entering an unfamiliar world that harbored no affection for them.

Initially, the Sultan welcomed them, assuming they were distant Muslims. Da Gama took advantage of this misinterpretation, extending his stay and learning as much as he could about their ways. Two Portuguese sailors even ventured ashore to attend what they believed to be a Christian mass at a church; in truth, it was a Hindu temple mistaken for a place of Christian worship. Such confusions would occur once more during the course of da Gama's journey.

When the Sultan discovered that the Portuguese were actually Christians, who were considered enemies of Islam, their welcome evaporated instantaneously. Tensions escalated further. 'These Christians may come bearing gifts, but they threaten only ruin,' remarked a local merchant, watching events unfold with growing concern. Da Gama’s men committed acts of theft against local fishermen, and the Portuguese responded by bombarding the town with their shipboard cannons. This display of force stunned a community accustomed to trading rather than conflict. The episode marked a decisive shift: rather than forging alliances through negotiation, the Portuguese established a precedent of violence that would characterize their interactions in the Indian Ocean. They soon departed, leaving behind animosity before they could establish any commercial ties.

This sequence of misunderstanding, violence, and retreat would replay itself time and again in their voyage.

In April at Mombasa, suspicion and hostility awaited them. The local Muslims had been informed of the events transpiring in Mozambique. When officials from Mombasa approached the São Gabriel to conduct an inspection, they realized the Portuguese were not merchants but armed marauders intent on destruction. An attempt was made to sever the ships' anchor lines, compelling them to run aground on the shore where capture was imminent. Da Gama ordered the torture of one Muslim pilot they had taken into custody in hopes of extracting information; their cries resounded throughout the harbor. In a state of terror, the Portuguese unleashed random gunfire upon the city before fleeing northward.

Yet, at Malindi, approximately two hundred miles up the coast, their fortunes shifted.

The Sultan of Malindi was embroiled in conflict with his counterpart in Mombasa. He extended an invitation to the Portuguese, hoping they could serve as allies against his adversaries. He provided supplies and, most crucially, a pilot knowledgeable about the route to India. This guide, often erroneously referred to as Ahmad ibn Majid, was historically recognized; however, he was actually a Gujarati Muslim with years of experience navigating the Indian Ocean. He understood the intricacies of the monsoon winds, ocean currents, and key landmarks.

Guided by this expert navigator on April 24, 1498, da Gama's fleet set sail across the Indian Ocean. The voyage took twenty-three days until May 20, 1498, when they spotted land, the Western Ghats emerging from India’s Malabar Coast. They had achieved their goal: traversing by sea from Europe to reach India. Anchoring near a small beach called Kappad, approximately six miles north of Calicut, a principal trading port along India's western coast, they had arrived.

Vasco da Gama became the first European to accomplish this feat. He was thirty-eight years old when he commanded four modest ships and around one hundred forty men. Scurvy and other illnesses were responsible for the deaths of thirty individuals during their voyage. His trade goods, while suitable for exchange with Africans, fell woefully short for a major Indian port. He lacked the credentials or gifts suitable for an encounter with an important ruler, as well as any understanding of Indian culture or the political landscape.

And it was against this backdrop that he would soon meet the Zamorin of Calicut.

THE MISUNDERSTANDING

Calicut, as it stood in 1498, was among the wealthiest cities in the world. It was situated along the Malabar Coast, teeming with goods such as pepper, ginger, cinnamon, cardamom, and cloves, which were stored in abundance in its warehouses. The harbor was bustling with a myriad of ships from various origins, including Arab dhows, Chinese junks, Persian ships, and East African ships. In Calicut, over a dozen languages could be heard in the market, creating an atmosphere filled with diverse cultures. Hindu temples adorned the cityscape with their intricate stone carvings. The Zamorin, signifying "Lord of the Sea," ruled a cosmopolitan commercial empire that made Calicut indispensable to the spice trade.

When news reached the Zamorin about strange ships docking at his port, he dispatched officials to investigate. Four boats approached the Portuguese fleet, and their oarsmen couldn't help but gawk at these unfamiliar vessels. Who were these people? Where did they come from? What brought them here?

Among those sent onshore was one of da Gama's convicts. He encountered two Muslim merchants from Tunis who spoke Spanish. The first words the Portuguese heard in India, as recounted by the voyage's chronicler, were: "May the Devil take thee! What brought you hither?"

The convict, likely terrified and deeply out of his element, conveyed da Gama's message to meet Christians and acquire spices. In this single sentence, he encapsulated Portugal's ambition to claim souls, seize riches, and believe these goals could coexist harmoniously.

On May 28th, da Gama ventured onto the city's ground in a palanquin surrounded by crowds of Indians who had never encountered Europeans. Hindus wore cloth around their waists, gold earrings, and ash marks on their foreheads. In stark contrast, the Portuguese were dressed in heavy attire and armor, finding it incredibly hot in the tropical climate. Everything was foreign to them: the smells of spices and incense, the sounds of new tongues, and the sight of temples which they mistook for Christian churches.

Da Gama was escorted into the royal palace to meet with the Zamorin. What ensued was a tragic clash of cultures and an economic mismatch.

Da Gama presented his gifts: four scarlet cloaks, six hats, coral branches, twelve almaçares (a specific cloth), seven brass vessels, sugar, oil, and honey, items that might awe an African chief but were deemed insulting to the Zamorin's officials. They laughed at the gifts, viewing them as a sign of disrespect. The merchants arriving in Calicut would typically bring gold, silver, fine cloths, and jewels. Even the poorest trader in Calicut's market offered better goods than what the Portuguese envoy brought.

The Zamorin displayed diplomatic grace by feigning pleasure with da Gama's gifts but was not fooled. These visitors were insignificant merchants; they had nothing worthwhile to trade for Malabar's valuable spices. The Muslim merchants of Calicut, seeing potential competitors, warned the Zamorin that the Portuguese would disrupt established trade networks.

Da Gama remained in Calicut for three months, his frustration growing as he tried to barter his goods, cheap textiles and brass trinkets, with no buyers in sight. He demanded permission to establish a Portuguese trading factory but received no commitment from the Zamorin. Da Gama wished to load his ships with pepper and return home in triumph; however, he lacked anything valuable to offer in exchange. The Muslim merchants actively sabotaged his efforts by spreading rumors and offering better terms to the Zamorin.

In late August, after much struggle, da Gama managed to secure a small amount of spices, likely through selling items from his ships or borrowing money from local traders. It was not the grand cargo he had envisioned but something nonetheless. The Zamorin granted him letters for the Portuguese king and allowed him to depart. The parting was tense; da Gama forged no alliances in India, secured no commercial treaties, and accomplished little more than announcing Portugal's presence and ambitions.

On August 29th, 1498, the fleet set sail for Portugal.

THE RETURN

The voyage back home was a living nightmare. For 132 days, they traversed the Indian Ocean, enduring a journey that was nearly five times as long as their outbound trip. The crew's misjudgment of the monsoon season significantly delayed their return. The winds that aided them during the departure now stood against them, making navigation extremely difficult. As time wore on, scurvy spread rapidly among the men, causing swollen gums and tooth loss. Their bodies became riddled with bruises, turning a deep purple hue. Many were too weak to stand, lying helplessly on the decks, unable to contribute physically or mentally. Some succumbed completely, passing away silently after a night of slumber. Upon reaching Malindi on the African coast in January 1499, approximately one-third of the crew had perished. The Sultan of Malindi, maintaining his amity, provided vital fresh produce that helped preserve the remaining survivors. Nevertheless, there were insufficient men left to navigate all four ships. Da Gama ordered the São Rafael set ablaze; Paulo, his brother and captain of that ship, was already gravely ill. Continuing along the African coastline, they stopped in Mozambique, rounded the Cape of Good Hope, and proceeded northward. More casualties occurred, and the ships began to leak, with provisions dwindling alarmingly. In the Azores, Paulo da Gama breathed his last into his brother's arms. While Vasco stayed by his side, Nicolau Coelho sailed ahead to Lisbon with the remaining vessels. Nicolau Coelho arrived in Lisbon in July 1499, nearly two years and two weeks after their departure. Vasco da Gama followed suit in September, having buried his brother in the Azores. Out of a hundred seventy men who had embarked, only about fifty-five made it back home. Two out of the four ships were irretrievably lost. The cargo of spices was meager. Their journey's impact extended beyond lost vessels and diminished returns; the lives that perished along the way bore witness to its toll. Each empty berth represented a son, father, or brother who would never see their homeland again, an unsettling reminder of the human cost associated with discovery.

Despite everything, it was enough to change the course of history entirely. The spices they brought back sold for sixty times the entire expedition's cost. In Lisbon, the scent of spices spilled across warehouse floors like a cascade of golden coins, a radiant treasure that surpassed mere numerical value. More importantly, da Gama had successfully proven the route's feasibility; he had reached India and returned home. He brought back invaluable knowledge, including charts, information on ports and rulers, and intelligence on the Muslim traders who controlled Indian Ocean commerce. King Manuel I showered him with titles and rewards. Da Gama was granted nobility status with the title "Dom," bestowed the prestigious "Admiral of the Seas of Arabia, Persia, India, and all the Orient," granted the town of Sines and its revenues as a hereditary possession, and received a substantial pension. He could now affix his family crest to his residence and even marry into a higher social class. In 1500, da Gama wed Catarina de Ataide, daughter of an influential household. They would have six sons and one daughter together. Settling into the life of a nobleman, he managed his estates, attended court events, and reveled in his newfound fame. This recognition and wealth bestowed upon da Gama mirrored the competitive imperial strategies of the era, akin to Spain's use of its silver fleets to exert influence and expand its empire. Much like the rich silver mines financing Spanish endeavors, Portugal's dominance in the spice trade underscored an economic rivalry that spurred European expansion across the globe.

Yet, King Manuel I remained unsatisfied with Portugal's control over the Indian Ocean route. Following da Gama's success, Manuel dispatched Pedro Álvares Cabral with a fleet of thirteen ships to assert Portuguese dominance. Cabral established a trading post in Calicut, but also ignited violence that resulted in several casualties. Muslim merchants retaliated against the newly established factory, prompting Cabral to bomb and massacre Arab ships in the harbor. By the time Cabral returned to Portugal, conflict had erupted between Portugal and Calicut. King Manuel sought someone ruthless to continue where Cabral left off, someone like Vasco da Gama.

THE RETURN OF THE ADMIRAL

The armada that set sail from Lisbon on February 10, 1502, was no ordinary fleet. It was an armada. Dom Pedro Cabral commanded twenty ships, ten robust naos and ten lighter caravels loaded with troops, artillery, and supplies. He had been appointed as Admiral after switching to the prestigious Order of Christ and received orders to impose Portuguese dominance over the Indian Ocean by force if necessary. This expedition was not about discovery but about using terror to achieve political goals. King Manuel I aimed to eliminate Muslim traders on the Malabar Coast, punish the Zamorin of Calicut for allowing Portuguese casualties, establish protected trading posts, and redirect India's wealth to Lisbon rather than Cairo or Venice. His mission was clear: instill fear of Portugal throughout the Indian Ocean. Cabral understood his mission completely. Upon reaching East Africa, Cabral's fleet began demanding tribute from local rulers. At Kilwa, a wealthy African city, he forced the Sultan to agree to pay annual gold tribute to the Portuguese crown or face destruction. The Sultan had no choice but to accept, as the Portuguese naval power was overwhelming. Moving on, at Sofala, another major trading port, they established a trading post and coerced local rulers into trading only with Portugal. Cabral's strategy was evident: wherever he went, stronger weapons secured his demands. Centuries-old systems of negotiation were now being reshaped by European naval superiority.

On September 29, 1502, the Miri incident occurred off the coast of Calicut. Cabral had the option to ransom an Arab ship captured near Calicut or to confiscate its cargo and release the passengers. Instead, he ordered them all burned alive. This brutal act of terrorism was intended to send a clear message: resist

Portuguese demands and face certain death, regardless of age, gender, or innocence. This ruthless behavior was not out of line with the Just War theology prevalent in the 16th century, which saw violence and terror as permissible means towards what was considered a righteous goal, spreading Christianity. The news of the Miri massacre spread quickly across the Indian Ocean. Upon arriving at Cannanore, where the local Raja had friendly relations with Portugal, Cabral was warmly welcomed. The Raja understood which side to support. Cabral negotiated favorable trading terms and established a factory in exchange for promises of cheap spices. At Calicut itself, Cabral orchestrated a campaign of calculated cruelty. He seized small fishing boats and their crews, ordinary fishermen with no part in the conflict between Portugal and Calicut, and executed them publicly. Then he mutilated their bodies and sent them along with a letter demanding that the Zamorin expel Muslim merchants and pay for Portuguese losses. The Zamorin refused, arguing that Portugal had already taken more wealth from Muslim traders than they had lost themselves. He pointed out that Calicut could not simply expel its large Muslim population. He contended that the violence was disproportionate and unjustified. Cabral responded by bombarding the city for two days, exploiting his armada's artillery advantage. The defending ships were no match for the Portuguese cannons in terms of firepower or naval tactics, and the city suffered severe damage and civilian casualties. After the bombardment ended, Cabral sailed to Cochin, a smaller port that had long resented Calicut's dominance. Here, he found willing allies, forming an alliance that would last decades and establishing a trading post with Portuguese protection. The Zamorin was not deterred and assembled a fleet of twenty-nine ships for a naval battle against the Portuguese in October 1502. It turned into a one-sided massacre as Cabral deployed his armada's concentrated firepower effectively. The technical superiority of European naval warfare was decisively demonstrated, leaving many enemy vessels sunk or burned. Cabral continued his campaign along the Malabar Coast after this victory. He attacked Muslim shipping, bombarded ports that resisted his demands, and established trading posts in friendly ports. By February 1503, when he finally departed for Portugal, he had fundamentally shifted the balance of power in the Indian Ocean. The Zamorin was humbled, and Portuguese trading posts were established at Cochin and Cannanore. Upon returning to Lisbon in October 1503, laden with spices and treasure, Cabral fulfilled his mission to punish Calicut and establish Portuguese control over the region. He was celebrated as a hero, but also created an empire with a reputation that would haunt future generations.

THE YEARS OF SHADOW

Vasco da Gama's second voyage should have cemented his position at the heart of Portuguese power. Instead, he spent the ensuing twenty years largely out of favor with royalty.

The reasons behind this were multifaceted. King Manuel I, despite acknowledging da Gama's triumphs, harbored resentment over various issues. Da Gama had not managed to completely subdue the Zamorin, and Calicut remained hostile towards Portugal. Furthermore, his uncle, Vicente Sodré, whom he had left in charge of a patrol fleet, disobeyed orders and perished in a shipwreck while pursuing personal gain rather than safeguarding Portuguese interests. These shortcomings, coupled with da Gama's haughty demeanor, soured his relationship with the king.

In 1505, when Manuel I decided to appoint the inaugural Viceroy of Portuguese India, a governor who would oversee all activities in the East, he deliberately bypassed da Gama for this prestigious role. The position went to Francisco de Almeida, a nobleman with more favorable court connections. Subsequent governors Afonso de Albuquerque and Lopo Soares de Albergaria were chosen from among other candidates. Da Gama, the individual who had forged the path to India and imposed Portuguese dominance through intimidation, was sidelined.

He retreated to his estates in Sines and later Évora, where he managed his lands, raised his family, attended Mass, became entangled in local disputes and lawsuits, feuded with the Order of Santiago over property rights, and eventually joined the Order of Christ in hopes of regaining royal favor. He was affluent, titled, and respected within his region. Yet, he remained outside the circle of power.

In 1518, when Ferdinand Magellan, a Portuguese navigator who had served in India, defected to Spain and proposed a voyage of discovery, da Gama threatened an identical endeavor. He wrote a letter to the king, suggesting that if Portugal did not value his services, Spain might be more willing to engage him. This was a bluff, and Manuel I called it; no further action ensued.

Upon Manuel I's death in 1521, da Gama found himself in his sixties. The new monarch, João III, was young and faced numerous challenges within Portuguese India. The colonial administration had become corrupt, with Portuguese officials enriching themselves at the crown's expense, mismanaging trade agreements, waging poorly coordinated military expeditions, and rendering the empire that da Gama had helped establish dysfunctional.

In 1524, João III concluded that drastic measures were necessary. He required someone with authority, experience, and a reputation for ruthlessness to address these issues in India, intimidating corrupt officials and reinstating order. This individual was none other than Vasco da Gama.

In February 1524, the king appointed da Gama as Viceroy of India, the highest position in Portugal's eastern empire. This honor was bestowed upon him alone, with only Francisco de Almeida having previously held the title of Viceroy. Da Gama was additionally named Count of Vidigueira, elevating him to the higher nobility, and his sons were promised prestigious positions. After two decades in obscurity, he was being summoned back into the spotlight.

He was sixty-four years old and had not set foot on a ship for nearly two decades; yet he remained Vasco da Gama and was determined to journey to India one final time.

THE LAST VOYAGE

The fleet that set sail from Lisbon back in April 1524 consisted of no less than fourteen vessels. Esteemed company joined their father on this endeavor; two of da Gama's sons, namely Estêvão and Paulo, accompanied him. The flagship was named Santa Catarina do Monte Sinai, a grand carrack embarking on its final voyage to India. This fleet did not merely represent military might but also signified generational change, sons poised to carry forward what their father had initiated.

This journey proved arduous from the outset. Out of the fourteen ships that initially set sail, anywhere between four and five failed to reach their destination. These were lost in tempestuous storms, scattered by unforgiving weather conditions, or vanished into the abyss that is the Atlantic Ocean. By September, when the surviving ships finally arrived at Goa, da Gama's health had begun to wane. The toll of a lifetime spent at sea, combined with tropical illnesses and the inevitable weight of age, had taken its toll on him. Yet he remained resolute, driven by his mission.

Goa, captured by Afonso de Albuquerque in 1510 and now serving as the capital of Portuguese India, was a city rife with stone fortresses, Christian churches constructed upon the ruins of Muslim mosques, settlers and subjects intermingling in a tapestry of cultures. Its bustling trade came hand-in-hand with brutal violence; corruption ran rampant within its walls.

The governor at that time, Duarte de Menezes, had lined his pockets and those of his friends with wealth unjustly taken from the crown. Goods destined for trade were often pilfered, tax revenues disappeared into thin air, military resources were squandered without purpose or direction, and justice was arbitrary and capricious. The Portuguese empire in India, built on a foundation of force and intimidation, was rotting from the inside out.

Da Gama tackled this state of affairs with the same unyielding hand he had wielded years prior. He dismissed officials deemed corrupt, scrutinized financial records, and ordered inquiries into malfeasance. It was abundantly clear that old ways were no longer acceptable. His very presence, as both Admiral of the Seas and conqueror of Miri, who had dared to threaten Calicut, instilled enough fear in all to know change was inevitable.

Yet his body betrayed him. In late November 1524, da Gama fell gravely ill, Portuguese accounts attributing it to rampant malaria prevalent in Goa and Cochin. Modern medicine suggests another systemic infection akin to anthrax could have been the culprit, exacerbated by years of tropical disease exposure. Whatever the exact cause, his condition worsened rapidly.

He was relocated to Cochin, possibly seeking a more propitious climate or better medical care. The city where he had established Portugal's first successful trading post now bore witness to his decline. He lay in a house close to the Church of Saint Anthony, wracked by fever and pain while his mind ebbed and flowed between consciousness and delirium. His confessor stayed by him, as did his sons and Portuguese officials who convened, already contemplating succession and its implications for their own standings.

Christmas Eve, December 24th, 1524, marked the end of Vasco da Gama's life. He was interred with honors within the Church of Saint Anthony in Cochin. The Viceroy of India, Count of Vidigueira, opening the East to European conquest, was laid to rest far from the Portugal that had been his homeland. It wasn't lost on contemporaries how he who had spent his life seeking a return to Portugal had ultimately found his final resting place in India.

His sons were promptly stripped of their titles and dispatched back to Portugal, as Henrique de Menezes assumed the governorship. Da Gama's brief reign over India ended as swiftly as it commenced.

Yet the body of Vasco da Gama did not rest in foreign soil. Fifteen years following his death, in 1539, his remains were exhumed and returned to Portugal. They were reinterred at Vidigueira in a grand mausoleum. Subsequently, due to shifting political landscapes and conflicts, they were relocated again to the Jerónimos Monastery in Belém, Lisbon, a structure partly funded by the wealth from India's trade that da Gama had pioneered. Or so it was believed; when the casket was opened in the 19th century, it held nothing but a skeleton and two skulls. It remains uncertain which, if any, belonged to the Admiral himself. Even in death, his identity remained elusive.

THE PRICE OF EMPIRE

In 1572, forty-eight years after Vasco da Gama's death, the renowned Portuguese poet Luís de Camões published Os Lusíadas (The Lusiads), a celebrated epic poem that extolled Portuguese exploration. At its core was da Gama, portrayed as an individual of mythical stature, who ventured beyond the known world to introduce Christian civilization to uncharted lands.

This piece of literature became Portugal's national epic, profoundly influencing how the Portuguese perceived their history and identity. It transformed da Gama into a symbol of courage, faith, and national pride, with streets named after him and statues erected in his honor. His expeditions were taught in schools as unqualified triumphs.

However, there was another narrative, one that Camões only hinted at indirectly and which later generations chose to overlook.

The route opened by Vasco da Gama became a thoroughfare for empires. Within a generation of his first voyage, Portuguese ships dominated the Indian Ocean, controlling key ports from Mozambique to Malacca and extracting tributes from African and Asian kingdoms. They monopolized the spice trade, which catapulted Lisbon into becoming the wealthiest city in Europe and enriched the Portuguese crown tremendously. But as Portugal extended its reach across the seas, other empires were expanding on land, crafting narratives of power and ambition that transcended oceans. The Mughals in India consolidated a formidable empire known for its cultural riches and administrative sophistication, while the Ottomans, with control over the eastern Mediterranean and parts of Europe, fashioned a realm marked by military prowess and architectural grandeur. These empires, formidable in their own right, offered alternative visions of imperial power, defined by their dominance on land rather than at sea. Beneath Portugal's maritime dominance lay a foundation of terror. The tactics employed by da Gama, shelling cities, executing prisoners, and burning ships with civilians aboard, became standard practice among the Portuguese. Those who followed in his footsteps carried out even more egregious acts. Afonso de Albuquerque seized Goa by annihilating its entire Muslim population, while Francisco de Almeida obliterated an Egyptian fleet, massacring thousands. Often, Portuguese captains subjected prisoners to torture, incinerated villages, and mistreated non-Christians as subhuman.

The justification for these atrocities was consistently the same: spreading Christianity and bringing civilization to so-called ignorant pagans. Yet, within this rhetoric of divine mission lurked a more selfish agenda, the extraction of wealth and resources through sheer violence and subjugation. The facade of a noble cause often served to mask the true intention of empire-building and exploitation, blurring the lines between ideology and economic gain.

This had a devastating impact on the peoples of the Indian Ocean. The centuries-old cosmopolitan trading system was shattered. Arab and Persian merchants, who had traded peacefully from Africa to China, became targets of Portuguese piracy. However, amidst the aggression and coercion, glimpses of cooperation and unity still persisted. In the bustling markets of Zanzibar, African, Arab, and Indian merchants would share stories of their travels and adventures, creating a cultural tapestry that defied the geopolitical tensions of the era. Musicians from different backgrounds often played together, blending their instruments and styles into harmonious performances, echoing the rich diversity of the region. Despite the looming threat of European domination, these small acts of cultural exchange illustrated a temporary harmony that existed before the full force of colonial disruptions. African kingdoms were coerced into servitude. Indian rulers faced the choice of submitting to Portuguese demands or risking bombardment. The Zamorin of Calicut, who had welcomed strangers in 1498, spent the following century battling against Portuguese attempts to annihilate his kingdom.

The Indian Ocean, once a realm of relative cultural exchange and commercial collaboration, became an arena of European imperial dominance. Da Gama's journeys marked the inception of European colonialism in Asia, a process that would endure for 4.5 centuries and fundamentally alter global power dynamics.

Yet, assessing the man himself presents a more complex picture. Da Gama was not a fiend who delighted in cruelty for its own sake. Contemporaries described him as aloof, calculating, and devoid of strong emotions. His destruction of the Miri was not an act of rage but rather a calculated move to instill terror in his adversaries. His onslaught on Calicut was not motivated by revenge but by a desire to demonstrate Portuguese prowess. Although his violence was chilling, it was a means to an end, not an expression of personal animosity, which may make it even more unsettling.

Moreover, he was not singular in his actions. The Portuguese crown explicitly instructed him to use force to establish trade monopolies and assert

Christian supremacy. Other European nations, Spain, England, France, and Holland, embraced similar tactics as they built their empires. The brutality exemplified by da Gama was not a sign of personal pathology but an embodiment of imperial policy.

What distinguished him was his efficiency rather than savagery. He was an extraordinary navigator who accomplished voyages deemed impossible. As a commander, he maintained discipline on expeditions that lasted for years. Da Gama also possessed strategic acumen, understanding how European naval superiority could overcome the numerical advantage of Asian forces. His relentless pursuit of objectives and indifference to causing suffering in achieving them made him precisely the sort of individual an empire demands.

EPILOGUE

In the Church of Santa Maria in Sines, where Vasco da Gama first gazed upon the world, there is a humble stone monument dedicated to him. It does not portray flames or conquest; rather, it depicts him on his knees, seeking divine guidance and solace. This image exudes tranquility and dignity, yet leaves much unspoken, inviting interpretation.

Vasco da Gama's legacy is multifaceted, embodying both his groundbreaking achievements as a pioneer in navigation and exploration and his role as an agent of violence and conquest. He stands as a testament to expanding the world's horizons while casting a dark shadow with acts of brutality that contributed to the rise of empires. As we ponder his dual legacy, we must ask ourselves: how do we balance the marvel of discovery with the cost of devastation? This conundrum invites us to reflect on the complexities of historical figures who both shaped and shattered worlds, urging an ongoing dialogue about the true price of progress.

When Gil Eanes sailed past Cape Bojador in 1434, he proved the existence of lands beyond by bringing back roses. Similarly, when Vasco da Gama ventured to India in 1498, he returned with spices but also left destruction and burning ships in his wake. Both individuals opened new paths and altered history, though the outcomes of their journeys were vastly different.

Da Gama's accomplishments were authentic; the sea voyage from Portugal to India was one of humankind's greatest feats of navigation, akin to the Apollo moon landings in its audacity, technical skill, and courage. The journey's success, despite immense obstacles such as vast distances, uncharted waters, and inaccurate maps, highlights Portuguese seafaring prowess and da Gama's command abilities.

However, Portugal could have chosen an alternative route, integrating with existing Indian Ocean trading systems rather than relying on conquest. Da Gama might have negotiated commercial treaties based on mutual benefit instead of subjugation through terror. Yet, the choices made by him, King Manuel I, and Portuguese leaders were rooted in the belief that non-Christians were inferior, business competition equated to warfare, and violence as a means to achieve goals.

On board the Miri, those who lost their lives in September 1502 were not soldiers, officials, or merchants but devout pilgrims returning from their most sacred journey. Their death served no military purpose; they were murdered to intimidate Muslim merchants into submission.

This brutal cost is what built Portugal's empire and laid the foundation for Europe's dominion over the Indian Ocean, financed by the blood of many. The wealth that boosted Portugal's golden age and the knowledge driving further exploration came at a great human price.

Vasco da Gama was well aware of these methods and their consequences; he firmly believed in imposing Christian European civilization on the world through any necessary means. The ends justified the means, and the souls saved and profits earned outweighed the broken bodies and cities reduced to ashes.

History has grappled with how to remember Vasco da Gama as a national hero in Portugal or as the bringer of centuries-long colonial oppression in India. Both perspectives are true yet incomplete.

Perhaps the most authentic view is acknowledging that opening the world did not necessarily make it better, but rather made it more interconnected, open to European power, and vulnerable to its violence. The Age of Discovery was also the Age of Conquest; explorers who ventured beyond the horizon brought back knowledge, spices, and the ways and reasons for building empires.

On Christmas Eve 1524, in Cochin's confines, a man lay dying. Vasco da Gama himself had traversed farther than most in his generation, altering world history with his journeys. He became wealthy, titled, and renowned; his name would endure long after royalty was forgotten.

In his final moments, as fever ravaged him and priests recited prayers, what did Vasco da Gama reflect upon? Did he envision the lives lost on the Miri or hear the tormented screams of captive prisoners? Did he recall the cities he had bombarded and the lives destroyed?

Or perhaps he only pondered his achievements, the insurmountable voyages completed, Portugal's wealth secured, and distant shores touched by Christianity.

The actual man, his thoughts, doubts, and private fears, is unknown because there is no record to reveal these aspects of him. History remembers him as a name and a series of actions with lasting effects for centuries; the true man behind them remains enigmatic.

What we do know is that he ventured beyond his world's edge, discovering not mythical creatures but people, civilizations, faiths, and ways of life distinct from his own. He, along with those following in his footsteps, subjected these peoples to European domination through technological superiority and systematic violence.

The routes he opened are still used today; the Suez Canal and modern shipping lanes follow paths he pioneered. The globalized world we inhabit owes much to him, yet so does the pattern of European colonialism that shaped centuries of exploitation and suffering. As we navigate our current world, the same trade routes, now bustling with container ships, echo the paths of spice-laden vessels of the past. The seamless flow of goods across these waters is a reminder of both technological triumph and historical consequence. It's worth questioning: What modern trade routes do we accept without scrutinizing their human cost? This inquiry connects da Gama's legacy with today's choices, prompting each of us to consider how our dependence on these lanes influences the balance between progress and its moral implications. The very lines that trace commerce today compel us to confront the ethical challenges they carry.

The spices he brought back have lost their value; they are no longer worth a kingdom's ransom. The empire he helped build has crumbled. Portugal is no longer a superpower, India is independent, and global power dynamics have shifted once again.

However, the enduring memory remains of a man who achieved both greatness and atrocities, flinging open doors without looking back. He stands as both hero to some and villain to others, a fitting ambiguity. Vasco da Gama was neither demon nor saint but something more complex: a product of his time shaped by cultural prejudices and ambitions, gifted with extraordinary abilities yet burdened with ordinary moral blindness. He did what his king commanded and what the world of his time rewarded.

The world we inhabit now, its betterment and worse, is partly his creation.

Pedro Álvares Cabral

April 22, 1500 – The Coast of an Unknown World

The ocean's foliage danced in unfamiliar currents, its movements whispering secrets untold beneath the surface. On the deck of his lead vessel, Pedro Álvares Cabral observed kelp patterns unlike any seen before, hinting at mysteries beneath the foam. The sea appeared green, distinct from the deep blue vastness of open waters. Above them, birds wheeled in unfamiliar flight paths. His pilot couldn't identify them. Off to the west, a solid form stood out, unmoving and clear, with no cloud. Land.

Land was not supposed to exist here. Cabral's armada of thirteen ships had set sail from Lisbon six weeks prior, following Vasco da Gama's route bound for India. They should have been traversing the Atlantic far and wide, skirting Africa to catch the winds toward the East. Yet something deviated them from their course. Perhaps it was the powerful currents that carried them further west than charted, a miscalculation in their navigation, or even a tactical maneuver to investigate the theories of land untold by tales. Cabral, with a racing heart, dared contemplate if the world were grander and queerer than any map had shown. At thirty-two years old, he was nobly born but less seasoned, commanding Portugal's mightiest fleet ever dispatched for trade. An army of fifteen hundred men sailed under his orders, their ships laden with goods meant for Indian spice markets. King Manuel I demanded haste in commerce, profits swift and sure, avoiding error at all costs. He had no tolerance for commanders who dallied or squandered resources on unforeseen ventures. Yet before them rose land where an endless sea was pledged. Cabral made a choice that would ripple through time. With trepidation yet curiosity, he decided to touch the new world for a brief span, just enough to discern its worth and nature before onward passage to India. He could not fathom that an unplanned turn had befallen him, a continent reborn from the sea's embrace. The discovery outclassed all spices of India in value. Cabral's fate would sing not of commerce but a serendipity that upended the globe, redefining humanity and sovereignty alike. As seaweed glided by, the distant silhouette sharpened into the frame of lands uncharted. Pedro Álvares Cabral, bound by duty yet unforeseen purpose, steered his vessels toward an era no cartographer had conceived.

THE MAKING OF A COURTIER

To truly comprehend how Pedro Álvares Cabral rose to lead Portugal's foremost fleet, it's essential to recognize that he was never intended for such a role from the onset. His narrative commences in this context.

Cabral's origins trace back to around 1467 or 1468 in Belmonte, a remote inland Portuguese town, far from any sea influence. He came from the prestigious Cabrals, a noble family of the highest status in Portugal. His father, Fernão Cabral, was alcaide-mor of Belmonte and served on the king's council. His mother, Isabel de Gouveia, had connections to influential families. The Cabrals' lineage spanned centuries, boasting vast estates and a strong sense of duty and privilege in military service to the crown.

Pedro was not destined for greatness as he was not the eldest son, inheriting merely his name and family ties. For noble youth like Cabral, service was the path forward, be it through military, diplomatic, or courtly roles. This transition into service marked the next phase of his journey, where success lay in securing powerful patrons, demonstrating unwavering loyalty, and being prepared to seize opportunities when they arose.

In his formative years, Cabral became part of King João II's household, a place where power converged and ambitions clashed. His role as a moço fidalgo, or junior courtier, involved ceremonial duties and an intricate education in the mechanisms of court life. Though not glamorous, attending ceremonies, delivering messages, and participating in hunts and tournaments provided invaluable knowledge about power dynamics.

In the late fifteenth century, the Portuguese court was focused on finding a sea route to India. For years, their expeditions mapped the West African coast and established trading posts. The ultimate goal was to circumnavigate Africa and access the lucrative spice trade. This shift to overseas exploration set the stage for Cabral's involvement. The stakes were immense. Spices like pepper, cinnamon, cloves, and nutmeg commanded much higher prices in Europe. Traditional overland routes were dominated by Muslim merchants and Italian city-states, who reaped substantial profits. Portugal yearned for a sea route to monopolize this wealth.

The turning point came in 1497, when Vasco da Gama embarked from Lisbon with four ships, explicitly tasked with reaching India. He achieved the impossible, rounding the Cape of Good Hope, navigating up the East African coast, and traversing the Indian Ocean with critical assistance from a Gujarati pilot, finally arriving at Calicut in May 1498. Upon his return to Portugal in 1499, his vessels overflowed with spices and gems, affirming that India could indeed be reached by sea.

King Manuel I ascended the throne in 1495 and promptly initiated plans for another expedition far grander than da Gama's, a massive commercial and military fleet aiming to solidify Portuguese dominance in the Indian Ocean, secure trading rights, and return with an astonishing cargo to astound European markets.

For this monumental mission, King Manuel sought more than merely skilled navigators or seasoned captains; he required a nobleman whose pedigree would intimidate Asian rulers, someone capable of diplomacy, earning respect, and representing Portugal with dignity. In essence, the perfect candidate was from an esteemed family background like Cabral's.

Pedro Álvares Cabral fit this bill impeccably. He bore ties to the high nobility, had court connections, and served the crown dutifully, albeit without exceptional distinction. What mattered most was his lack of powerful adversaries among court factions; he was competent, loyal, and unlikely to spark political unrest.

However, what Cabral lacked was tangible experience at sea. He had probably sailed before, as many Portuguese nobles did possess some ship knowledge. But he had never spearheaded a major naval expedition, braved open waters, or orchestrated the complex logistics of an overseas fleet far from home.

None of this seemed to deter King Manuel; birth trumped skill in his eyes. Cabral would have expert pilots and navigators. His role was to embody the crown's interests, make weighty decisions, and advance Portuguese influence. On March 9, 1500, King Manuel formally appointed Cabral as 'capitão-mor', captain-major, of a fleet bound for India. He was told to chart da Gama's route, build ties with the Hindu rulers of the Malabar Coast, secure trading rights, set up a fortified factory, and return with precious cargo to justify the expedition's cost.

The instructions provided were detailed, practical, and singularly focused on India; no exploration of the Atlantic was mentioned, no authorization to investigate uncharted shores, nor any hint that anything other than swiftness and commercial triumph held importance. Cabral's mission was clear: reach India, establish Portuguese presence, and return triumphant.

What transpired instead would astonish all, most of all Pedro Álvares Cabral himself.

THE FLEET OF FORTUNE

The armada congregated in Lisbon's estuary during the dawn of 1501, showcasing Portugal's mightiest naval force yet. It comprised thirteen vessels: grand naus capable of carrying massive amounts of freight and agile caravels designed for speed. The crew tallied fifteen hundred souls - mariners, troopers, traders, clergymen, and specialists. Their cargo bays brimmed with copper, seashells, fleeces, wine, and a fortune in silver coins to barter for spices. Such an endeavor was exorbitantly costly; King Manuel's investment reached around 2 million réis, nearly 2% of Portugal's annual revenue at the time. This was not simply exploratory but a business proposition under royal patronage aimed at immense returns and dominion over the Indian Ocean trade.

Leading the armada was Cabral aboard a large nau, possibly ranging between four hundred and six hundred tons, equipped with state-of-the-art navigational instruments and numerous cannons. The fleet included seasoned captains such as Pero Escobar, who had charted the path around Africa's bulge for Bartolomeu Dias in 1488, along with veterans from da Gama's journey acquainted with the sea route. Among them was Dias himself on a smaller caravel - his presence would prove crucial beyond anticipations.

Dias stood as an icon; he was the first European to traverse the southern tip of Africa, affirming that a maritime path to India was viable. His voyage from 1487 to 1488 unveiled the Cape of Good Hope and paved the route for da Gama's later journey. Now in his fifties, Dias embarked again with the wisdom accumulated over the years.

The fleet boasted an unusually high number of nobles compared to a typical commercial mission - Nicolau Coelho, Aires Gomes da Silva, Simão de Miranda, and Sancho de Tovar were among them, seeking prosperity and status. Their inclusion highlighted the dual nature of this expedition: it was as much a trading enterprise as a military one.

On March 9, 1500, Lisbon witnessed the largest fleet ever dispatched by Portugal to India. Its triumph would enrich the kingdom; its failure would devastate the crown and disgrace Portugal before other European powers. A grand ceremony marked their departure. Priests anointed each ship while Cabral accepted the expedition's standard from the king - a symbol of his sovereign's mandate that he carried on behalf of the realm. Sailors bade farewell to kin, knowing they might never return due to the horrifying mortality rate in Indian waters, estimated at thirty to forty percent. Disease outbreaks, storms, battles, and accidents were constant perils.

The fleet glided south along Africa's coastline, following the established path, pausing at the Cape Verde Islands for replenishment before venturing southwest into the open sea with the intention of circumnavigating Africa and striking towards India. This strategy was rooted in hard-earned navigational knowledge; prevailing winds and ocean currents followed predictable trajectories. Vessels bound for the Cape of Good Hope traditionally benefited from an eastward course post a considerable westward departure, capitalizing on favorable trade winds while sidestepping hostile coastal currents.

However, the fleet failed to meet expectations. Some historians theorize this was intentional; Cabral received secret instructions to scout uncharted lands in the western Atlantic, while others attribute it to navigational mishaps. The truth likely lies in between; perhaps Cabral was charting an exploratory course or was simply carried off course by stronger-than-expected currents and breezes. Regardless of the cause, by mid-April, they found themselves far west of their intended location. And on April 22, 1500, a lookout's gaze fell upon what puzzled Cabral - signs of land: seaweed, birds, then unmistakably, soil.

FIRST CONTACT

The coastline that appeared over the horizon was unlike any seen by Portuguese explorers before. Towering cliffs clad in thick vegetation. Beaches composed of snow-white sands. Rivers gracefully empty into the sea. And in the distant background, majestic mountains rise in vibrant blue towards the sky.

Cabral approached cautiously due to the dangers of unfamiliar shores, which could harbor hidden reefs or hostile tribes, as well as potential diseases detrimental to his crew. He dispatched smaller vessels ahead of him to sound the depths and search for suitable anchorage. By nightfall, the fleet found refuge in an embracing natural harbor, naming it Porto Seguro ,Safe Haven.

The next day, they came across local inhabitants. These were Tupiniquim people, a part of the wider Tupi-speaking groups that inhabited the Brazilian coast. They arrived in canoes, their voices speaking a language unknown to the

Portuguese, and their gestures pointing towards land. Pero Vaz de Caminha, an esteemed chronicler who accompanied Cabral and documented everything meticulously, described them as follows: "Their skin is brown with hints of redness; they have pleasant faces and well-shaped noses. They go about completely naked without any covering for their body parts or their faces."

This absence of clothing fascinated and perplexed the Portuguese, given their previous interactions with Muslim and African peoples who wore distinct attire indicative of social status. The Tupiniquim's lack thereof conveyed to the Portuguese a sense of innocence, akin to purity before the fall of man. Caminha, in his letter to King Manuel, frequently highlighted this: these people seemed to embody a natural grace untouched by civilization, yet remained oblivious to Christian truths.

The initial interactions were peaceful; the Tupiniquim provided food, such as fish and exotic fruits, as well as unusual birds. The Portuguese reciprocated with bells and trinkets. Their communication was primarily gestural, lacking verbal understanding but conveying basic amicability. It appeared that the indigenous people harbored curiosity rather than hostility towards the strangers in their midst. They marveled at the ships' construction, admiring the iron fittings and the canvas sails.

Cabral sent men ashore for exploration, discovering a landscape brimming with abundance. Rivers teeming with fish. Dense forests rich in timber resources. Soil that appeared fertile to their eyes. Parrots were so abundant that their calls overshadowed human voices. Living among these wonders were the Tupiniquim inhabitants, who cultivated crops such as manioc and engaged in fishing and hunting, leading to a paradise-like existence, as the Portuguese perceived it.

However, Cabral now faced a dilemma. His mission was to reach India, and with each passing day on this unexpected coast, his primary goal seemed further out of reach. Supplies were depleting rapidly. The favorable winds for crossing the Indian Ocean had a limited season, and if he tarried too long, he risked missing the ideal window altogether, forcing him potentially into perilous wintering along the African or Indian coasts.

Yet, this land also represented potential value. Its location suggested it might fall within Portugal's sphere of influence as defined by the Treaty of Tordesillas in 1494. This treaty divided much of the non-Christian world between Spain and Portugal, with a line running approximately 370 leagues west of the Cape Verde Islands. If this newly discovered land lay on Portugal's side, especially considering its approximate location about 510 leagues west of the same islands, it would legally belong to the Portuguese crown.

To assert their claim and meet the requirements for legal proof, Cabral ordered an imposing wooden cross erected on the shore, marking the land as formally Portuguese. A priest conducted Mass while both Portuguese soldiers and Tupiniquim watched in mutual confusion. This ritual held no meaning for the indigenous people but was crucial theatrically for establishing jurisdiction that would hold weight within European legal systems.

Next, Cabral tasked Pero Vaz de Caminha with writing a detailed account of their discovery to be sent back to King Manuel. This letter became one of the most pivotal documents in Brazilian history, providing firsthand insights into the land's nature, its inhabitants, and the Portuguese interpretation of this encounter. The tone was remarkably positive; Caminha highlighted the fertility of the soil, the healthy climate, and the peacefulness and innocence of those they encountered. He suggested that this land offered great potential for colonization and conversion.

To ensure that no detail was overlooked in formally documenting their claim, Cabral dispatched one of his ships back to Lisbon with these accounts. His mission had expanded beyond mere exploration; by acknowledging this new find while remaining uncertain of its significance relative to his original orders, he sought guidance from the king on how to proceed next.

After spending nine days along Brazil's coastline, Cabral gave the order for the fleet to resume its journey towards India. They departed on May 2, 1500, leaving behind not only the wooden cross but also presumably bewildered indigenous inhabitants whom they had named Ilha de Vera Cruz, Island of the True Cross. This would later be renamed Terra de Santa Cruz and eventually Brazil, in reference to brazilwood, a valuable commodity that would become an early export.

Cabral steered his fleet eastward towards Africa and India, fully aware now that his voyage was merely beginning. He could not have foreseen how this unplanned stopover would end up being far more significant to Portugal than the pursuit of spices had ever been.

INTO THE ABYSS

The South Atlantic voyage was extremely challenging.

Over three weeks, Cabral's fleet kept sailing eastward amid increasingly difficult circumstances. The vessels were laden with supplies for their journey to India, including provisions, water, trading items, and weaponry. They appeared low in the sea, making them susceptible to large waves. In addition, the crews had already dwindled due to illnesses contracted in Cape Verde and Brazil; they continued to work in rotating shifts to control the sails and maintain the course.

On the 24th of May, 1500, disaster struck.

A tempest emerged from the southeast, a colossal weather system typical of South Atlantic winters. Within hours, the winds intensified into gales, while towering waves engulfed the ships' masts. It rained heavily, reducing visibility to nearly zero. The fleet dispersed as each captain fought for survival against overwhelming odds.

Historical accounts depicted the fear experienced that night. Sailors tied themselves to the rigging in a desperate attempt not to be swept overboard. Water seeped through hatches, while cargo shifted in the holds, threatening the ship's fate. Seamen prayed aloud, convinced they were on the verge of death.

At dawn, Cabral assessed his remaining fleet and found four ships missing.

Among them was Bartolomeu Dias' caravel, the man who discovered the Cape of Good Hope and had charted a pioneering route around Africa, which had tragically perished. His ship sank with all hands aboard, leaving behind approximately fifty lost souls in the icy waters of the South Atlantic.

This loss deeply affected the fleet's morale. If even Dias, their most experienced navigator, could meet such an end in these waters, what were the prospects for others? Cabral commanded that the remaining ships continue onward despite doubts sowing among them as to whether this expedition was jinxed.

The fleet sailed to the East African coast and finally set foot on land near Sofala, in present-day Mozambique. They were battered, diminished, and shaken but still determined to reach India as per their mission.

They continued north along Africa's coastline, visiting ports to trade, gather intelligence, and repair storm damage. At Mozambique Island, they encountered Muslim merchants who harbored ill will towards the Portuguese ships. Word of da Gama's voyage had spread across the Indian Ocean trade network; the Muslims saw the Portuguese as aggressive, unreliable, and prone to using force to secure commercial advantages.

At Kilwa, a significant East African port for trading, Cabral attempted to foster good relations with the local sultan by sending ambassadors bearing gifts and proposals for a commercial agreement. The sultan, Emir Ibrahim, was cordial but non-committal; accepting Portuguese demands would alienate his Muslim business associates, while refusal risked Portuguese aggression. He played for time, providing vague promises without making any firm commitments.

Cabral understood the sultan's delaying tactics and chose not to press further. His main goal remained India, and he was keen on preserving resources for future challenges in this uncharted region.

In July 1500, they crossed the Indian Ocean following da Gama's route. They made landfall at Calicut, a powerful city-state situated on the Malabar Coast and central to the spice trade.

This marked what Cabral had been preparing for since his departure from Lisbon; Brazil, the storm in the South Atlantic, and the African coastal voyage were all leading up to this moment. He aimed to establish Portuguese commercial dominance over India and secure spices that justified the entire venture.

However, his arrival coincided with a political and economic minefield that would ultimately result in catastrophe.

THE CALICUT MASSACRE

Calicut in 1500 was a bustling metropolis renowned for its cosmopolitan atmosphere and potential wealth along the Malabar Coast. At the helm of this thriving port, the Zamorin oversaw an intricate commercial network comprising Hindu merchants, Muslim traders from Arabia and Persia, Chinese merchants, and various other factors from across the Indian Ocean. The prosperity of Calicut hinged upon maintaining its cosmopolitan openness, ensuring a neutral ground where diverse groups could engage in trade safely.

Enter Cabral with his arsenal, crosses, and unwavering belief that Portuguese interests deserved precedence.

The Zamorin entertained Cabral with courtesy, as was customary; foreign merchants were welcomed, provided warehouse space, and allowed to conduct their business. However, Cabral's demands transcended standard commercial agreements. He sought exclusive trading rights, a fortified factory independent of local authorities, and preferential customs rates, all guaranteed by a treaty backed by Portuguese naval power.

The Zamorin, an experienced ruler, recognized that acquiescing to Portuguese demands would incense Muslim merchants who controlled most of Calicut's trade. These influential merchants boasted connections across the Indian Ocean, substantial capital, and age-old commercial ties. They were not to be taken lightly or offended casually. Yet rebuffing the Portuguese might invite violence, as accounts of da Gama's 1498 visit remind us, when cities were bombarded and ships seized.

The Zamorin employed stalling tactics akin to those of the Sultan of Kilwa, extending hospitality and engaging in drawn-out negotiations while promising much but delivering little. Meanwhile, Muslim merchants in Calicut organized resistance, aware that Portuguese presence threatened their entire commercial standing. If Portugal secured a fortified factory with monopolistic rights, these merchants risked being marginalized or expelled.

Tensions escalated throughout the late 1500s, culminating in skirmishes between Portuguese sailors and Muslim merchants in Calicut's markets. Cabral frequently lamented the Zamorin's failure to honor commitments, while the ruler deflected, citing complex local conditions and promising to finalize agreements in the near future.

Then, on December 16, 1500, violence erupted.

A faction of Muslim merchants launched an attack on the Portuguese factory in Calicut, with conflicting accounts suggesting whether the Portuguese provoked or were independent targets. What is clear is that the assault was fierce and devastating; fifty Portuguese, including the factory owner Aires Correia, were killed, trade goods seized or destroyed.

Cabral responded instantaneously with brutal force. He ordered his fleet to shell Calicut for hours, causing widespread destruction and loss of life. Following this onslaught, Cabral seized a dozen Arab merchant vessels, taking their crews prisoner before setting them ablaze.

This aggression was designed to terrorize the Zamorin and Muslim merchants into submission, aligning with strategies employed by Portuguese commanders in Morocco and along the African coast. However, it marked a catastrophic blunder; the destruction rendered peaceable trade between Portugal and Calicut impossible. The Zamorin, deeply humiliated and incensed, became an eternal adversary of Portugal. Across the Indian Ocean, word spread about Portuguese treachery and brutality. Consequently, Cabral's fleet found itself severed from its intended markets.

In pursuit of alternative trade, Cabral journeyed south to Cochin, a smaller port ruled by a distinct raja. Here, he encountered a more amenable audience; the Raja of Cochin, rival of the Zamorin, discerned potential in Portuguese prowess for local political advantage. An alliance was forged, not grounded in commercial logic but rather on the dynamics of local power struggles.

Yet, this accord came with its own set of challenges. Cabral had to accept prices dictated by the Raja, conduct business under Cochin's terms, and contend with the reality that Muslim merchants would actively sabotage Portuguese interests throughout the region.

THE HOMECOMING

In January 1501, Cabral embarked on the homeward journey to Portugal, leading a fleet that had dwindled to only five or six ships - though the exact number is uncertain. Losses incurred during storms, detaching vessels for other tasks, and leaving one ship behind as an enduring fixture in Cochin significantly diminished his company. The cargo holds were laden with spices such as pepper, cinnamon, and ginger, along with other precious goods. Despite all adversities, the voyage could still prove profitable if these wares made it back to Lisbon.

The return journey was fraught but unremarkable; Cabral's fleet replenished supplies and navigated through East Africa before rounding the Cape of Good Hope with minor incidents and then sailed northward into the Atlantic. By June 1501, the Portuguese coast came into view.

Fifteen months had passed since Cabral's departure. He faced colossal losses, roughly halving his ships and around eight hundred men perishing. His exploration of a new land in the western Atlantic was brief but promising, despite nearly being obliterated by a massive storm. Moreover, he ignited a war in Calicut that would influence Portuguese-Indian relations for many years. However, he returned with spices worth far less than King Manuel's investment in the expedition.

Upon his return to Lisbon, Cabral received a varied response. On one hand, he had fulfilled his mission by reaching India, establishing a foothold at Cochin, and bringing home spices. The discovery of Brazil was intriguing, but its potential was not yet fully acknowledged; it would later emerge as a groundbreaking event. On the other hand, the losses were immense - half the fleet vanished, many lives were lost, significant diplomatic and military dilemmas arose in India, and profits, though considerable, fell short of expectations.

King Manuel meticulously assessed Cabral's performance. As a shrewd political strategist, he recognized that exploration and conquest entail substantial risks and costs, often resulting in failure. While Cabral had not met the criteria for failure - returning with spices and new geographical knowledge - neither did his achievements surpass greatness.

In 1502, Manuel appointed Vasco da Gama to lead an expansive fleet of twenty ships carrying 1,500 men and abundant weaponry to subjugate the Zamorin and assert Portuguese dominance through forceful occupation. Da Gama, having previously commanded a successful voyage to India, was entrusted with this ambitious escalation.

Cabral did not receive command of this new fleet but was instead recognized for his service with financial compensation, effectively retiring him from active maritime exploration at the age of thirty-four, marking the end of his career as an explorer.

THE LIFE AFTER GLORY

For many individuals, leading the voyage that uncovered Brazil would have been an exceptional achievement of a lifetime. However, for Cabral, this feat merely formed part of an extended period of private life following his return to Belmonte.

He returned home and lived like a nobleman from the provinces. In 1503, he married D. Isabel de Castro, and together they had at least four children: two sons and two daughters. These children were raised according to the customs of the Portuguese nobility - the boys prepared for careers in military or religious service, while the girls would marry well, securing alliances and property.

Cabral was involved in local governance, resolving disputes, and managing his properties. This daily routine reflected the ordinary life of someone from his social class; comfortable yet unremarkable. He held the position of fidalgo on the king's council, which carried certain privileges and responsibilities. There was nothing in his everyday life to suggest that he had ventured across the Atlantic, discovered a new continent, or endured hardships resulting in a significant loss of life.

Did Cabral ever regret retiring from the sea and its possibilities? The sources remain silent. He left no memoirs or letters revealing his inner thoughts. We learn about his later life from bureaucratic records such as property transactions, royal appointments, and legal documents. Occasionally, historical accounts mention him before he blends back into the unrecorded routine of provincial life.

In 1518, he joined the royal council, which brought prestige but not much real power. He may have occasionally attended council meetings in Lisbon, though he mainly stayed at his estates.

His death is believed to have occurred around 1520, possibly in Santarém, where he maintained a residence. The exact date is unknown. His remains were first buried in a chapel he had established and later transferred to the family church in Belmonte. There is no known epitaph or contemporary portrait of him, nor are there any sculptures or royal monuments commemorating his achievements. Within one generation, many Portuguese forgot his name. When they recalled the discovery of Brazil, it was often attributed to other commanders or considered a collective effort rather than the achievement of a single individual.

This obscurity is striking, considering that Cabral claimed territory that would become Portugal's most valuable colony, anchoring an empire across an entire hemisphere. Yet, he died in relative anonymity with his achievement overshadowed by more dramatic conquests and celebrated commanders.

Why was this so? Partly because the discovery of Brazil was unintentional, lacking epic battles or conflicts with indigenous empires. The Tupiniquim people Cabral encountered lived in small village societies without large-scale states, monumental architecture, or known resources of precious metals. To the Portuguese in 1500, Brazil seemed less valuable than other discoveries, such as India's spices and Africa's gold. It appeared to be just another stretch of land inhabited by people living in what looked like a primitive state.

Partly because of Cabral's character, he was neither boastful nor one who sought favorable accounts or political recognition. He fulfilled his duties, accepted rewards, and retreated to private life. While noble in its way, this modesty did little to create legends.

And partly because of timing – at the beginning of the sixteenth century, Portugal was focused on India, the spice trade, and building the Estado da Índia. Brazil became a secondary concern that only gained significance decades later, when Portugal recognized its agricultural potential, sugar production capacity, and strategic importance in the Atlantic system.

By the time Brazil's value to Portugal was acknowledged, Cabral had passed away, forgotten by many. The discovery garnered recognition, but the person responsible for it remained largely unknown.

WHAT HE FOUND

To fully grasp Cabral's voyage's significance, one must consider its long-lasting impact rather than focusing solely on his lifetime achievements. He claimed territory inhabited by numerous indigenous people, whose existence remains brief and fleeting in historical records.

Vera Cruz, the region named by Cabral, was not a barren wilderness awaiting European settlers; it was home to millions of people – estimated at 2-8 million – living in hundreds of unique societies with distinct languages, cultures, and political frameworks. Among those who encountered Cabral were the

Tupiniquim, just one of many groups.

These societies possessed intricate structures and knowledge of their environments that eluded Cabral's comprehension. They had an advanced understanding of their surroundings, complex social hierarchies, and rich oral traditions. For instance, they practiced controlled burning in the Amazon, creating 'terra preta,' a type of anthropogenic dark earth that enhanced soil fertility and facilitated dense agricultural production. By doing so, they cultivated productive landscapes by shaping natural environments over generations of human intervention; dozens of plant species were nurtured, many of which became globally significant crops such as manioc, sweet potatoes, peanuts, and cacao.

However, certain shortcomings rendered them susceptible to European colonization: the lack of immunity to Old World diseases such as smallpox, measles, influenza, and typhus; limited metallurgy, resulting in inferior weaponry compared to European steel; and the absence of draft animals, which hindered agricultural productivity and military mobility. Most societies were organized at the village or tribal level, without centralized states capable of mounting effective resistance to invasion.

Initially, the Portuguese showed little interest in colonizing Brazil. For the first three decades after Cabral's voyage, their presence was sporadic, with occasional ships stopping to collect brazilwood – a tree whose bark produced valuable red dye. Early traders established temporary relations with indigenous groups, who would harvest brazilwood in exchange for metal tools, cloth, and other goods.

However, by the 1530s, Portugal altered its approach. The crown began granting territory to Portuguese nobles who were expected to colonize their lands independently. This capitania's system facilitated permanent settlement. Colonists arrived, introducing sugar cultivation techniques developed on Portugal's Atlantic islands. Indigenous people were either enslaved or forced inland, while African slaves began arriving in greater numbers.

By the mid-sixteenth century, Brazil underwent a significant transformation. Coastal forests were cleared for sugar plantations; indigenous populations dwindled due to disease and violence; European settlement expanded inwardly. A hybrid society emerged – Portuguese, indigenous, and African – creating something previously unseen.

None of this was foreseen during Cabral's brief nine-day encounter in 1500. He could not have envisioned the monumental changes that would follow his landing. He observed a coastline, friendly natives, and potential territory for the crown before filing his report and moving on.

Yet his inadvertent landing marked the beginning of processes that reverberated across centuries. The convergence of three continents' worth of peoples led to the destruction of indigenous societies, the emergence of the largest slave-holding society in the Americas, the development of plantation agriculture shaping global capitalism, and the formation of Brazil as a nation with its own language and culture. This chain of events not only shaped Brazil but also left a legacy that continues to influence today's global trade practices, a direct line from early plantation economies to the complex networks and trade behaviors witnessed in our modern globalized world.

All these developments can be traced back to that unplanned encounter in April 1500.

THE VERDICT OF HISTORY

How do we assess Pedro Álvares Cabral? From his time, he was a decent, if unremarkable, leader who successfully carried out a demanding mission under tough conditions. He discovered new lands, endured a devastating storm, established a Portuguese foothold in India, and brought back valuable treasures. He adhered to his orders, and while the outcomes were mixed, that's just how exploration worked during the Age of Discovery.

Looking through today's lens, things get more complex. He claimed regions inhabited by countless people without their consent or even awareness. His actions initiated colonization, leading to atrocities like genocide via illness and violence, enslavement of indigenous and African peoples, and the ruin of cultures. It matters less that he didn't commit personal crimes during his brief stay in Brazil; what's significant is the havoc his actions triggered.

Still, it's unfair to pin everything on Cabral himself. He was only there for nine days. He erected a cross, held Mass, and filed reports. He didn't massacre indigenous peoples or establish plantations or import slaves. The horrors of colonization emerged later, as other individuals made different decisions.

This raises a deeper question: How accountable are discoverers for what they stumble upon? Can we distinguish between claiming land and the repercussions of that claim?

Perhaps it's about embracing multiple truths at once. Cabral was an individual who followed his job based on the beliefs of his time. He also served as an imperial agent who unintentionally caused immense suffering. He wasn't purely heroic or villainous, but a historical figure whose significance exceeded personal intentions or moral worth. This reflection prompts us to ponder modern choices that might lead to unforeseen devastation and to consider present-day policies or actions that could have negative impacts in the future. By combining our judgments of past events with forward-looking considerations, we can deepen our understanding of today's responsibilities for creating a fairer, more empathetic world.

What is clear is that his discovery had transformative effects he never envisioned. Brazil became Portugal's most significant and wealthiest colony, eventually surpassing Portugal itself in population and economic influence. Brazilian sugar reshaped global trade patterns, while later gold discoveries funded Portuguese royal extravagance and fueled European development. The slave trade brought about the largest movement of Africans in history, with around five million ending up on Brazilian plantations between the sixteenth and nineteenth centuries.

All this, the abundance and the horror, the cultural blending and the atrocities, the forging of a new civilization and the destruction of others, traces back to that day in April 1500 when Cabral's crew noticed unusual seaweed patterns and strange bird flights.

THE PORTUGUESE CALCULATION

Understanding Cabral's journey fully requires insight into Portugal's goals in 1500 and why the discovery of Brazil initially seemed secondary to their objectives.

In 1500, Portugal was a modest kingdom with an approximate population of 1 million, about the same size as present-day Rhode Island, yet it sought control over trade routes spanning three continents. This audacious feat was made possible by decades of Atlantic exploration, during which they developed specific technological and organizational advantages.

However, colonization demanded resources, ships, settlers, soldiers, supplies, and every real invested in Brazil meant diverting funds from India. Every settler who went to Brazil was one less in Goa or Malacca. Despite these challenges, Portugal attempted to manage both theaters simultaneously, stretching its limited resources across vast distances.

The result was an empire characterized by paradox, a power that was both impressive and fragile, wealthy yet bankrupt, with a global reach but vulnerabilities at every point. Portuguese influence in Asia peaked early in the sixteenth century but gradually declined as Dutch and English competition intensified. The expansion of Brazilian colonization was slow due to metropolitan constraints and indigenous resistance until the late sixteenth-century sugar boom made it so lucrative that it could no longer be disregarded.

Cabral's expedition marked a turning point in these historical trends. He was dispatched to reinforce Portuguese authority in India but stumbled upon the territory that would eventually surpass its significance. Neither he nor his contemporaries recognized this irony; they viewed Brazil as a minor distraction compared to Asia. Only in retrospect did the discovery of Brazil emerge as the most consequential aspect of his voyage.

THE PERSONAL COST

What emotions ran through Pedro Álvares Cabral during the aftermath of his voyage? It's hard to say for certain, but the historical records hint at a man grappling with a stark contrast between what he anticipated and the situation upon his return.

He had overseen the grandest expedition that Portugal had ever sponsored. He expected to return as a hero, boasting riches and celebrated as the one who cemented da Gama's legacy. However, instead of triumph, he faced losses, entanglements, and mixed outcomes. Half of his crew perished, along with half of his fleet. Yes, he discovered new lands, but their value was uncertain; in Calicut, battles led to long-standing enmities. His successor, Vasco da Gama, was called upon to rectify matters with brute force.

Cabral, who was steeped in the values of chivalry and success, must have found it taxing to accept such an outcome. Despite fulfilling his duties, adhering to orders, and surviving hardships that claimed stronger men, he was effectively set aside for those whose mettle seemed more ferocious.

Though letters he penned post-expedition are lost, one can envision their content: rationales, accounts of mishaps, pleas for another chance, assurances of lessons learned. It's a defensive tone, indicative of an individual seeking to salvage his standing and career.

King Manuel's answers, though courteous and indecisive, must have been crushing. Acknowledgment of service, promises of recompense, yet no more expeditions awaited him. His role as an explorer ended; go home, lead a quiet life, cherish what you have.

So Cabral returned home, wedded, begetting children, managing holdings, and sitting on governing boards. He led the life befitting his lineage, a secure, esteemed existence, far removed from the ocean that had momentarily thrust him into history's limelight.

Did he regale his offspring with tales of Brazil and India? Did he recount the storm claiming Bartolomeu Dias' life, Calicut's siege, or Tupiniquim's bare figure inspecting his vessels with awe? Or did silence prevail as he recognized that his fleeting brush with fame overshadowed his post-expedition ordinariness?

The records propose the latter. There is no indication that Cabral sought to be hailed as a discoverer or profited from his feat. Instead, he faded into the minor nobility as a fidalgo among countless others, his distinction merely known by those familiar with his story.

This peaceful retirement may reflect on Cabral's character; he was not a seeker of glory nor driven to become legendary. He reconciled with reality even when it fell short and accommodated himself to a humbler state.

There's nobility in such acceptance, yet there is also melancholy in a life punctuated by one significant event, followed by years devoid of drama.

THE SHIPS THAT DID NOT RETURN

Amidst the most chilling aspects of Cabral's journey are the ships that vanished, the vessels that ventured into storms or obscurity and failed to return to Portugal. One such incident involved Bartolomeu Dias's caravel, which met its end in a South Atlantic storm, taking with it one of Portugal's greatest navigators and approximately 50 men. Their deaths were swift, a ship sinking in frigid, tumultuous waters would swiftly claim everyone on board. Among these nameless faces was João Castanho, a young sailor from Porto, an experienced hand with dreams of returning home triumphant, his pockets filled with riches. Like countless others, João's story ended tragically, erased by the sea's unforgiving grasp. However, four other ships disappeared under more enigmatic circumstances.

One vessel, helmed by Diogo Dias (Bartolomeu's brother), became separated from the fleet during the tempest and never reconnected with it. This ship endured extraordinary trials after surviving the storm, including being stranded in the Indian Ocean for months, enduring a shipwreck, capture by Muslim merchants, and escaping against all odds. Yet, Diogo Dias never made it back to Portugal; his ship vanished during the return journey, swallowed up either by the enormity of the South Atlantic or the African coast.

Another ship mysteriously disappeared after departing from Brazil. It set sail from Porto Seguro and was never seen again, leaving no wreckage or survivors in its wake. It is assumed that it met a tragic end somewhere between Brazil and Portugal, possibly succumbing to storms, reef crashes, or simply being consumed by the sea.

These losses were not uncommon; each major Portuguese expedition suffered ship losses. The Indian run served as a grave for many. Storms in the South Atlantic and near the Cape of Good Hope claimed the lives of numerous sailors. Disease also ravaged crews in tropical waters, while shipworms gnawed through hulls and navigational errors led to wrecks along uncharted shores. In essence, the sea posed an enemy more lethal than any human adversary.

What stands out is how nonchalantly these losses were accepted. Portuguese chronicles casually mentioned missing ships without even recording the names of commanders or the number of men lost. It seemed as though the prevailing attitude was: this is the cost of empire, men will die, and ships will sink; we shall merely build more ships and recruit additional hands. Recently, a fragmented diary entry circulating among sailors expressed the sentiment: "Today, we lost João to the sea. He was my closest friend on this perilous voyage. Minds like his deserve better than the depths of the ocean, yet here we are, disregarded by those who chart our destiny from afar." This ruthless treatment of their own people, not merely those subjugated in colonization, characterized Portuguese expansion. The monarchy regarded Portuguese lives as disposable resources to be invested in endeavors that might yield profits or simply disappear into the Atlantic without a trace.

The men who sailed with Cabral were aware of this reality. They knowingly signed on despite knowing the mortality rates hovered between 30 and 50 percent, a fact well known to them. Yet, driven by poverty, ambition, religious fervor, or sheer desperation, they embarked regardless. What other options did they have? To stay in Portugal and slowly starve, or sail to India with a chance of returning home wealthy? For many, it was no choice at all.

Hence, they boarded leaky ships that consumed food crawling with weevils, endured months of storms and disease, the monotony of which could drive one mad. And if fortune smiled upon them, they might return home with enough money to live comfortably; though more often than not, death awaited a drowning, a disease, conflicts over trade rights, or simply disappearing into the vast expanse of the ocean.

Their names are mostly lost to history, appearing in crew manifests as mere statistics: "sailors, forty-eight" or "soldiers, thirty-two." Only those of high-ranking officers, nobles, and men of influence are given individual recognition within the chronicles. The common sailors who braved the ships, who fought against storms, who battled to save their vessel from foundering they remain mere shadows in statistics.

Yet they were the backbone that made the empire possible. Without them, no ships would have set sail. Without their labor, suffering, and sacrifices, Portugal would not have had access to Asian wealth. Cabral commanded them, but it was their hands, hearts, and lives that enabled his voyage to unfold. And most of them perished in doing so.

THE INDIGENOUS CATASTROPHE

While Cabral returned to Portugal and enjoyed his peaceful retirement, events unfolded in Brazil that he would never witness or even be aware of.

The indigenous people who welcomed the Portuguese with curiosity and amity in April 1500 were facing an impending disaster beyond imagination. Over 100 years, their population might decline by as much as 90%. Entire tribes vanished into oblivion, languages perished, and societies that thrived for millennia were decimated so thoroughly that later generations would hardly recall their existence. Yet, amidst this catastrophic loss, some indigenous communities demonstrated remarkable resilience and adaptability. They adapted to new circumstances, preserving essential aspects of their cultural heritage. Their stories of endurance, creativity, and vitality continue to play a significant role in influencing contemporary Brazilian culture. From language and art to spiritual practices and medicinal knowledge, the enduring legacy of these communities supports a nuanced understanding of survival alongside tragedy.

The primary cause of this devastation was illness. The Portuguese vessels carried smallpox, measles, influenza, and other deadly pathogens to which the indigenous population had no immunity. These diseases spread with alarming speed, surpassing even the pace of Portuguese colonization. A single ship's brief stop at the coast could unleash a chain of pathogens that would journey inland through native trade routes, claiming the lives of those who had never encountered Europeans.

The first major epidemic likely struck within a decade after Cabral's arrival. The accounts from that period described indigenous populations devastated by mysterious maladies causing fever, pustules, and death. Survivors recounted tales of entire villages being wiped out, corpses lying unburied as there were insufficient survivors to perform funeral rites, and social structures collapsing with the demise of elders and leaders.

The Portuguese were cognizant of the immense loss of indigenous life but viewed it as a divine sign, clearing the land for Christian colonization. Few expressed sympathy or acknowledged their role in perpetuating the disease. The prevailing belief was that these unfortunate deaths ultimately paved the way for Portuguese colonization and the spread of Christianity.

In addition to illness, deliberate violence played a part in the decline of the indigenous population. As Portuguese settlements expanded, colonists enslaved indigenous people to work on plantations under brutal and often fatal conditions. Resistance was met with punitive expeditions that obliterated villages and slew those who dared to fight back.

Some indigenous groups allied with the Portuguese against their traditional rivals, hoping to exploit European military might for their own benefit. While these alliances sometimes provided short-term advantages, they all ultimately faced the same grim fate. Portuguese colonization was relentless and eventually dispossessed all indigenous peoples, irrespective of whether they had sided with or opposed them.

By 1600, the coastal indigenous population had dwindled significantly from its 1500 levels. Survivors were displaced inland, enslaved, or assimilated into the hybrid colonial society sprouting along the coast. Their descendants persisted, but no longer as autonomous peoples governing their territories autonomously.

Cabral bore no direct culpability for this catastrophe; he spent merely nine days on Brazilian shores and took no life. Yet, his claim to the territory on behalf of Portugal initiated a chain of events. He marked the land with the cross that sanctioned colonization, documented his discoveries that piqued Portuguese interest, and opened the floodgates for conquest, disease, and slavery.

Could he have foreseen these consequences? It is unlikely in detail. Yet the pattern was set; Portuguese colonization in the Atlantic islands and along the African coast showcased the ramifications of European territorial claims and exploitation. Indigenous populations dwindled or were enslaved, settlers seized control, and local societies bent under colonial interests.

Cabral, as an educated nobleman familiar with Portugal's colonial expeditions, certainly grasped the general blueprint. Based on precedents from the Atlantic islands and the West African coast, where indigenous communities suffered grievous losses due to disease, forced labor, and displacement, he could have anticipated societal disruptions. However, he likely did not predict the scale of devastation that would unfold following his discovery.

Whether such ignorance absolves him is a question for each reader to answer personally.

THE DOCUMENT THAT SURVIVED

Among all the documents from Cabral's voyage, one stands out for its immense value: Pero Vaz de Caminha's letter to King Manuel I describing their arrival in Brazil. Caminha served as the official scribe aboard the fleet, documenting significant events and transactions. This particular letter, dispatched from a ship upon returning to Lisbon, spans approximately 8,500 words. It offers detailed accounts and displays remarkable consideration for indigenous people.

The letter describes the physical attributes of the Tupiniquim people, their nakedness, gestures of friendship, weapons (bows and arrows), lack of metal tools, villages, canoes, as well as their apparent naivety about Christian beliefs. Caminha even offers a recommendation that would have grave consequences in the future: he believed the best outcome for these lands was to save their inhabitants.

This phrase encapsulates the Portuguese belief that justified centuries of forced religious conversion, cultural destruction, and colonial domination. Indigenous peoples were seen as souls to be saved rather than individuals with valid ways of life. Their religions were considered ignorant and in need of correction, while their social structures were viewed as primitive and to be replaced by Christian civilization. Furthermore, they believed the land was vacant despite being inhabited, awaiting appropriate use by Christians.

Caminha's letter also reveals confusion on the Portuguese side about what they had discovered. He described the land as "very large," but remained uncertain whether it was an island or a continent and questioned its strategic or economic potential. The letter appears to be written by someone intrigued by something noteworthy, yet without clear importance.

Unfortunately, King Manuel's reaction to this letter remains unrecorded; however, his subsequent actions suggest that he regarded Brazil as a minor discovery unworthy of immediate investment. No colonization expedition was launched, nor were additional ships sent to explore the coastline. Instead, the information was stored away, and attention remained focused on India.

The letter eventually faded into obscurity within the royal archives for centuries, only to be rediscovered in 1817. Although its significance wasn't immediately recognized, today it is regarded as one of the foundational documents of Brazilian history – the first European account detailing Brazil and its indigenous inhabitants.

However, for much of its existence, this document was simply a forgotten report, buried within dusty archives - its author now deceased, its subject matter deemed insignificant.

This obscurity parallels Cabral's historical significance. The discovery that would ultimately prove crucial was initially regarded as the least important. The landing, which altered an entire continent's demographic and cultural landscape, was treated as nothing more than a minor sidestep from more pressing matters. Only through hindsight does it become apparent that April 1500 marked a pivotal moment in history.

THE EMPIRE HE ENABLED

By approximately 1520, after Cabral's passing, Brazil held little significance for Portugal; however, within a single generation, it would become integral to both their prosperity and sense of identity.

This transition began with sugar cultivation. Portuguese settlers utilized techniques from Madeira and São Tomé and recognized that Brazil's climate and soil were perfect for growing sugar. At the time, sugar was an extremely lucrative commodity in high demand throughout Europe. A successful plantation could make its owner wealthy within a few years. Yet, cultivating sugar required immense amounts of labor, which proved to be dangerous and led to the death of many workers at alarming rates. Indigenous people, who were already succumbing to disease, couldn't provide sufficient labor. Consequently, they resorted to African slavery as the solution, an outrageous decision by any moral standard.

From the 1570s onward, Portuguese ships began transporting increasing numbers of enslaved Africans to Brazil. By 1600, around 15,000 Africans had been forcibly brought to Brazilian plantations. By 1700, the number had risen to approximately 600,000. When slavery was finally abolished in Brazil in 1888, roughly five million Africans had been transported there, making Brazil the largest slave society in the Americas.

This massive forced migration gave rise to a new hybrid society that emerged from the violent collision of three continents, neither European, indigenous, nor African. Brazilian Portuguese developed distinct characteristics influenced by both indigenous and African languages. Brazilian religion merged Catholic practices with beliefs drawn from African and indigenous cultures. Consequently, Brazilian culture became a complex blend of diverse influences held together by the Portuguese language and the Catholic Church.

Portugal itself grew increasingly reliant on Brazil's wealth. Revenue from sugar funded royal expenditures, military campaigns, and Lisbon's lavish court life. The discovery of gold in Brazil's interior during the 1690s dramatically increased Portugal's wealth, allowing for the construction of extravagant Baroque churches and palaces that still define much of Lisbon's architecture.

By the eighteenth century, their roles had reversed; Brazil was no longer a dependent colony of Portugal, but Portugal relied heavily on Brazil. When Napoleon invaded Portugal in 1807, the Portuguese royal family sought refuge in Rio de Janeiro, effectively turning Brazil into the center of the Portuguese Empire. After the Napoleonic Wars, upon returning to Lisbon, Brazil declared its independence in 1822 under a Portuguese prince. This unique situation allowed for an independent empire while maintaining dynastic continuity.

None of these developments was imaginable back in 1500 when Cabral spent nine days along the Brazilian coast. The trajectory that unfolded from accidental discovery to colonial dependence to imperial upheaval was shaped by contingencies, opportunities, and decisions made by thousands over several centuries. Yet it all originated from that initial claim the cross Cabral planted, the Mass his priests celebrated, and the letter Caminha wrote. The legal act of discovery established a framework for everything else that transpired. Portugal claimed Brazil due to Cabral's declaration that it belonged to Portugal, and this was accepted by other European powers as legitimate.

This illustrates the strange, lamentable logic behind colonial law. A man could simply spend nine days somewhere, erect a religious symbol, write an account, and thereby establish his nation's 'right' to conquer, settle, and exploit the territory and its inhabitants for countless generations. The people who lived there, their ownership, rights, and very humanity held no significance in this transaction. European legal fictions erased their presence and authority. A parallel can be drawn with contemporary struggles of indigenous communities against modern land claims, where historical treaties and unjust land laws undermine their sovereignty. In countries like Canada, the United States, and Australia, legal battles over ancestral lands reflect colonial practices, demonstrating how 'nine-day sovereignty' still affects us today and making it crucial to condemn such actions.

Cabral did not create this system; he operated within frameworks shaped by papal bulls, royal treaties, and centuries-old legal and theological traditions.

However, he was the catalyst that made these abstractions a concrete reality in Brazil.

THE SEAWEED REMEMBERS

April 22, 1500. The coast of an unknown world.

Pedro Álvares Cabral stands on his ship's deck, gazing at the drifting algae, contemplating their significance. Could this be an island? A peninsula? Perhaps even a new landmass? The knowledge and experience to fully comprehend this sight elude him.

He opts to explore further. It's a logical choice for any naval commander faced with unfamiliar terrain. Dutifully following his monarch's directives of diligence and service, he deduces that a few days are needed to ascertain the nature of this coastline before proceeding on their voyage to India as originally planned.

The days that follow would alter history's course in ways unimaginable at the moment.

Not immediately nor evidently. The repercussions manifest over decades, long after Cabral has passed. Yet, the foundation was laid during this unintended discovery and its subsequent documentation.

Twenty years hence, Cabral finds himself in his manor house in Belmonte, a nobleman overseeing familial estates. Does he ponder Brazil? Wonder what became of the land he discovered or the people he encountered? Realize that his fleeting exploration had far greater implications than all his time spent in India?

Unlikely. Without insight into colonial progress and oblivious to the demographic crises unfolding on Brazilian shores or the economic shifts pivotal to Portugal's future wealth, there was no reason for him to suspect the magnitude of his discovery. It remained just another episode in a career that concluded with disappointment.

Yet, history judged otherwise. Cabral is mostly remembered for Brazil. His troubles in Calicut, his trade mission to India, and his survival from catastrophic storms are relegated to footnotes. What resonates across centuries is not his conflicts or missions but that serendipitous landfall.

The seaweed was mistaken. The birds are unfamiliar. Yet a noble man faithfully carrying out orders stumbled upon a continent.

Everything that followed the rise and fall of empires, the cultural amalgamations and devastations, the accumulation of wealth alongside suffering, all trace back to his decision to investigate this unusual coastline.

Not predestined nor fated. Hundreds of subsequent choices influenced outcomes. Yet, Cabral opened up possibilities with his choice to explore.

He was ignorant of the consequences he could not foresee. Neither excused for what he unknowingly set in motion, nor absolved simply due to ignorance. He acted; actions had effects that rippled across centuries and continents.

This is the burden of historical agency to act without full knowledge, yet remain accountable. To make decisions unaware of their ramifications, yet those ramifications shape countless lives. To follow orders, fulfill duties, serve monarchs, and partake in transformations not intended nor controllable.

Pedro Álvares Cabral fulfilled his duty, explored uncharted land, claimed it for his king, and resumed his journey. Competent and thorough, he was utterly unaware of the profound impact his actions would have.

The seaweed continued its drift. The coast emerged on the horizon. And history veered toward an unforeseen future. Discovery often reveals more than sought after, altering everything in ways not fully grasped by the discoverer.

Centuries later, we ponder a moment initially perceived as ordinary yet revolutionarily significant, executed with competence but causing upheaval, marked by one man as duty yet experienced by multitudes as distress. Cabral departs the shore, a cross marks the beach attracting Tupiniquim curiosity, and ships prepare to depart. Meanwhile, in the unfolding possibilities of what's to come, an entire world awaits birth, transformation, and devastation beyond anyone's present imagination.

The seaweed was wrong; that was the first sign. All else followed suit.

Gaspar Corte Real

Summer 1501. Somewhere in the North Atlantic. 50° North Latitude.

The ice floated eerily. Gaspar Corte-Real gripped the tiller with determination. His eyes were fixed on the ethereal white mass emerging from the thick fog. The air nipped at his lips, leaving them numb with icy moisture. Before him was an iceberg. Its true magnitude hid beneath the sea, its form shaped by ceaseless wind and water. Rubble and fractured ice clung desperately to the hull. They created a chilling symphony of cracking bones. In the shadowy depths below deck, fifty-seven indigenous captives cowered in fear. They were bound, trembling, their homeland forever lost. Around him, thirty Portuguese sailors battled the rigging with hands stiffened by frost. Frost adorned their beards. Months at sea had drained the vibrancy from their eyes. Ahead lay only a veil of fog, drifting ice, and an unforgiving ocean. It broke the spirits of all who dared confront it. Corte-Real wrestled with the tiller, aiming to steer clear of the menacing iceberg. The wind was obstinate and biting. It pushed against him with relentless force. His caravel, laden with human cargo and battered by the trials of the sea, responded hesitantly. The other two ships were lost in a storm days before. They were swallowed up by an endless expanse of gray. He could only speculate on their fate. Had they endured, retreated, or succumbed beneath the waves? At forty-seven, born into nobility and seasoned in navigation, he knew better than to challenge the northern seas so late in the year. Yet, ambition, not caution, drove him forward. A burning desire for discovery was etched in Portuguese veins. He yearned to etch his name in history. The dream of a Northwest Passage to Asia would surpass all other triumphs. Instead, he encountered only ice, fog, and a coastline so unforgiving that even its natives waged daily battles for survival. Now, weighed down by captives and valuable maps, he was adrift in an ocean that tolerated no mistakes. The caravel never returned to Portugal. Corte-Real vanished into the North Atlantic, leaving behind nothing but silence. No wreckage surfaced. No survivor recounted the tale of those final moments. What force could eradicate a fleet and leave not even a whisper upon the waves? He disappeared alongside his captives, crew, and the answers that tormented Portuguese explorers. What discoveries had he made? How far had he ventured? What secrets did he unearth about the northern path to Asia? Within a year, his brother Miguel set sail north in search of him. Miguel vanished just as completely. Two brothers, two ill-fated expeditions, and over a hundred souls were swallowed by the North Atlantic. Their fate remains an enigma. This is their story, a tale of an era when ambition for glory and riches led men to risk everything against an unyielding sea.

THE THIRD SON

Gaspar Corte-Real, born around 1450, came from a family that symbolized Portugal's aspirations in that era. His father, João Vaz Corte-Real, had risen through the ranks to become captain-donatário of Terceira Island in the Azores, one of the most prestigious positions in the Portuguese maritime empire. This role was not a mere inheritance but was granted by the king as a token of his service and loyalty. As such, the Corte-Reals were ambitious climbers; in Portugal's burgeoning Atlantic economy, fortune awaited those daring enough to explore beyond known boundaries.

The name itself, Corte-Real, meaning "royal court," spoke volumes about their ambitions despite being minor nobles, or fidalgos, linked to the great families of Portugal. They were respectable but lacked true power, seeing opportunities in the Atlantic Islands. On the mainland, land was scarce and power concentrated among ancient noble houses. However, in the Azores, Madeira, and the Cape Verde Islands, fortunes could be made through agriculture, shipping, and trade. The Corte-Reals seized these opportunities eagerly.

João Vaz Corte-Real might have been involved in voyages to Greenland or the North Atlantic in the 1470s; however, the historical record is fragmentary, leaving his participation unclear. Portuguese documentation from this period is incomplete, with claims of early voyages to North America often intertwined with national legends. Nonetheless, it's clear that João Vaz was an adept administrator and possibly a seasoned navigator, the very kind the Portuguese crown needed to govern its Atlantic territories. He governed Terceira competently, fostering agriculture, encouraging settlement, and laying the foundations for the Azores' prosperity.

Gaspar grew up amidst such expansion and possibilities. Being the third son, he was set to inherit little under Portugal's inheritance laws, which favored the eldest son. His options were limited: the church, military service, or the sea. Gaspar chose the sea, drawn by its allure even as it beckoned him with danger. By his twenties, he had participated in Atlantic voyages, learning navigation and seafaring skills and grasping the complex interplay of winds and currents governing ocean travel.

The late fifteenth-century Atlantic was not entirely uncharted but remained perilous. Portuguese navigators had been treading these waters for generations, charting routes to the Azores and Madeira, venturing down the African coast, and mastering the trade winds that could span vast distances. Yet every voyage bore inherent risks: ships were lost in storms, crews succumbed to scurvy or dysentery, and navigational errors led vessels into despair. The fatality rates were high enough for sailors to draft wills before setting sail and attend mass for spiritual solace.

Despite these perils, the rewards of successful voyages were immense. During Gaspar's youth, Portugal morphed from a minor kingdom into a burgeoning maritime empire. The African coast yielded treasures of gold, ivory, and slaves. The Atlantic Islands flourished with sugar and wine production. Beyond these riches lay tantalizing prospects. There were direct routes to Asia's spice markets. Alliances with fabled realms like Prester John's beckoned against Islam. Undiscovered lands could hold untold wealth. For an ambitious figure like Gaspar Corte-Real, these possibilities were irresistible. In this seafaring life, Gaspar often carried a small pendant, a gift from his mother, as a talisman against the dangers of the unknown. Sometimes, standing on the deck and gazing at the vast horizon, he whispered a quiet prayer for protection and success, hoping that his name would one day echo through the annals of history.

1495. Terceira, Azores.

By midlife, Gaspar had forged what could be considered an unremarkable but respectable career path. As a seasoned sailor, he had participated in multiple Atlantic crossings, though the exact routes and activities were not well-documented. His reputation was strong enough to earn him favor from King Manuel I of Portugal, who granted him land and privileges, suggesting that valuable service had been rendered. He further elevated his standing by marrying into an influential Azorean family, integrating himself within the islands' elite circle. He had secured all that could reasonably be expected from someone of his background.

Despite this, he lacked the accolades. There were no grand discoveries to his name nor any new lands conquered. His fame did not align with that of explorers like Bartolomeu Dias or Vasco da Gama, who were making monumental strides, such as rounding the Cape of Good Hope and plotting expeditions to India. At forty-five, Gaspar was beyond the typical age for embarking on perilous quests; he watched as younger navigators claimed what he had dreamed of achieving.

Gaspar's fortunes shifted with a combination of geopolitical changes and family aspirations. In 1497, Vasco da Gama proved that a direct route to India could be taken via the southern seas, fueling the Portuguese ambition for new trading routes. This breakthrough intensified the quest for finding even more lucrative alternatives. The once-neglected northern route, known as the Northwest Passage, suddenly became a focal point for Portuguese strategists seeking to circumnavigate or traverse North America and reach the Pacific Ocean. At the same time, Spain's successful expeditions to the New World, including Columbus's discovery of the Americas, added pressure on Portugal to expand its territorial and trade interests. The potential of a Northwest Passage offered both a competitive edge against Spain and access to untapped Asian markets, aligning with Portugal's strategic goals of extending its empire and controlling significant maritime routes.

Around the same time, John Cabot had successfully navigated the North Atlantic under English patronage, demonstrating that lands existed across the ocean and that systematic exploration of those waters was feasible. This stirred the Portuguese crown's competitive spirit, prompting a desire to assert its own claims in these regions.

Gaspar Corte-Real presented several qualities that made him an appealing candidate for exploring northern latitudes. His extensive experience in Atlantic navigation provided a solid foundation. Familiarity with the Azores as a departure point for northern expeditions was a plus, possibly bolstered by his family's historical ties to North Atlantic exploration. His age allowed for mature judgment and experience while maintaining enough vigor for the physical demands of prolonged voyages. Lastly, he demonstrated an ambition that drove him to embrace risks others might shy away from.

In either 1499 or 1500, the exact year varies, Gaspar secured royal approval to explore northern waters at his own expense. This marked a departure from crown-funded expeditions; Gaspar was spearheading an early form of venture capitalism. He pooled personal resources and gathered support from family and associates, staking everything on the potential for groundbreaking discoveries. The stakes were high; success promised immense wealth and prestige, while failure risked financial ruin as he bore all financial risks alone.

Preparation for this undertaking took considerable time. Ships needed reinforcement against ice and overstocking provisions for extended trips through sparse resupply zones. He assembled a team willing to venture into harsh northern climates far from any Portuguese settlement. The endeavor was strongly backed by the Corte-Real family, utilizing their influence among Azorean merchants and shipowners.

The first voyage likely departed in summer 1500, with the number of ships involved unclear, potentially three or fewer. They sailed northwest from the Azores towards Greenland and eastern North America, guided by contemporary geographical theories that held Asia extended much further east than it truly is. Their goal was to find a relatively swift passage to Asia's coast via high northern latitudes, despite being misinformed about the actual geography.

Details of Gaspar's discoveries during this voyage remain elusive. He reportedly made landfall in some northern location in Greenland, Newfoundland, and Labrador, and upon returning to Portugal in late 1500, his accounts encouraged further exploration. King Manuel I was sufficiently impressed to grant him additional privileges for a subsequent journey.

Yet the first voyage served as reconnaissance only; it was on the second expedition in 1501 that Gaspar overextended himself.

INTO THE ICE

May 1501. Lisbon.

Three caravels lay anchored within the Tagus River, their hulls recently sealed, their cargo holds filled with provisions sufficient for six months or more. At the helm was Gaspar Corte-Real, accompanied by his brother, Miguel, as an investor and supporter, though it does not appear that Miguel joined the expedition in person. Although the exact number of men remains undocumented, a typical trio of caravels might carry around seventy to ninety souls in total, comprising captains, pilots, sailors, soldiers, artisans, and perhaps even a priest to uphold spiritual observances.

These caravels were not grand vessels; they measured roughly sixty feet in length, each hoisting single masts with lateen sails that could be angled to catch breezes from various directions. They were constructed for the Atlantic's demands: shallow drafts to navigate uncharted shores, robust frames to withstand fierce seas, and uncomplicated riggings suitable for a small crew. Inside, the conditions were less accommodating. Blankets soaked up the oily tar that seeped from the deck, leaving sailors to endure nights of discomfort. While water and cold infiltrated every corner, biscuits crawled with maggots, a grim reality that twisted the stomachs of the men gathered in the cramped quarters below deck. However, they lacked preparation for ice conditions. No European ship of this era was truly equipped for the trials Gaspar would encounter in northern latitudes.

Their provisions would have followed standard Atlantic fare: casks filled with water and wine, salted pork and fish, hard biscuits teeming with weevils, dried beans and onions, cheese, and vinegar. The wine served a practical purpose, as water stored in wooden barrels soured swiftly, while wine retained its drinkability for far longer. The biscuit, despite its unpleasant nature, provided necessary sustenance for labor-intensive tasks. Salt pork provided protein but contributed to scurvy because it lacked fresh nutrients.

Aboard were also items deemed suitable for trade: copper and brass trinkets, fabrics, knives, and beads. These goods often preceded coercion or violence, rather than representing neutral exchanges, and they typically formed part of a strategy to forge relations with indigenous peoples. And weaponry was present too: crossbows, swords, pikes, armor for the soldiers, powder, and shot if firearms were aboard. Such armaments served not solely for defense. Portuguese expeditions in Africa had set a precedent by combining trade with raiding, acquiring some slaves through purchase while seizing others outright. There is no reason to assume Gaspar's northern voyage operated under any different modus operandi.

The departure would have adhered to familiar customs: a mass said aboard ship or at Lisbon's waterfront chapel, blessings bestowed by priests and possibly by representatives of King Manuel, and farewells exchanged with family members aware of the likelihood of never seeing these men again. Then came the practicalities: the ships were pulled away from the docks with oars and towing lines, sails raised once clear of the harbor, and set a southwest course to round Cape St. Vincent before veering northwest towards the Azores.

The journey from Lisbon to the Azores was routine for seasoned Atlantic sailors. A stopover at Terceira would have been for final provisions and perhaps to recruit additional crew. Then northwest they sailed into open waters, heading for latitudes where prevailing winds would carry them toward Greenland and North America's coast.

Weeks passed as they crossed the ocean. Shipboard life settled into a routine: watches of four hours on and four off, constant vigilance over sail and rigging, meals eaten from wooden bowls while bracing against the vessel's motion, prayers at dawn and dusk, and the perpetual adjustment of course as winds shifted. The water transitioned from the deep blue of the mid-Atlantic to the grey-green hues of northern waters. Cold air prevailed. Seabirds became more plentiful, suggesting nearby land. And then came ice.

First, as scattered fragments drifting close to the hull. Then there are larger sections that require cautious navigation to avoid. Then, through fields of slush that caravels had to force their way, ice scraped against planking with sounds that were grating on the ears. The temperature plummeted further. Sailors who sailed comfortably in shirts now layered themselves in every piece of clothing available. Rigging stiffened from ice's touch. Water splashed onto the deck and froze almost instantly, creating a slippery hazard for the planks.

The coastline they finally reached bore no resemblance to Portugal's coasts or even those explored by African-bound Portuguese sailors. It was harsh and barren, dominated by rocks and ice. Stunted vegetation clung tenaciously where it could on this unforgiving land. Inlets and bays were clogged with ice even in summer, and the few beaches consisted of stones rather than sand. This landscape had been shaped by glaciation sculpted by ice sheets that had only recently retreated northward.

Yet, inhabitants thrived here, the indigenous people of these shores. They might be Beothuk in Newfoundland or Inuit in Labrador/Greenland, having developed remarkable adaptations to survive this brutal climate. Hunting seals and whales, fishing, and gathering sparse plant life sustained them while maintaining social structures that allowed communities to endure despite the cold that could kill outsiders within hours.

Portuguese observers viewed these people through the lens of their own cultural norms, branding indigenous Americans as primitive savages suitable for enslavement or conversion. Yet, these observers failed to recognize not only the resilience and ingenuity of the indigenous peoples, but also their agency in shaping and sustaining their societies amid adversity. The sophisticated survival technologies they developed, such as clothing tailored from skins and tools including the Inuit kayak, boats, and sleds adapted for ice and snow, reflected both a deep understanding of the harsh environment and a purposeful adaptation to its demands. These innovations not only supported complex social structures among the Beothuk and other groups, but also demonstrated deliberate strategies of resistance, adaptation, and cultural continuity in the face of external pressures. However, these achievements and forms of self-determination remained invisible to those defining civilization primarily by stone architecture and written language, thus overlooking the active roles indigenous peoples played in maintaining and transforming their communities.

To them, these people represented potential slaves, vulnerable to capture like Africans, who could be sold in Portugal, offsetting exploration costs. This was merely an extension of existing patterns; Portuguese ships had been transporting African slaves since the 1440s, with established legal and moral frameworks endorsing slavery. Extending this to the North Atlantic required no new justifications.

Captures likely occurred over several weeks as the expedition navigated along the coast, landing parties encountering indigenous groups and seizing them by force or trickery. For the Beothuk or Inuit, seizure meant sudden iron stench of chains, cacophony of alien voices, and cries from family torn apart. Historical accounts suggest around fifty-seven people, men, women, and possibly children. Shackled aboard caravels in holds unfit for human cargo, stripped of belongings, they transformed into property.

The trauma experienced was absolute: seized by strangers speaking an unintelligible language, confined on unfamiliar vessels to a fate unknown, suffering terror and grief from separation, enduring ill-treated conditions, all recorded only indirectly through Portuguese accounts treating them as mere goods.

To Gaspar and his crews, these captures represented success, and slaves equated profit. They served as evidence of discovery and living proof of new lands for potential resource exploitation. King Manuel would interrogate the captives to learn about their homelands, assessing whether territories warranted further Portuguese interest.

However, first they had to return to Portugal, a point where the expedition's fate irreversibly darkened.

THE SEPARATION

Late Summer 1501. The North Atlantic. Approximately 50° North.

The three ships sailed close to one another when weather permitted them, apart only a few miles during daylight hours. This approach gave them mutual support as they covered more ground by surveying different parts of the coast. But this method also had its risks; if they got separated in fog or storm and couldn't find each other again in such a vast sea, it could be disastrous.

The late summer storm that hit them wasn't unusually strong by North Atlantic standards, but it was fierce enough. The American coast often sees violent weather, with low-pressure systems moving eastward that can raise waves up to 40 feet high and produce winds capable of tearing sails and masts. This area is notorious for its harsh conditions, with water temperatures often plunging to about 10°C, greatly increasing the risk of hypothermia. Historical data suggests that storms at this latitude occur on average twice per month during late summer. Sailors referred to this section as the "roaring forties," between forty and fifty degrees latitude, where storms are most brutal. However, Gaspar Corte-Real was further north in those days, where conditions were even worse.

In hours, the ships were separated by the storm's wrath. Visibility dropped to nearly zero as heavy rain turned everything into a grey chaos. Each ship fought for its survival, sailors gripping whatever they could as waves from all directions tossed their vessels violently. The Portuguese crews had experience, but it did little to help against such ferocious seas. It was luck as much as skill that determined who would survive.

When the storm finally cleared after hours or even days, accounts aren't clear; the other two ships were nowhere to be seen. Whether they had sunk, turned back, fought their way to different positions, or just vanished beyond sight was uncertain. Standard procedure is to search the last known position, but determining one's location after such a storm can be nearly impossible without modern instruments.

Gaspar Corte-Real faced a tough decision: continue looking for his missing companions, risking his own ship in increasingly treacherous conditions and burning through precious supplies? Or head back home to Portugal, hoping the others had done the same so they could reunite there. Whatever his reasoning was, he chose to sail home. Perhaps he believed the other captains would do the same. Maybe his provisions were running low, or his ship and crew needed repairs. We will never know for sure.

Two of the three caravels returned to Portugal in early October 1501, but Gaspar's ship did not. They brought between forty and fifty men who were exhausted, battered, but alive with tales of exploration, encounters with natives, captives as evidence, and geographical data collected. But overshadowing their triumphant reports was the terrible news: Gaspar Corte-Real had vanished into the North Atlantic as autumn turned to winter.

The surviving captains would have recounted how they got separated, endured the storm, and failed to find Gaspar's ship again. They might have expressed hopes he'd found a safe harbor somewhere along the coast, wintering there until spring returns. But seasoned sailors knew the reality: one caravel alone after storms, low on supplies, in waters filled with ice, sailing further into violent autumn weather had little chance of survival.

King Manuel received this news with mixed feelings. The expedition brought significant discoveries that greatly expanded Portuguese knowledge of northern geography. The captives could be interrogated for valuable information about resources and routes. However, losing Gaspar Corte-Real, a skilled navigator and royal favorite, was a heavy blow. Manuel ordered the remaining survivors questioned thoroughly, demanding every detail of what Gaspar had discovered and his ship's last known location.

The interrogations provided tantalizing clues to Portuguese geographers. The lands explored by Gaspar were extensive; whether they belonged to Asia or represented a new continent blocking access to it remained a mystery. The indigenous populations were numerous, with no signs of complex political structures, suggesting territories ripe for conquest. Coasts teemed with fish and potentially other resources. And somewhere in those cold northern waters lay the fate of Gaspar's caravel: either seeking refuge until spring or resting at the bottom of the sea.

Miguel Corte-Real, brother to Gaspar, found it hard to stomach the prospect of leaving his sibling behind without making an attempt at saving him. Or perhaps he could not stand the disgrace that came with abandoning his kin unaided. It's also possible he saw a chance to finish what Gaspar started and reap the accolades for himself. Whatever drove Miguel, he wasted no time in plotting a rescue mission.

King Manuel backed this plan, offering ships and official sanction. The Portuguese monarchy had already invested in the Corte-Real family's ventures towards the north and stood ready to protect that investment. If Gaspar were still out there, rescuing him would uphold Portugal's claims on his discoveries. If not, perhaps Miguel could salvage maps, logs, and other knowledge for future expeditions' use. Of course, if Miguel did not return either... such are the perils of venturing into uncharted waters.

The preparations for this expedition took considerable time. The treacherous winter storms that made northern journeys untenable gave Miguel ample opportunity to gather ships and crews, pore over reports from returning captains, and strategize his search efforts. He intended to set sail for the last confirmed location of Gaspar's caravel and execute a methodical search, checking harbors and bays where a distressed ship might have taken refuge, scouring for remnants or survivors, and questioning local inhabitants about sightings of European vessels or men.

May 1502. Lisbon.

Miguel's journey commenced from Lisbon during late spring in 1502 aboard three ships, mirroring the route that Gaspar had traversed a year prior. The crews consisted of seasoned Atlantic mariners, including some veterans of the 1501 voyage, who could serve as navigational guides for the expedition. These vessels were laden not only with supplies sufficient for extended voyages in northern waters but also with trade goods and armaments necessary to interact with local inhabitants.

The outward journey seemed devoid of noteworthy incidents, according to surviving accounts that mention neither turbulent weather nor other unusual obstacles. By early summer, Miguel's fleet had arrived at the coasts which Gaspar had charted in his previous year's travels. The search proceeded methodically, with the ships surveying the shoreline and dispatching boats into every cove deemed worthy of exploration. They also investigated islands and bays while seeking information from indigenous groups about European maritime incursions. However, communication was evidently an immense challenge without interpreters.

Despite a thorough investigation, no remnants of Gaspar's expedition were unearthed, no debris, survivors, or records that might have hinted at their whereabouts. The only potential clue could have been artifacts seemingly originating from stranded cargo, though these items couldn't be conclusively tied to Gaspar's vessel with certainty. It seemed as though the sea and ice had fully consumed both the ship and its crew, leaving behind no trace of their ultimate destiny.

Throughout the summer, Miguel persisted in his search, extending beyond the reaches that Gaspar had ventured into, probing unexplored coastlines in a forlorn hope that his brother might have journeyed further than anticipated. However, as autumn approached and climatic conditions deteriorated, he faced the same dilemma as Gaspar: to persevere with the risk of being trapped by winter gales or retreat to Portugal.

Miguel's choice diverged from his siblings, or perhaps he encountered a different fate while making the same decision. He resolved to press forward in the search, potentially wintering in the Americas or venturing further westward in pursuit of untapped discoveries. The historical archives offer no closure as Miguel Corte-Real never returned to Portuguese shores. His expedition, like Gaspar's, vanished into obscurity.

THE HAUNTING

October 1502. Lisbon.

The initial voyage back to Lisbon in October 1502 was undertaken alone, carrying news that had devastating consequences for the Corte-Real family. Miguel's expedition had become separated while exploring the northern coastlines, leading to a disheartening reunion at their predetermined meeting point, with Miguel's ship nowhere in sight. The other captains waited as long as resources and weather allowed before reluctantly returning to Portugal, leaving behind Miguel, who faced an uncertain fate.

This pattern of events bore a chilling resemblance to the experiences of two expeditions over two years, involving two Corte-Real brothers vanishing in similar circumstances. The coincidence suggested something more systematic than mere misfortune. It was possible that the North Atlantic, particularly at those latitudes, posed too great a danger with its thick ice, violent storms, and poorly charted coasts. Perhaps both brothers made the same critical error: advancing too far, too late in the season, without adequately considering how quickly conditions could deteriorate in the northern autumn. Alternatively, there might have been a particular hazard, such as dangerous currents or reefs, that led to their ships' demise.

Yet, it was also conceivable that the disappearances were not entirely coincidental. The indigenous peoples who had been captured and enslaved by the Portuguese held valid reasons for animosity towards these ships. If either Gaspar or Miguel sought provisions or attempted further captures, they might have faced violent resistance from groups that had been previously raided. These groups would be prepared to defend themselves and could even organize ambushes. A caravel's crew, if dispersed ashore for provisioning or exploration, risked being overwhelmed by determined attackers. Similarly, the ship itself, left with only a skeleton crew responsible for anchor watch, might have been captured or destroyed before the landing party could return.

However, the Portuguese accounts seemed to dismiss this possibility entirely. They regarded the disappearances as maritime disasters caused by weather and navigation mishaps. Nevertheless, the absence of evidence did not equate to evidence of absence, particularly in a period where the North Atlantic witnessed violent and exploitative interactions between Europeans and indigenous peoples, likely sparking retaliatory actions. The fact that both Corte-Real brothers vanished while engaged in slaving raids may not have been coincidental.

The impact on the Corte-Real family was immense. Their patriarch, João

Vaz, had lost two sons in consecutive years, both pursuing the same northern routes and vanishing without a trace or explanation. The expeditions were heavily invested with resources, ships, provisions, crew wages, and trade goods, resulting in catastrophic losses of capital and personnel without yielding anticipated returns such as slaves, geographical intelligence, or valuable resources from new territories.

King Manuel prohibited a third rescue expedition. Vasco Corte-Real petitioned for permission to search for his lost siblings, but was denied. Two expeditions had already been lost, and attempting another would only compound the tragedy, with no reasonable hope of success. The North Atlantic proved too dangerous, its secrets too well guarded, leading Portugal to explore these regions through channels other than Corte-Real ventures.

The king's decision, while pragmatic, offered cold comfort to the family left in perpetual uncertainty about the brothers' fate, whether they died quickly in storms or more slowly from starvation and exposure, whether captured by indigenous peoples, or if either could have survived to understand their fate. This ongoing uncertainty was perhaps worse than confirmed death, leaving the family suspended in permanent hope and grief.

The captives brought back from the 1501 expedition became mere historical footnotes, their individual fates unrecorded. They were primarily mentioned in Portuguese documents as evidence of Gaspar's discoveries, rather than as individuals with names and stories. They likely ended up as slaves in Lisbon, dispersed among various owners, and their knowledge of northern geography extracted through interrogation before being absorbed into the anonymous mass of enslaved people working on Portuguese households and estates.

The information these captives could have provided is lost forever. They possessed firsthand knowledge of the coasts explored by Gaspar from a perspective unique to those who lived there. They understood seasonal patterns, resources, and dangers in ways no European could acquire through brief summer visits. This included valuable information about routes, populations, and territories that Portuguese geographers desperately sought. However, their knowledge was filtered through interpreters barely understanding the languages being spoken, recorded by scribes only interested in details relevant to future Portuguese exploitation, and valued only insofar as it served Portuguese interests.

The irony is biting: the individuals best positioned to explain what happened to Gaspar Corte-Real were probably his captors. They might have known whether his ship reached their coasts after the separation, if he attempted further raids, or if he foundered somewhere identifiable. Yet, their voices scattered, their potential testimony lost to history.

The Portuguese crown gradually abandoned intensive exploration of the North Atlantic. The southern route to Asia through the Cape of Good Hope was proving profitable beyond imagination. Vasco da Gama's second voyage in 1502 returned with cargo worth sixty times the expedition's cost. In contrast, the northern route, if it existed at all, demanded traversing ice-choked waters and hostile territories without obvious resources. The cost-benefit calculation did not favor additional investment.

However, other European powers eventually resumed the search for the Northwest Passage. English, French, and Dutch expeditions sailed through these waters over the following centuries, suffering comparable losses and mysteries. Henry Hudson vanished in the bay that bears his name, while John Franklin's entire expedition was lost, with one hundred and twenty-nine men perishing in the Canadian Arctic. The North Atlantic claimed hundreds of ships and thousands of lives before finally yielding its navigational secrets.

Nonetheless, the Corte-Real brothers were among the first Europeans to fall victim to this particular geographical obsession. Their disappearances helped establish the North Atlantic's reputation as one of the most dangerous waters on Earth, a place where experienced navigators could vanish without a trace. Ice, storms, and distance created barriers that Renaissance technology could not reliably overcome. And so, the ocean keeps its secrets, a refrain that echoes in maritime lore and reminds us of the unforgiving nature of these uncharted waters.

THE TRACES

What truly did Gaspar Corte-Real find? The puzzle has puzzled historians for centuries due to incomplete records, national tales spun, and, unfortunately, Gaspar's absence to tell his tale himself. It is strongly believed that he explored the eastern coast of Canada, perhaps most famously Newfoundland, or maybe Labrador, possibly stretching as far as Greenland. Yet pinpointing the exact location proves impossible without access to detailed maps or accounts, which were lost with his ship.

The indigenous peoples he may have encountered were likely Beothuk in Newfoundland, a community that would be fully annihilated by European colonization over the following centuries. The last known Beothuk woman, Shanawdithit, passed away in 1829, along with the remnants of her people's language and culture. Should Gaspar's captives have included any Beothuk, they would have been among those who would experience an onslaught that would continue for nearly three hundred years; yet the groundwork was laid during these initial raiding parties.

Some scholars argue that Gaspar ventured farther west than is generally acknowledged, potentially reaching Hudson Bay or the interior of Canada via rivers that flow into the Atlantic. These theories rest on scant evidence, Portuguese documents' vague descriptions, ambiguous references to "great rivers" and "large lands," and claims by later English and French explorers of finding European artifacts in supposedly uninhabited territories. However, extraordinary assertions require robust proof, which is notably absent here.

What is without doubt is that his expeditions contributed to the evolving European understanding of geography, though this knowledge remained fragmented and muddled. Maps from the early 1500s demonstrate the North Atlantic coastline with increasing accuracy, integrating intelligence from expeditions by Portuguese, Spanish, English, and French. Certain maps feature labels such as "Land of Corte-Real," indicating that Gaspar's revelations were highly regarded by contemporaneous cartographers.

Perhaps the most lasting impact might be more metaphorical than geographical. The tragedies faced by Corte-Real taught Portuguese overseers that northern exploration was perilous beyond imagination, necessitating substantial resources and incurring losses Portugal could not afford, given its commitments to Africa and Asia. This realization steered Portuguese expansion southward toward Brazil and the South Atlantic, leaving the North Atlantic open for other European nations who would also endure their own trials of discovery and loss.

1520. Terceira, Azores.

Vasco Corte-Real's remaining years were spent as an elder statesman on his family's property in Terceira, holding the position of captain-donatário similar to that of his father. Nearly two decades had elapsed since his siblings disappeared. King Manuel was no longer alive; João III now ruled Portugal. The Portuguese dominion had greatly expanded, and Brazil's colonization was flourishing, African trading posts provided steady income, and Asian spices flooded Lisbon's markets. The initial era of exploration, marked by the uncertainty of each voyage, had given way to a period of exploiting established routes.

Yet, questions persisted in Vasco's mind. He would often wander along Terceira's coasts at dusk, watching waves crash against volcanic outcrops, envisioning his lost brothers adrift within that expansive sea. Did they perish swiftly, shipwrecked amidst storms or encased by ice? Or did they succumb to starvation and disease aboard a vessel where rescue seemed evermore elusive?

The agony of uncertainty was its own torment. Vasco petitioned repeatedly for permission to search for his kin but was consistently denied. He conversed with any returnees from northern voyages, hoping for scraps of information that might bring closure. He commissioned masses in their memory, not knowing the fate of their souls, offering prayers into the void. The Atlantic claimed Gaspar and Miguel without offering their remains for proper burial.

This was the cost of exploration during an era of discovery: not merely the lives lost, commonplace in seafaring, but the vanishings, the unanswered mysteries, and the families left forever in limbo. For each hero who returned with tales and treasures, there were others swallowed by the sea, their fates untraceable, their deeds forgotten except to loved ones.

Vasco passed away in 1532, thirty years after his siblings' vanishing without resolution. The Corte-Real lineage continued in Portuguese nobility but dwindled in influence, no longer associated with bold ventures but with tragic absence. The northern shores that had consumed Gaspar and Miguel remained largely uncharted by Portuguese vessels, instead falling to English and French mariners who chased similar dreams and faced comparable fates.

Those indigenous people taken into bondage and transported to Portugal left behind no accounts of their ordeal. Their names were not recorded; their origins were vaguely cited as "the domain of Corte-Real." They died unclaimed and unmourned, save for distant families who might never have known the fate of their kin seized by force. The symmetry is grim: Portuguese households mourning explorers lost at sea, indigenous communities grieving for relatives taken without consent, all victims of a system rife with exploitation.

The fifty-seven captives who arrived in Portugal aboard surviving 1501 ships exemplify one of the earliest instances of European enslavement of North American aboriginals. Decades before mass slave raids decimated populations along the Atlantic coast, they established a cycle repeated over and over: European arrivals establishing initial rapport, followed by the seizure of individuals for sale. The trauma engendered by such encounters fueled mistrust toward Europeans, spoiling future interactions.

Had Gaspar's ship made landfall after their separation, had he attempted to disembark and carry out more raids, he would have confronted people already scarred by Portuguese aggression. Initial captures set the stage for resistance. A plausible scenario, though unproven, suggests that a caravel crew, around thirty men, would be ill-prepared once ashore, outnumbered and vulnerable against indigenous fighters familiar with their terrain and armaments, capable of inflicting heavy casualties. This scenario could account for the complete disappearance of a ship's company, overpowered, with their vessel potentially seized or destroyed, and survivors assimilated into local populations, leaving no trace.

Yet, this is speculation. The sea offered a simpler explanation: the vessel encountered storms, ice, or leaks leading to its sinking. Such losses were common during the Age of Exploration, with the North Atlantic particularly hostile, and ambitious commanders pushed caravels beyond their limits, leading to catastrophic failures. No conspiracy nor violence was necessary for the disappearance; the inherent dangers of early seafaring, coupled with northern conditions, sufficed to explain the vanishing.

THE COLD MATHEMATICS

The economics of exploration in the early 1500s operated on a cold calculus that prioritized discovery over survival. Success was measured by the return of trade goods, geographical knowledge, or territorial claims; these gains justified losses such as ships and lives. Unsuccessful ventures that only ensured safe return without new discoveries were considered unprofitable investments. This system encouraged reckless risk-taking behavior, rewarding those who ventured beyond reasonable limits while penalizing caution. However, not all voices agreed with this relentless pursuit of profit and expansion. Some figures, like Friar Antonio, a contemporary moralist and critic, dared to question the prevailing logic. In his writings and sermons, he openly challenged the morality of such ventures, cautioning that the quest for wealth could blind men to the sanctity of life and to the divine laws that guide ethical conduct. His dissenting perspective offered a much-needed counterbalance, highlighting the era's ethical tensions and the human cost of mercantile and imperial ambition.

A speculative question arises: Had the Portuguese crown perceived northern cod as having the same allure as Asian spices, would they have continued their investment? Such a shift in focus might have altered Portuguese ambition and economic priorities, potentially changing the course of European exploration in the North Atlantic.

Operating within this system was Gaspar Corte-Real. At forty-seven years old, when he embarked on his first northern expedition, he felt the weight of age and the urgency to secure his legacy before opportunities slipped away. Witnessing younger counterparts reap rewards for their discoveries fueled his determination to make a name for himself, elevating his family's status to that of minor nobility.

Consequently, Corte-Real took bold steps: venturing late in the season when prudence dictated retreat, navigating icy waters other captains avoided, and overloading his ship with captives who promised profit but compromised seaworthiness. Each decision aimed to boost potential value while amplifying the risk of catastrophic failure.

Miguel's subsequent rescue mission operated under the same calculus. Prudent individuals would have acknowledged Gaspar's loss, mourned it, and moved on. However, such restraint did not build empires or family fortunes. Miguel saw an opportunity in his brother's tragedy: to search for him while also fulfilling Gaspar's exploration goals, claiming discoveries, and showcasing familial bravery and loyalty. Success would vindicate the Corte-Real name; failure meant accepting its consequences without living to see them.

This psychology of exploration, the risky pursuit of glory, refusal to accept limits, relentless pushing despite overwhelming caution, was the hallmark of the Age of Discovery. Yet, it was also its downfall. We remember the survivors who got lucky or made better choices. For every successful explorer like da Gama or Magellan, there were numerous Corte-Reals: capable men lost in quests that exceeded their grasp.

King Manuel's decision not to authorize a third Corte-Real expedition reflected a different calculation. Two ships and two experienced captains were lost, with no tangible returns except for limited, immediate geographical information. From the crown's perspective, the northern route did not warrant further investment when southern routes to Africa and Asia produced actual profits. The North Atlantic could wait for other nations to explore at their expense.

From an economic standpoint, this decision was rational. However, it also signified defeat, acknowledging that Portuguese power had its limits; not every ocean could be conquered, nor every mystery solved. For a kingdom constructing a global empire, this realization posed a significant psychological adjustment: the world proved larger, more complex, and riskier than early triumphs suggested. Easy conquests were replaced with challenging frontiers resistant to exploration and exploitation.

1534. Porto Santo, Madeira Archipelago.

An aged Portuguese mariner found himself at a riverside inn, indulging in wine and regaling groups of young landlubbers with tales from his adventures on the open seas. He claimed he had been part of Miguel Corte-Real's ill-fated voyage, one of the few souls to return after their leader vanished into the vast unknown. The younger men listened intently, yet with a healthy dose of skepticism, as they were regaled with stories of towering glaciers bigger than parish churches, aboriginal tribes adorned in peculiar paints, impenetrable forests devoid of sunlight, and frigid temperatures capable of stealing one's breath.

One of the young listeners, driven by both curiosity and concern, asked the pressing question, "What befell Miguel Corte-Real?" The seasoned sailor took his time to refill his cup before offering an account steeped in contemplation. "In the thick fog, we were separated. Miguel yearned to explore a river mouth, hoping it might lead inland to untold treasures. We waited anxiously at our prearranged meeting point for weeks, but there was no sign of him. Disagreements arose among the captains; some advocated searching while others clamored for returning before the tempests of autumn arrived. In the end, we were forced to abandon hope and retreat lest the merciless winter grasp us."

"Do you suppose he survived?" inquired one of the onlookers.

With a nonchalant shrug, the sailor replied, "The North Atlantic is indifferent to our beliefs or wishes; it claims what it desires. Both Miguel and Gaspar knew the perils they courted, yet they pursued their quests unabashedly. The ocean granted them their wish, but at an unknowable cost."

Though undoubtedly embellished and possibly partly fictionalized over time, the essence of the narrative remained unaltered: brothers Corte-Real had disappeared into the unforgiving North Atlantic, a tale of ambition clashing with the harsh realities of early exploration.

As time marched on, their vanishing became the stuff of maritime legend in Portugal, serving as cautionary parables to aspiring navigators. They embodied lessons learned from past blunders, cautionary tales meant to remind that unbridled ambition could lead to ruin, overestimating one's abilities, and disregarding prudence could have dire consequences. These stories were more than mere entertainment; they offered valuable insights into Portuguese maritime capabilities and served as a beacon to avoid past mistakes.

Yet in their transformation into legend, complexity was lost. Gaspar and Miguel became archetypes: the reckless adventurer and the loyal brother, respectively, consumed by an insatiable hunger for glory and each other's devotion, yet both undone by their unchecked pride. But beneath these tales lay individuals with inner conflicts and external pressures; their decisions were shaped not just by a yearning for discovery but also by financial necessity and familial commitments.

History often strips away such intricacies to craft simplified narratives that future generations can understand, reducing the multifaceted lives of historical figures into easily digestible accounts.

THE INHERITANCE

The disappearances associated with Corte-Real have left lasting impacts beyond mere tales of old. Details accrued by surviving 1501 expeditions influenced Portuguese maps and shaped subsequent ventures from various European powers. Cartographers included "Terra de Corte-Real" in their depictions of the North Atlantic, marking territories as explored despite uncertainties. In an era where early 1500s cartography was a lucrative business, these maps were exchanged, copied, and even pilfered by merchants, sailors, and competing governments. Today, the legacy of these expeditions persists in modern place names that bear Portuguese roots, offering a concrete reminder of the historical mapping stakes and the ongoing influence of these early voyages.

The English were particularly engrossed. Following John Cabot's arrival in Newfoundland in 1497, English traders were keen to tap into the region's bountiful fisheries. However, Portuguese assertions based on Corte-Real's voyages posed diplomatic hurdles for England, leading to decades of tension. Eventually, the practical difficulties of Portugal asserting control over such distant lands, coupled with a growing British naval prowess, rendered their claims increasingly irrelevant. Yet, during the early 1500s, which explorer first reached these northern lands was a matter of serious contention among European courts.

Although Newfoundland's burgeoning fishing industry did not directly stem from the Corte-Reals, their voyages undeniably bolstered the broader knowledge expansion of North Atlantic geography. By the 1520s, Portuguese, Spanish, French, and British fleets fished the Grand Banks on a massive scale, extracting seemingly inexhaustible cod supplies. These fishing endeavors set the stage for Europe's sustained presence in North America, leading to seasonal outposts that eventually evolved into permanent settlements.

The local indigenous inhabitants paid dearly for this European interest. The Beothuk people of Newfoundland, who may have included some of Gaspar's captives, were systematically decimated by European settlement activities. Disease outbreaks, violence, displacement from their ancestral hunting and fishing grounds, and targeted killings reduced their numbers drastically from a presumed few thousand in 1500 to complete annihilation by 1829. The Corte-Real raids marked the onset of a three-century-long genocide, with Europeans perceiving indigenous peoples not as rights-bearing individuals but as obstacles or resources to be exploited.

Although the Mi'kmaq, Innu, and Inuit further north faced better survival odds against European incursion, their experiences came at great cost. They found ways to integrate metal tools and weapons into their economies while preserving cultural heritage and territorial claims. Yet, diseases introduced by Europeans, such as smallpox, measles, and influenza, caused devastating epidemics among populations with no resistance, resulting in massive population declines of 50-90% in certain areas. Entire communities were obliterated, and social frameworks crumbled under such immense mortality rates over short periods.

None of this suffering could be directly traced back to Gaspar Corte-Real; he was merely one explorer among many whose impact ended with his early disappearance. Yet, his expeditions were part of the broader European expansion that brought calamity to indigenous peoples across the Americas. From an indigenous viewpoint, the distinctions historians make about individual explorers and their relative guilt lose significance compared to the overall impact: Europeans arrived, and native societies ceased to exist.

1580. Lisbon.

Portugal found itself in turmoil following the demise of King Sebastian at the hands of Morocco's forces in 1578. This untimely death sparked a succession crisis, leading to Philip II of Spain ascending the Portuguese throne. The epoch of Portugal's exploration was coming to an end as its dominion faced exhaustion and rivalry from more potent adversaries. What had once rendered Portugal affluent, the path to India, Brazilian colonization, and the lucrative African coastal trade now fell into other hands or dwindled in profitability compared to past eras.

Amidst these tumultuous times, a venerable figure spent countless hours within the Casa da Índia's archives, meticulously scrutinizing documents related to the Corte-Real expeditions. He claimed descent from Vasco Corte-Real and endeavored to substantiate ancestral claims to lands purportedly discovered by Gaspar. If Portugal could not itself lay claim to these regions, perhaps it might pawn its dubious title rights to English or French entrepreneurs willing to invest in distant shores' entitlements.

However, the documents proved exasperatingly incomplete. No maps had survived from Gaspar's voyages. The captains' reports were vague, prioritizing indigenous peoples and resources over precise geographical details. Even the accounts of captives taken in 1501 were murky because translators barely understood their languages. It was nearly impossible to piece together where Gaspar had ventured with such scant information.

The old man diligently copied what little he could find, aiming to construct a case for Corte-Real's territorial claims. Yet, even as he labored, he sensed the futility of his endeavor. The world had shifted its gaze elsewhere. The North Atlantic regions explored by Gaspar were now staked by nations capable of enforcing their sovereignty. Claims rooted in expeditions from eight decades prior, marked by tragedy and obscurity, wielded little weight amidst late 1500s power plays.

With his task completed, the documents neatly rolled, he exited the archives into Lisbon's vibrant commercial hubbub. Ships from various European nations filled the harbor, while merchants haggled in a cacophony of languages over intercontinental trade deals. Though Lisbon remained affluent, it no longer reigned supreme as the epicenter of maritime commerce. The world had broadened and intensified competition, leaving Portugal grappling to maintain its influence.

The Corte-Real expeditions, which were once coveted and brimming with promise, now lie relegated to historical footnotes. The brothers who perished in pursuit of icy glory remained only in the memory of their descendants, clinging to diminishing claims and recollections. The indigenous peoples encountered and enslaved by them had been entirely forgotten, their names obliterated, their plight only noted as a testament to discovery. The sea that claimed both explorers and victims kept its mysteries intact, indifferent to human yearning for recognition or dominion.

THE LEGACY OF ABSENCE

What we remember about Gaspar Corte-Real is mainly absence - absence of detailed records, absence of maps to document his discoveries, absence of witnesses to tell his story. His body was never found or buried, leaving a void for his family and unresolved questions from his disappearance. He exists in historical awareness largely as an enigma, a cautionary tale, or simply the name connected with lands he might have been the first European to explore. Yet, beneath those icy waters, perhaps lies an old astrolabe, a solitary relic hinting at a navigator's final journey and giving form to his absence.

Absence itself can be significant. The lack of news about the Corte-Real brothers underscores the thoughtless violence of the Age of Exploration, how easily people and ships could vanish, and how narrow the margin for error was in early modern navigation. Their disappearance reminds us that behind every successful expedition we celebrate, there are unsuccessful ones we've forgotten - voyagers who made slight mistakes or were unfortunate and perished, crews who died far from home without any witnesses or mourners.

The indigenous captives taken by Gaspar represent another type of absence: the erasure of colonialism's victims in historical consciousness. We do not know their names or stories, nor what happened to most of them - whether they died on board, in Portuguese slavery, or found a way to survive and adapt. Their absence from history is not random but inherent within systems that valued European lives while treating indigenous peoples as objects rather than subjects.

The Corte-Real expeditions achieved their immediate goals: expanding European knowledge of geography, capturing and transporting indigenous people for enslavement, and establishing Portuguese claims on northern territories. But they fell short in their larger objectives: finding the Northwest Passage, discovering routes to Asian wealth, and securing lasting prominence for the Corte-Real family. The cost of these mixed outcomes was devastating: two ships lost, two captains dead, over a hundred people killed or enslaved, families torn apart across both continents.

Was it worth it? From whose perspective? There's no straightforward answer. The knowledge gained led to European colonization of North America, with varying impacts depending on one's side in the colonial dynamic. The precedents set - of European dominance over indigenous lands, of indigenous peoples as suitable for enslavement, and northern seas being navigable yet dangerous - shaped subsequent Atlantic history. These consequences spread out in ways Gaspar could not have foreseen or likely cared about.

In the end, what remains is ice, fog, and the cold arithmetic of exploration. Ships venturing into the unknown, decisions made with incomplete information, ambitions clashing with reality. The ocean doesn't care; it takes what it wants while keeping its secrets. The North Atlantic was as merciless in 1501 as it had been millions of years before humans learned to sail. Gaspar Corte-Real opposed it with Renaissance technology and medieval dreams. The ocean prevailed, much like it usually does.

Beneath the frigid waters lie remnants of his caravel, scattered across the ocean floor, corroded by centuries of salt and currents but still there - marking where another navigator ventured too far and paid the ultimate price. The wooden hull has long decayed. Iron fittings have rusted away. Human bones have been dispersed by ocean currents or consumed by marine life. Yet, the site remains, even if it's never found - a grave for people whose names we know and those we've forgotten, Europeans and indigenous Americans united in death as they were not in life.

The fog lifts eventually, and the ice thaws with the seasons. Ships now traverse these waters equipped with technology that would seem like magic to Gaspar Corte-Real: engines independent of wind, navigation systems accurate within meters, and communication capable of calling for help over thousands of miles. The ocean has been mapped, measured, understood - yet never truly conquered or made safe from its ability to claim the careless or unlucky.

Gaspar Corte-Real's legacy mirrors that of all explorers who vanished: a reminder that the unknown comes at a cost, that knowledge is often bought with lives, and that empty spaces on maps were filled in, at great expense, by people whose names are now mostly forgotten. He was neither hero nor villain but something more complex and relatable - an ambitious man making calculated risks, sometimes prevailing and other times failing, ultimately overcome by forces beyond his control.

The stone monuments erected by other Portuguese explorers on African shores have mostly endured, signifying the empire's expansion. But Gaspar left no markers in the northern ice or enduring symbols of his journey. All that remains are charts, possibly inaccurate, accounts from captains who did not return, legends that grew over centuries until they bore little resemblance to the real person. And silence. The permanent and absolute silence as a ship sinks beneath the waves, leaving no trace, evidence, or answers. Just absence, a blankness stretching across history like the Arctic fog and, in its place, the enduring question: what stories lie frozen in the depths, waiting to be discovered, if ever the ice should break?

Duarte Pacheco Pereira

September 1498, Somewhere on the Coast of Brazil

The vessel surged violently. Duarte Pacheco Pereira clutched the railing tightly, his knuckles etching against the wood seasoned with salt spray. Through the vaporous veil of morning, an enigmatic shoreline gradually emerged from obscurity. Vast verdant forests cascaded onto pristine beaches of ivory sand. Murky rivers meandered into the choppy sea. Above, unseen avian creatures traced patterns in the sky and unleashed their raucous calls. These mingled with the untamed sounds of this land. As the enigmatic landscapes unfurled before him, Pacheco recalled a faded map, incomplete, whispering of lands yet to be understood. A thrill of discovery mingled with the perils of the unknown. This was not the continent he knew as Africa or India. No landmark aligned with the meticulously drafted maps secured within his quarters.

He had been commissioned for an expedition into the uncharted. King Manuel's instructions were explicit yet shrouded in mystery: traverse the waters beyond the Pope's demarcation, report back on your findings, and maintain silence. Pacheco embarked upon this clandestine journey with a fleet of three caravels and a handpicked ensemble of mute sailors. They ventured across seas spoken of in hushed tones within taverns but absent from official documents. As he stood at the helm now, an illicit landfall unfolded before him. Instantaneously, it dawned on him: this revelation would be swallowed by secrecy, consigned to the vaults of state rather than heralded as a victory.

In due course, the date of discovery would spark controversy among chroniclers. The voyage itself would vanish from the annals of history. His name was purged from the record by royal scribes. Soldier, scholar, seafarer, and secret agent, Duarte Pacheco Pereira was the vanguard European to map and meticulously scrutinize the South American coasts. Though official accounts remained silent on his deeds, cryptic references in archives suggest that internal logs documented his travels. A persistent silence enveloped his discoveries: "No word from royal scribes, no place in the chronicles," as some documents might have recorded. Years hence, he would allude to his journey in a veiled manuscript, one ill-fated to fade into oblivion. Meanwhile, others would bask in the limelight of recognition. "Cabral's discovery," they declared in exuberant tones, celebrating those who stood in the radiance of charts not wholly theirs. They availed themselves of his charts and wisdom while politics and clandestine motives relegated his accomplishments under the radiance of those who followed.

This narrative is of an intellect overshadowed, of discoveries swallowed by epochs, and of a life extraordinary yet eclipsed by the ceaseless mechanisms of empire.

THE MAKING OF A RENAISSANCE MIND

Duarte Pacheco Pereira's birth, estimated around 1460, took place in the scenic Santarém, situated on cliffs overlooking the Tagus River, some forty miles from Lisbon. At this time, Portugal was a small, impoverished nation fixated on its singular ambitions. In contrast to Spain and Venice, which were preoccupied with territorial conquests and Mediterranean trade, respectively, Portugal directed their gaze towards the unexplored Atlantic. This unique trajectory was driven by religious zeal, economic necessity, and an insatiable curiosity about the world. As a child, Pacheco often watched the ships docking on the Tagus River, wondering about their distant voyages and the secrets they might carry. The spirit of Prince Henry the Navigator, whose legacy had profoundly permeated Portuguese culture even six years posthumously to Pacheco's birth, fueled this venture. Navigation transcended mere technical skill; it became a patriotic duty, a mathematical art pursued with fervent dedication.

Though little is documented about his youth, indicators suggest a life of privilege and learning for Pacheco. His family possessed the means and influence to secure him a position at court, although not amongst the highest nobility. He received the conventional education befitting a Portuguese gentleman: fluency in Latin, mathematics, rhetoric, and martial skills. However, his thirst for knowledge surpassed these confines. In astronomy and mapmaking, he was tutored by experts versed in both ancient doctrines and the findings of explorers returning from Africa. Pacheco mastered the computation of latitude via solar and celestial observations. He was innovative in cartography, not simply replicating existing maps, but creating new ones based on evidence. He came to understand that the practical arts of navigation and mathematics were interwoven to broaden human comprehension of the world.

By his early twenties, Pacheco had embarked on his seafaring career. Initially found along the West African coast, he played a part in Portugal's expansion efforts there. This endeavor was marked by both brutality and profitability; Portuguese trading posts traded European goods for gold, ivory, and malagueta pepper, with the slave trade becoming an increasingly prominent component. Pacheco's involvement in this system, as evidenced by his later writings, reflected the viewpoints of his era regarding African peoples as potential converts, trade partners, and slaves while acknowledging their humanity yet subordinating it to Portuguese interests. He was engaged in skirmishes against resistant African kingdoms and gained experience with the peculiar challenges of tropical weather, coastal currents, and ship maintenance in a humid climate that quickly decayed wood and rope.

The West African coastline served as his unparalleled educational arena. It imparted lessons no academy could offer. The myth of Ptolemy's endless Africa was debunked by Portuguese cartographers who revised world maps each season. A pivotal journey along the coast led to a transformative insight for Pacheco. During an expedition near the Cape Verde Islands, he navigated treacherous waters and encountered a local fishing village. Their maps and star charts were far more precise for navigation than any European map. In a modest hut, the village elder, with eyes full of wisdom and hands tracing the stars on parchment, shared his insight: "The earth speaks through these constellations. Know them, and the oceans will guide you home." These words, when compared with his formal education, revealed the superiority of lived experience over inherited knowledge. Pacheco realized the world was far greater, stranger, and more complex than anyone had dared to conceive.

THE SOLDIER-SCHOLAR AT WAR

The war began in 1495, as Pacheco led a Portuguese military expedition to North Africa. His journey took him across treacherous waters into the heart of conflict. Fortified cities stood precariously against Moroccan forces. Pacheco's valor was most notably displayed during the siege at Azamor and the defense of

Graciosa. Wounded more than once, his skill and courage left a lasting impression upon those who bore witness.

Contemporary records herald him as a warrior whose audacity knew no bounds, yet was tempered by an acute awareness of tactics. He could be seen rallying his men in dire straits and, with mathematical precision, directing the artillery's fire. Pacheco's exploits were not merely confined to physical feats; he also possessed an intellect that thrived upon challenges.

Upon King Manuel I's ascension in 1495, Duarte Pacheco Pereira emerged as one of the most distinguished figures. He was surrounded by a multitude of talents. He was more than a scholar navigating the sea or a soldier with a penchant for numbers. Instead, he embodied the Renaissance ideal, where intellect and action were indissolubly intertwined.

THE SECRET VOYAGE

The Treaty of Tordesillas, signed in 1494, unwittingly sowed the seeds of doubt about the world's true shape. The treaty divided the as-yet undiscovered lands between Spain and Portugal, drawing an invisible knife across the Atlantic, situated 370 leagues west of the Cape Verde Islands. All lands to the east belonged to Portugal, while everything to the west was Spain's. Portuguese navigators held suspicions that their own observations contradicted this division.

King Manuel I needed certainty about the existence of land beyond the designated line, which could be claimed before Spain's awareness. Thus, before Vasco da Gama embarked on his voyage to India in July 1497, another clandestine expedition was dispatched westward. Duarte Pacheco Pereira took command of this covert mission, with instructions to investigate any potential Portuguese territory that might have been overlooked in the treaty's calculations.

The exact departure date remains uncertain, but evidence points to late 1497 or early 1498. With three caravels equipped for extended voyages across the Atlantic and armed with the finest instruments available, Pacheco set sail west-southwest. His orders were clear yet undocumented: search for land along a specific latitude, map any coastline discovered, assess local resources and populations, and return swiftly to report directly to the King.

Weeks of uneventful crossing passed before land appeared exactly where Portuguese calculations had predicted it would. On 13 October, the logbook notes revealed, the coast was lined with crimson trees that added a vivid contrast to the grey ocean. Pacheco meticulously charted this new coast, observing its natural harbors, river mouths, and indigenous peoples from a cautious distance. Two days later, on 15 October, he noted a safe harbor, ideally suited for future anchorage. He noted Brazil wood, a valuable dye source that thrived abundantly. His meticulous recording of latitudes provided precise maps that would later guide future Portuguese explorers.

Pacheco instantly recognized the significance of his findings, realizing they were a vast continental landmass rather than an isolated island. As he contemplated the strategic implications of his discovery, it became evident that the detailed charts he was compiling not only broadened geographical knowledge but also strengthened Portugal's maritime navigation and claims in the New World.

However, Pacheco understood the importance of secrecy. Spain had strong claims to the western Atlantic. Proving Portugal's right to this land demanded careful calculations and mapping. It was safer to keep his discovery quiet until Portugal could defend it if necessary.

After completing his mission and returning to Lisbon around 1498 or early 1499, Pacheco reported directly to King Manuel I, leaving no official records of their meeting. Our knowledge comes from later references in his writings and the pattern of subsequent Portuguese expeditions.

Pedro Álvares Cabral, on a voyage to India in 1500, ostensibly sought favorable winds by sailing far west. Instead, he "discovered" Brazil in April 1500, landing near where Pacheco had previously explored, guided by charts that revealed precisely where land could be found. One ship returned immediately to Portugal to report the discovery, securing official recognition for Portugal's claim.

Whether this was a coincidence or an orchestrated finding shaped by Pacheco's reconnaissance remains unclear. The fiction of accidental discovery suited Portuguese interests perfectly as it suggested they had stumbled upon territory rather than deliberately sought it out.

Duarte Pacheco Pereira's name disappeared from the historical accounts, his discovery locked away in royal archives. Cabral received fame, accolades, and a place in history, while Pacheco quietly resumed his duties, erased from the record of his greatest triumph by the nation he had served so loyally. The scent of Brazil wood, sharp and vibrant, haunted him when thoughts of eclipsed honor lingered in quiet moments. He was left grasping at forgotten memories that would never touch the prominence they deserved, yet those sensory echoes persisted in the stillness, a reminder of the silent cost of obscurity.

THE GUINEA COMMANDER

If Pacheco felt resentful about being overlooked, his writings do not reflect it. He carried on as if nothing had happened. In 1503, King Manuel appointed him commander of São Jorge da Mina fortress on the Gold Coast in West Africa, the most significant Portuguese trading post in Guinea. The position signified great trust placed in him. Mina profited immensely from its gold trade with inland African kingdoms and also served as a slave-trading center that Pacheco managed efficiently. During this period, correspondence reveals his management of construction projects, negotiations with merchants and rulers in Africa, disciplining Portuguese soldiers who abused their authority, and meticulous records of gold and enslaved individuals passing through the fortress.

He successfully defended it, too. The African kingdoms in that region had complicated relations with the Portuguese, characterized by alternating cooperation and conflict depending on political situations and economic interests. In 1505, a significant African force attacked Mina, intending to destroy the fortress and drive out the Portuguese. Pacheco personally oversaw the defense. He employed artillery effectively, organized sorties that disrupted the attackers' camps, and maintained morale among his soldiers who were outnumbered at least tenfold. Reports describe him fighting on the walls with scars from North African campaigns, refusing to hide in the fortress while his men faced danger.

The siege failed, leaving the attacking force with heavy casualties after they were unable to breach the fortress's defenses. Pacheco informed Lisbon about their victory and requested reinforcements, supplies, and adjustments in Portuguese strategy for the region. King Manuel granted him titles and money as a token of appreciation for his service, even though the specifics of that service remained unknown to the public at large.

However, Mina had devastating effects on Pacheco's health. The climate was extremely lethal for Europeans. Diseases frequently ravaged the garrison, and Pacheco contracted recurring fevers, likely malaria, which weakened him and contributed to his chronic health issues later in life. Isolation was profound due to months passing between ships from Portugal, limiting communication to infrequent letters that took six months to reach their destinations. The responsibility of overseeing the slave trade must have had some psychological impact, despite Pacheco's ability to rationalize it within the theological and economic beliefs of his time.

Pacheco returned to Portugal in 1506, his body weakened by fevers and hardships. Despite his exemplary record, the glory of his deeds faded as new explorers attracted the king's attention. The route to India discovered by Vasco da Gama brought ships full of spices, overshadowing Pacheco's Atlantic voyages and their modest profits.

THE GREAT BOOK NO ONE READ

Duarte Pacheco Pereira is believed to have begun his magnum opus around 1505, possibly even during his time in West Africa or, at the very least, following his return to Portugal. He entitled it "Esmeraldo de Situ Orbis" ("The Emerald of the World"), a compendium combining his decades at sea, years dedicated to astronomical observation, and thorough studies of both classical authors and contemporary explorers. This work exemplified Renaissance humanism in practice, integrating Ptolemaic geography with direct observation, mathematical theory with sailing skills, and ancient wisdom with the novel insights arising from Portuguese expansion. Pacheco highlighted a notable problem he solved using this fusion of disciplines: during one voyage, he was tasked with determining his ship's position at noon. By carefully observing the sun and comparing his readings with tables he had meticulously crafted, Pacheco calculated the latitude accurately, allowing his crew to navigate confidently through uncharted waters. Specifically, he employed the technique of measuring the sun's altitude with a quadrant or astrolabe at its zenith, recording the angle above the horizon. He then consulted ephemerides to correlate this measurement with the day of the year, which enabled him to establish their precise latitude. This procedure was detailed in the 'Esmeraldo' as not only a theoretical exercise but a means to practically orient the vessel, adjust course, and avoid hazards such as uncharted reefs or dangerous currents. The crew's successful progress depended on these calculations, turning abstract knowledge into dependable navigational practice. By recording both the process and its application, Pacheco's treatise ensured that future Portuguese navigators could replicate his methods, thereby preserving Portugal's navigational advantage.

The "Esmeraldo" was structured into four books, with only Book I and part of Book II surviving. Book I covered theoretical geography, celestial navigation, latitude calculations, the usage of navigational tools, and the mathematical foundations of cartography. Pacheco elucidated these concepts clearly, targeting educated readers who still required practical guidance. He respectfully challenged classical authors whenever their claims conflicted with Portuguese observations, but did so to expand ancient knowledge rather than discard it.

Book II offered precise navigational instructions for voyaging along the West African coast from the Strait of Gibraltar to the Cape of Good Hope. It included detailed latitude measurements for numerous coastal features, descriptions of currents and prevailing winds, directions on where to find resources like fresh water and supplies, warnings about perilous reefs and hostile tribes, and insights into potential trading opportunities. This amalgamation represented decades' worth of Portuguese expertise, some gleaned from Pacheco's own experiences, much of it compiled from other pilots.

Discreetly within this practical guide for navigation were references to his covert voyage. Pacheco described the lands beyond the Atlantic, hinting at a "great continent" west of Africa, a region regularly sailed through by Portuguese fleets. He provided no detailed maps or specific directions on how to reach it; instead, he shared just enough information to assert personal knowledge while avoiding any revelation that might compromise state secrets. These passages were seemingly intentional ambiguities that communicated truth without divulging sensitive details.

The manuscript never saw the light of day during Pacheco's life. He presented it to King Manuel, presumably hoping for royal support in its publication. However, it vanished into the royal archives, where it was classified as a document too sensitive for public release due to its valuable navigational intelligence. Portugal's edge in Atlantic and African navigation relied on keeping precise geographical knowledge private from competitors. If Pacheco's work had been published, it would have gifted this advantage to any literate navigator of rival nations who could purchase the book.

The manuscript remained unused. Although some Portuguese pilots and royal officials privately circulated copies, the "Esmeraldo" failed to reach a broader audience. Over time, Pacheco's observations became outdated as new explorers provided more accurate information. His astronomical tables were eventually replaced by even more precise versions. It wasn't until nearly four centuries after its original creation that a scholar published the "Esmeraldo" in 1892; it was then regarded not merely as a historical artifact but also as a testament to early Portuguese knowledge, although it was no longer applicable for practical navigation.

BETRAYAL AND VINDICATION IN INDIA

In 1504, driven by a quest for redemption or perhaps simply unable to resist the royal summons, Duarte Pacheco Pereira set sail once again for India. The Portuguese position in Asia was precarious at best. Although Vasco da Gama had successfully opened up the route, maintaining Portuguese dominance in the face of sophisticated Asian kingdoms and well-established Arab trading networks demanded relentless military efforts. It was under these circumstances that the Viceroy in India sought out seasoned military commanders. Despite his age and declining health, Pacheco stepped forward to offer his services.

Upon arrival in Cochin, the chief Portuguese stronghold on the Malabar Coast, Pacheco found chaos reigning supreme. Relations between the Portuguese and local rulers were marked by unpredictable oscillation between alliances and warfare, heavily influenced by shifting commercial interests and dynastic politics. The Zamorin of Calicut, a ruler of immense power in the region, perceived the Portuguese as disruptive interlopers threatening centuries-old trading traditions. In March 1504, this resentment culminated in the Zamorin assembling a massive fleet and forming an alliance with Arab merchants to attack the Portuguese factory at Cochin, aiming to obliterate their presence in India once and for all.

Commanding the defense was Duarte Pacheco Pereira, backed by about 150 Portuguese soldiers and artillery. They were bolstered by several hundred Indian allies from the Kingdom of Cochin, which had allied with Portugal against the Zamorin. Yet, they faced an army estimated at 50,000 to 80,000 men, a number that, although possibly exaggerated by contemporaries, signified overwhelming numerical superiority.

Pacheco managed to hold Cochin for three months by strategically positioning his artillery to defend the fortress approaches, effectively targeting siege positions with his limited manpower, exploiting divisions within the besieging forces, and maintaining iron discipline among his soldiers. Contemporaries noted that Pacheco seemed to be everywhere at once, directing cannon fire, leading counterattacks, treating wounded men, rationing out dwindling supplies, and projecting unwavering confidence even in the direst circumstances. His knowledge of astronomy proved invaluable on the battlefield, guiding him to launch raids at optimal times based on his precise calculations of moonrise and sunrise. Moreover, his experience in North Africa influenced his defensive tactics, allowing him to hold fortified positions against much larger forces.

The siege ultimately failed, largely due to Pacheco's defensive maneuvers.

Portuguese reinforcements arrived, and the attacking army, already reeling from Portuguese resistance and internal political strife, retreated. Pacheco had single-handedly saved Portuguese India. His victory at Cochin ensured that Portugal maintained its foothold in Asia, paving the way for further expansion that would eventually make Portugal a dominant force in the Indian Ocean for a century to come.

The Portuguese Viceroy, Francisco de Almeida, sang Pacheco's praises in his dispatches to King Manuel, describing the defense as nothing short of miraculous and Pacheco's leadership exemplary. Rewards and honors seemed all but guaranteed. Pacheco was set to return to Portugal basking in glory, with his long service finally receiving the recognition it deserved.

Then came betrayal. Political factions in Goa and Lisbon, either envious of Pacheco's success or harboring old grudges from his years of service, accused him of embezzlement and misconduct during his tenure as commander at Mina years prior. These accusations were almost certainly fabricated or grossly exaggerated, merely another facet of the endless court intrigues that characterized Portuguese imperial administration. Yet, they found an audience. Upon returning to Portugal in 1505, Duarte Pacheco Pereira was arrested, tried, and imprisoned. His property was confiscated, his reputation tarnished. The courtroom resounded with voices proclaiming, "Here stands the man who sought to fill his coffers with royal treasure while hiding behind the pretense of imperial duty." In a private letter to a trusted confidant, Pacheco lamented, "I fear that the truth may be buried under the weight of accusations, for I have served with honor in every distant shore and fortress." This personal insight into Pacheco's own voice lays bare the profound injustice of his situation, allowing us to feel his turmoil. The legal proceedings dragged on for years, but despite Pacheco's defense, which relied on documentation and testimony from former subordinates, he lacked the political backing needed to counter the royal disfavor that hung over him. Although eventually acquitted of the most serious charges, the exoneration came too late and in relative silence to fully rehabilitate his reputation. It was around 1519 that Pacheco was finally released from prison, but by then his health was ruined, his finances decimated, and his name was associated more with scandal than with his decades of service.

THE FORGOTTEN END

Duarte Pacheco Pereira's twilight years passed unremarked. He eked out a living off a meager royal stipend, insufficient recompense for his lifelong dedication. His writings collected dust, unrecognized; his achievements overshadowed by others' claims to fame. Portugal's focus shifted to the newly discovered lands of Asia and Brazil, their eyes blinded to those who had paved the way beforehand.

His departure from life in 1533, possibly within Lisbon's walls, may have been marked by penury. No grand send-off recognized his existence; no royal proclamation mourned his passing. This man, who charted territories in Brazil before Cabral set foot there, defended Cochin against tremendous adversities and composed one of the period's foremost geographical treatises. This servant of Portugal for four decades across continents merely faded from history as invisibly as his clandestine voyage had been erased from records.

For generations, he was all but forgotten. Portuguese historians occasionally recalled him, yet frequently mixed up details or confused him with others entirely. The Esmeraldo lay like a forsaken manuscript in the archives, unvisited. It wasn't until scholars of the nineteenth century meticulously pieced together the early days of Portuguese exploration that Pacheco's name resurfaced, painstakingly reconstructed from fragmented notes, ancient cartography, and the subtle imprints he had left.

The evidence for his 1498 voyage to Brazil remains suggestive yet persuasive: he recorded in the Esmeraldo that he witnessed lands beyond the Atlantic firsthand. Subsequent expeditions followed trajectories that made sense only with the presupposition of preceding reconnaissance efforts. The veil of secrecy adhered to Portuguese policy regarding critical geographical intelligence. Present-day scholars generally concur that Pacheco reached Brazil before Cabral, though debates continue concerning exact dates and the scope of his exploration.

Nevertheless, being the first to achieve this feat matters less than what was ultimately lost. Duarte Pacheco Pereira encapsulated Portugal's most cherished qualities: intellect, valor, expertise, and steadfast loyalty. He sacrificed his well-being, prosperity, and even his honor for a nation that employed him and then discarded him. His narrative unveils the concealed aspects of empire, uncelebrated pioneers rather than renowned explorers, the diligent laborers and soldiers whose sacrifices shaped history, only to be left with nothing.

THE SHAPE OF FORGOTTEN GLORY

Picture a different September morning in 1498, the year when Duarte Pacheco Pereira embarked upon an unforeseen journey. From aboard his ship, he gazes upon an unfamiliar coastline that emerges slowly from the fog. It is at this precise moment that he senses not only the concealed nature of this discovery and his name being omitted from historical accounts, but also the silent foundation his work will lay for others' future triumphs. His experience embodies a recurring motif in the construction of empires: individuals whose contributions are deliberately obscured by state secrets or political expediency. As we reflect on Pacheco's narrative, can we truly account for all the 'unsung pioneers' who might still be overshadowed by modern structures and ambitions? How many more stories remain untold, guiding present-day endeavors under the veil of anonymity? This question invites us to think about today's silent trailblazers who build paths for future triumphs, echoing the hidden legacy Pacheco unknowingly left behind.

Pacheco may harbor a fleeting sense of injustice, yet it is more plausible that he experiences the paradox at the heart of empire-building, finding solace and pride in problem-solving, even as his efforts are deliberately overshadowed by Portugal's burgeoning global influence.

The moment passes. He instructs his crew to prepare for disembarkation. They will spend months meticulously surveying the coastline and gathering intelligence that will guide Portugal's ambitions for generations. Each observation, calculation, and meticulous annotation on his maps is a testament to his relentless expertise. Yet, none of this bears his name. All becomes state secrets, accessible only to those granted royal permission.

Upon returning to Lisbon, he personally presents his report to King Manuel, engaging in a conversation that spans perhaps an hour. The king queries him with precision, seeking details on what was discovered, its location, resources available, and methods of acquisition and defense. The king expresses gratitude. The meeting concludes. Official chroniclers remain uninformed of this exchange. Pacheco is reimbursed for expenses and resumes his duties, the magnitude of his achievement concealed as a secret he carries for decades, only to hint at it obliquely within an unpublished book, long after everyone involved has passed.

Such brilliance is consumed, not by failure but through bureaucracy and secrets quietly buried. Those capable are utilized and discarded with each new challenge. Duarte Pacheco Pereira devoted his knowledge, health, youth, courage, and life to Portugal. In return, he faced poverty, imprisonment, and centuries of obscurity until historians finally unearthed his story.

The Brazilian shore is bathed in increasing morning light. With unwavering composure, Pacheco bides orders, concealing any hint of bitterness. The sails unfurl, the ships alter course northward, embarking on an exhaustive exploration destined to bring glory to others. And Duarte Pacheco Pereira, deserving of remembrance, fades quietly into obscurity as his achievements slip from the nation's consciousness. Yet, as we peruse the pages of history, one cannot help but ponder: how many more stories like Pacheco's lie hidden in forgotten archives, awaiting their rightful place within the annals of discovery?

Tristão da Cunha

March 1506. Somewhere in the South Atlantic.

As the ship undulated upon tumultuous waves, Tristão da Cunha emerged as a frequently overlooked yet crucial pillar of an empire, the man who ensured the ships continued to sail while others basked in the limelight.' The compass swirled in disarray as he gazed at an isle manifested from the fog, a place no map dared acknowledge. This island rose majestically from the sea, a solitary black volcano veiled by mist. No chart documented it, nor had any pilot uttered its name. For three weeks, his fleet had hunted near the Cape of Good Hope, that treacherous threshold to the Indies and its coveted treasures. However, they stumbled upon this mountain, isolated in the vastness, enveloped in clouds, and enshrouded in greenery hinting at water yet offering no refuge. The waves loomed menacingly, the wind howled ferociously. Tristão da Cunha, an experienced navigator and knight entrusted by King Manuel, confronted a decision whose reverberations would span centuries.

He bypassed it. The island would ultimately bear his name. Yet on that March dawn, as tempests gathered and his fleet dispersed, Tristão da Cunha opted for practicality over triumph. He meticulously recorded the island's coordinates, sketched its outline, and veered eastward. The crew remembered this moment with a blend of perplexity and alleviation. They anticipated him to seize the land, erect a beacon, or confront the treacherous surf head-on. Instead, he altered course without vacillation. To many observers, it appeared as an opportunity squandered. Time would vindicate his choice. Nevertheless, history allowed almost every aspect of his life to fade quietly into obscurity.

THE HIDDEN GIANT

Tristão da Cunha's tale is rich in irony. He commanded expeditions with thousands of soldiers across perilous seas. He negotiated with influential leaders and fought in pivotal battles that influenced the spice trade, significantly boosting Portugal's wealth. Monarchs confided in him, rivals dreaded him, and even his adversaries held him in esteem. Nonetheless, today he is primarily known for a chain of distant volcanic islands in the South Atlantic, often regarded as "the most remote inhabited archipelago worldwide."

His legacy has gradually yet wholly diminished. Vasco da Gama swiftly gained fame upon reaching India. Afonso de Albuquerque earned the title "Caesar of the East" for his conquests. Tristão, instead, operated behind the scenes during Portugal's peak expansion period. Rather than making groundbreaking discoveries, he established structures and maintained the empire's functioning. His principal accomplishments lay in administration, diplomacy, and strategy. These fields seldom result in enduring renown.

To fully comprehend the era of Portuguese exploration, one must consider individuals like Tristão da Cunha not only as the pioneering explorers but also as the crucial second group. The initial cohort comprised visionaries who made novel discoveries. The subsequent group, akin to Tristão, sustained these prospects and facilitated the empire's expansion.

SON OF THE FIFTEENTH CENTURY

Tristão da Cunha's genesis transpired circa 1460 in a Portugal that was just beginning to define itself. The realm he stepped into as an infant bore scant resemblance to what he'd contribute towards molding by adulthood's dawn. In the 1460s, Portugal was a petite, impoverished kingdom teetering on Europe's fringes, hemmed between Castile and the vast Atlantic expanse, with possibly a million citizens and economically tethered to fishing, winemaking, and salt exports. Lisbon's royal court was modest by English or French standards, the nobility land-scarce yet yearning for prospects. Its most notable feats weren't on European soil, where it barely made waves; rather, they lay in the uncharted waters beyond Cape Bojador – a psychological barrier along West Africa's coast that had long daunted European seafarers until Prince Henry's intrepid navigators breached it in 1434. This backdrop underscores how Portugal's limited resources and its nobility's aspirations spurred its nascent maritime endeavors.

Born into the unremarkable yet respectable da Cunha lineage, Tristão's life began under the watchful eye of his father Nuno, who served the Crown and owned land near Covilhã, far from any tide. They were noble but not affluent; their standing was respectable but lacked real influence. Progress relied on royal patronage rather than birthright alone. Tristão's education mirrored that of every young Portuguese fidalgo: horsemanship, weaponry, Latin, mathematics, and a crumbling chivalric code in an era when gunpowder and commerce reigned supreme.

Yet this was also Portugal under João II, who ascended the throne in 1481 and embarked on a thirteen-year crusade against old feudal powers. He sowed the seeds for a new class of service-oriented nobles, whose prosperity hinged upon loyalty and Atlantic conquests. João II saw an opportunity that eluded his contemporaries - the future was not in fortified countryside castles but in seaports bolstered by cannons; professional mariners and soldiers rather than feudal levies.

Tristão da Cunha matured amid this seismic shift. By his twenties, Portuguese caravels were charting Africa's coastlines. In 1488, Bartolomeu Dias rounded the Cape of Good Hope, proving that a sea route to India was feasible. Christopher Columbus, spurned by João II in 1492, found the Caribbean, mistaking it for the East Indies. Vasco da Gama fulfilled this vision in 1498, reaching Calicut and returning with spices that fetched sixty times their cost, igniting the birth of an empire. As these milestones unfurled, Tristão often contemplated the relentless pace of exploration, pondering his place within this grand narrative - aware that each discovery rewrote the map of known worlds and his own destiny.

Tristão's perspective was not that of a mere chronicler but of an active participant in this unfolding epoch. By the 1490s, he was part of the royal court, tied to the future Manuel I's inner circle while he was still Duke of Beja. Upon João II's untimely death and Manuel's ascent in 1495, Tristão was already ensconced within the sphere of confidences for the new sovereign. This timing would be instrumental in shaping his destiny.

THE ARCHITECTURE OF EMPIRE

Manuel inherited a kingdom poised on the brink of change and held a vision for its fulfillment. His predecessor set in motion key initiatives. These included forging alliances with African kingdoms, venturing into Atlantic exploration, forming partnerships with Italian traders and financiers, and piecing together geographical knowledge. These efforts would morph blank map sections into seafaring paths. However, Manuel's reign marked the pivotal moment when Portugal erupted as a preeminent force in global commerce.

The plan hinged on audacious simplicity yet aimed high. By eschewing traditional land-based spice routes, monopolized by Venice, the Ottoman Empire, and various Arab and Indian middlemen, Portugal sought direct ties to source markets. This meant bypassing Venetian merchants, who bought from Egyptian intermediaries, who dealt with Indian traders. Portugal wanted pepper, cinnamon, cloves, and nutmeg. The goal was to eliminate middlemen, undercut competitors, and monopolize the market. The result would be tremendous profits. However, the risks involved were equally formidable.

Navigating to India required an arduous voyage across thousands of miles of open sea, without the comfort of nearby harbors. These waters were treacherous, prone to storms that could disperse fleets. Calm periods, while providing relief, also posed a threat as ships risked being stranded for extended periods with diminishing supplies. Once in India, Portuguese captains had to navigate local politics by negotiating with rulers and establishing and defending trading outposts against both European and Asian adversaries. Success entailed not only exceptional navigational skills but also the talents of diplomats, soldiers, merchants, and administrators, all of whom were adeptly coordinated across vast distances. The communication challenges were equally significant; messages from Lisbon might go unreplied to for years.

To summarize the logistical stakes, consider the 'three perils' that navigators faced: the immense distance, the threat of disease in unfamiliar climates, and the diplomatic challenges of engaging with various local rulers. These factors kept the margins razor-thin and demanded individuals like Tristão da Cunha, whose unique organizational prowess was pivotal to achieving these monumental ambitions.

THE FIRST MISSION

In 1506, King Manuel I entrusted Tristão da Cunha with an ambitious task. He was to command a fleet of fifteen ships and over 1,500 men on a voyage to India. This marked the largest such expedition Portugal had sent to India at the time. The mission entailed several objectives. Portugal sought to strengthen its influence in India following Vasco da Gama's contact. Tristão was to secure alliances with local rulers against Muslim adversaries. He would escort a diplomatic envoy to Ethiopia, a Christian kingdom considered vital for encircling Islam. He would also establish fortifications at key sites to consolidate control over the Indian Ocean.

Setting sail from Lisbon in April 1506, the fleet embodied an immense investment by the Portuguese crown. This reflected a profound faith in Tristão's capabilities. With fifteen vessels, numerous cannons, hundreds of soldiers, cargo for trade, diplomatic gifts, and letters of introduction, the mission carried high stakes. The voyage also benefited from the navigational knowledge of many generations who had deciphered the Atlantic's winds and currents. By then, the route should have been relatively familiar territory.

Da Gama had completed this journey, as did Pedro Álvares Cabral, who discovered Brazil in 1500 en route. The path was well-recognized: traveling south along Africa's coastline, then southwest to catch specific winds that would pass around the Cape of Good Hope and into the Indian Ocean before heading northeast. However, understanding the route did not guarantee safety; it remained perilous.

Around three weeks into their voyage, Tristão's fleet encountered an island later named after him in the South Atlantic. By choosing to bypass landing on this island despite its lack of immediate strategic value, a decision likely driven by dangerous sea conditions and the expedition's distant objectives, Tristão demonstrated his professionalism, prioritizing mission goals over personal glory. Yet, the black volcano rising from the sea would linger in the shadows of his legacy, a spectral presence that would remain imprinted on maps and in maritime lore.

Shortly thereafter, south of the Cape of Good Hope, the fleet faced extreme tempests that were synonymous with sailor nightmares. The accounts are fragmented but emphasize the ordeal: ships separated amidst darkness, massive waves, sails torn apart by gale-force winds, and an unsettling realization of their isolation without knowledge of any other survivors.

Among them, Afonso de Albuquerque's ship sought refuge in Mozambique for months, awaiting stragglers. Another, misdirected, ventured to explore Madagascar's coast. Several reached East African harbors, while Tristão pushed forward toward India with a portion of his fleet intact.

This scenario reflected the reality of sixteenth-century oceanic navigation: fleets were scattered, ships were lost, and captains made independent decisions over extended periods without orders, relying on their judgment and pilots' expertise. The fact that most of Tristão's fleet eventually reached India, reuniting at various ports over months, was in itself a testament to the Portuguese navigational capabilities, which allowed individual commanders to navigate vast oceans even after losing contact with their expedition.

However, this meant Tristão arrived in India with diminished forces and an altered schedule. The grand entrance he might have made with fifteen ships behind him never materialized. Instead, Tristão's arrival was piecemeal and scattered across multiple groups over time, complicating the demonstration of Portuguese power in a region where local rulers were growing increasingly cautious towards these foreigners who came bearing both trade goods and cannons.

THE POLITICAL MAZE OF MALABAR

The India discovered by Tristão in 1506 was quite diverse, comprising multiple kingdoms, sultanates, and bustling trade cities, each with its own ambitions, conflicts, and perspectives on the Portuguese arrival. Focusing on the Malabar Coast, where the Portuguese sought to establish themselves, we find a highly intricate landscape. The Zamorin of Calicut ruled over the most significant and influential port within the region, while the Raja of Cochin, who was at odds with the Zamorin, formed an alliance with the Portuguese as a strategic countermeasure against Calicut's dominance. Further south, the Kingdom of Quilon controlled another major port, further complicating matters. Moreover, across these territories lived Arab trading communities, which had long dominated the Indian Ocean spice trade and viewed the Portuguese as a threat.

Portugal's approach was centered on exploiting these rivalries for its advantage. They supported Cochin against Calicut, offered military aid to local rulers in exchange for trading benefits, and the right to construct fortresses. Leveraging their superior naval strength, they intimidated rivals and guarded allies. Gradually, they built up a network of bases through strategic alliances and treaties, securing control over the shipping lanes through which spices were dispatched to Europe.

Tristão's mission was part of this strategy but unfolded during an especially charged time. The relationship between the Portuguese and the Zamorin had deteriorated completely, with Portuguese attacks on Muslim ships having further strained relations. These attacks upset many local leaders who relied on that trade. Additionally, Portugal's small presence in Cochin, perhaps a few hundred soldiers and merchants, existed under constant apprehension, surrounded by populations far exceeding their numbers and potentially hostile at any moment.

It was into this precarious situation that Tristão da Cunha entered, tasked with bolstering Portuguese influence but lacking the strength to impose his will unilaterally. He would have to navigate a web of diplomacy where each choice had uncertain outcomes, and a misstep could ignite a conflict that Portugal's small foothold in India couldn't withstand.

THE BATTLE FOR RESPECT

One of Tristão's initial actions was indeed an attack, which may initially seem counterintuitive, especially considering his role as a diplomat. However, this choice reflected his astute understanding of how power dynamics unfolded in the Indian Ocean region. The Portuguese were significantly outnumbered, and their coastal trading posts relied heavily on local support for survival. If they had been perceived as weak, they would have been easy to overpower. Therefore, it was crucial for the Portuguese to be seen as a formidable force.

During late 1506 and early 1507, Tristão led Portuguese forces against Muslim trading vessels and carried out raids on coastal settlements allied with the Zamorin. These operations were not extensive conflicts but rather strategic demonstrations of power intended to showcase the capabilities of Portuguese ships equipped with cannons and disciplined crews. The Portuguese captured ships, seized their cargo, and sent a clear message: resisting them would lead to economic devastation, while cooperation could yield significant profits.

This approach of controlled violence was pivotal to Portugal's success in Asia. Unlike the Spanish in the Americas, who expanded their territories using relatively small forces by exploiting existing rivalries and introducing diseases that decimated indigenous populations, the Portuguese adopted a different strategy in Asia. Instead of large-scale territorial conquests, they constructed a maritime empire consisting of a network of fortified trading posts. These strategic locations controlled vital chokepoints, allowing the Portuguese to dominate trade without physically controlling the land. However, maintaining this empire required an unbreakable reputation for invincibility at sea. Consequently, Portuguese commanders could never afford to leave any challenge unanswered.

Tristão fully grasped these dynamics. His attacks were not mere acts of piracy but carefully calculated moves within a broader strategic framework. Each captured ship and each burnt warehouse reinforced the notion that engaging with or rejecting the Portuguese would come at a great cost. This strategy thrived on thin margins and was sustained by the myth that Portuguese power far exceeded its actual capacity, thus further enhancing their perceived invincibility.

THE ETHIOPIAN DREAM

In overseeing military operations on the Malabar Coast, Tristão also bore responsibility for advancing one of Manuel I's most passionate and ultimately ill-fated endeavors: forging ties with the Christian kingdom of Ethiopia. This fixation stemmed from an enduring European fantasy. For generations, tales had spread across Europe about a mighty Christian ruler named Prester John who reigned over a vast empire in the East, supposedly ready to ally against Islam. By the late 15th century, Portuguese explorers had pinpointed Ethiopia as the most plausible site of this fabled kingdom.

Manuel I, envisioning himself as a monarch commissioned to rally Christendom against the Muslim realm, believed that aligning with Ethiopia could fundamentally shift the power dynamics. Should Portuguese forces in the Indian Ocean harmonize efforts with Ethiopian armies in the Red Sea and Arabian Peninsula, they might jointly encircle and ultimately disrupt the Muslim trade networks linking the Mediterranean to Asia. This was a grand vision that significantly overrated both Portugal's capabilities and Ethiopia's willingness to forge such an alliance. Yet, it fueled Portuguese policy for many years.

Tristão's role in this endeavor was to support his cousin, a Portuguese envoy who journeyed from India to Ethiopia to negotiate with the Ethiopian emperor. From its inception, the mission encountered numerous obstacles. Navigating from the Indian coast to Ethiopia demanded traversing areas controlled by hostile Muslim authorities. Moreover, the Ethiopian court lay deep within Africa's Horn, accessible only through arduous expeditions across deserts and mountains. Compounding these challenges, the Portuguese had scant knowledge of Ethiopian governance, language, or culture.

Though the mission fell short of achieving its diplomatic goals, it did provide Portugal with a more precise understanding of Ethiopia than Europe had previously. Nonetheless, the failure was not attributable to Tristão; he fulfilled his duties by providing ships, supplies, and military protection for the envoy. The unattainability of the overarching strategy highlighted the Portuguese court's fantasies rather than any inadequacy on Tristão's part in executing his assignment.

BUILDER OF BASES

Tristão's enduring impact on Portuguese dominance in Asia lay not in his battles or negotiations, but in the construction and reinforcement of fortresses along India's shores. Though these strongholds may seem modest, with their basic stone walls and few structures, they were transformative in the dynamics of Indian Ocean commerce.

These fortifications enabled King Manuel I to extend military influence across vast oceanic distances, a feat unmatched by other European powers. Each fort served as a critical resupply hub for ships, barracks for soldiers, and an assertion of Portuguese sovereignty even in distant waters. Furthermore, their connectivity through trade routes created a web of communication, reinforcement, and supply that turned these outposts into a cohesive network.

Tristão grasped the strategic value of this system. He ensured that forts were adequately supported, staffed, and strategically located to regulate key trade routes. By negotiating with regional leaders, he secured permission to construct and maintain such bases. Moreover, he established the governance frameworks necessary for fort operations, including logistics, troop rotations, and communication systems to Lisbon.

This work may lack grandeur or heroism, but it was crucial in laying the foundation of the empire. Without these forts, Portuguese vessels would have lacked safe havens during storms, repair facilities, and supply depots. The absence of Tristão's organizational efforts would have left these bastions to disintegrate into chaos, undermining their garrisons and defenses. Thus, the Estado da Índia, which monopolized Indian Ocean trade for over a century, owed its existence to the groundwork set by men like Tristão.

THE PRICE OF DISTANCE

Life in Portuguese India was a harrowing experience. The climate proved fatal for European bodies unaccustomed to tropical heat and disease. Cases of dysentery, malaria, and other illnesses took a devastating toll on soldiers and sailors alike. Within the stifling confines of Goa's barracks, the air was heavy with perspiration and illness - a constant reminder of lurking tropical maladies. Ships languished in warm waters while supplies decayed. The psychological strain of being so far from home, the uncertainty of receiving any reply to their letters sent to Portugal for up to eighteen months, and the pressure of being vastly outnumbered by populations resenting their presence all took their toll on these men in ways no battle ever could.

As commander, Tristão bore an immense burden. Every decision he made carried life-or-death implications. Dispatching a ship to patrol certain waters might result in its loss at sea with all hands during storms. Stationing troops in one fort while leaving another vulnerable invited attacks. Trusting the wrong local allies could result in disastrous consequences. Unlike commanders closer to home who could seek counsel or adjust strategies as circumstances changed, Tristão operated independently, offering both freedom and terror. Portuguese commanders in Asia had to act like mini-sovereigns, exercising sovereign judgment because they lacked alternative support.

The few remaining personal letters and reports from this period reflect the strain these commanders experienced. They often expressed feelings of isolation, frustration with Lisbon's lack of support, and a desperate need for additional ships, manpower, and supplies. These writings detailed disease-ravaging garrisons and the ever-present anxiety of local rulers switching allegiances overnight. Despite such hardships, Portuguese commanders wrote about their determination to persevere - knowing that failure would not only bring personal disgrace but also lead to the collapse of Portugal's significant investments in this enterprise.

Tristão felt all these pressures too. However, unlike some peers whose stress led them to act erratically and violently, he maintained a reputation for composure. His decision-making was generally sound; negotiations were conducted with patience and respect, a rarity among Portuguese commanders who frequently treated Asian rulers with disdain. Moreover, his administration was efficient, ensuring that the resources Portugal allocated were used effectively rather than squandered through corruption or incompetence.

THE POISONED CUP

In 1507, after an approximately year-long stay in India, Tristão received instructions to return to Portugal. The precise reasons for this decision are unclear, but presumably involved standard Portuguese commander rotations, typically serving two or three years before returning home, as well as King Manuel I's wish to personally discuss Asian affairs with one of his most trusted agents. Preparations were made for the homeward journey, a passage expected to be just as perilous as the outbound trip but offering the prospect of reunion with family, familiar surroundings, and an escape from debilitating heat and constant disease risks.

However, before Tristão could embark on this return voyage, misfortune struck in both personal and political dimensions. Historical records mention his serious illness without providing detailed specifics. Some accounts speculate on poisoning as the cause, while others suggest tropical fever. Regardless, it is evident that for an extended period, he teetered between life and death, his anticipated departure postponed indefinitely.

Tristão recovered, but the sickness weakened him, potentially affecting his health in the long run. This ordeal also illustrated the vulnerability of Portuguese influence in Asia - the fragility of their power laid bare by the incapacitating potential of a single ailment. Similarly, a single tempest could devastate entire fleets, and a misstep in diplomacy might alienate allies into adversaries. The survival and sustenance of Portuguese dominance in the Indian Ocean hinged on luck, expertise, and unwavering perseverance. Should any one of these components falter or be removed, the whole edifice risked implausible collapse.

RETURN TO A CHANGED WORLD

Tristão finally returned to Lisbon in 1508, two years after embarking on his journey. The Portugal he returned to had experienced continuous prosperity from the trade with Asia. Ships full of spices would arrive regularly, their valuable cargo selling for prices that justified the risks and costs of such long voyages. The royal treasury had never been fuller. Manuel I, having earned the nickname "the Fortunate," a title reflecting both the remarkable success of Portugal and the king's ability to manage the empire it had created.

Yet Tristão did not retire upon his return. His skills were too valuable, and his loyalty had been proven time and again. Instead, King Manuel continued to employ him in positions of trust and responsibility. He served as a royal advisor and participated in diplomatic missions. In 1514, he was appointed to another significant assignment that would define the latter part of his career: the governorship of Portuguese India.

This position held immense power but also carried a crushing sense of responsibility. As governor, Tristão was responsible for all Portuguese activities spanning from the Cape of Good Hope to the Spice Islands. He was expected to expand Portuguese influence, protect their interests, manage complex diplomatic relations with numerous Asian rulers, and do so while facing inadequate resources, a climate that proved deadly to Europeans, and a communication system that could leave him without news from Lisbon for years at a time.

Tristão's second term in India, spanning 1515 to 1516, was characterized by the same qualities as his first: competence, steadiness, and a strategic vision that prioritized sustainable growth over grandiose gestures. He fortified existing fortresses, established new trade relationships, and worked diligently towards consolidating Portuguese control over trade routes linking the Malabar Coast to the Spice Islands. Though he did not secure any major victories or make dramatic discoveries, what mattered most was that he kept the system operational.

THE ALBUQUERQUE SHADOW

Tristão's governorship coincided with the later years of Afonso de Albuquerque, whose feats would come to embody the archetype of a daring and ruthless conqueror in Portuguese memory. Having captured Goa in 1510, Albuquerque established a permanent Portuguese stronghold in India. In 1511, he seized Malacca, granting Portugal control over the strategic straits through which much of Asia's trade flowed. He even attempted to capture Aden, situated at the entrance to the Red Sea, in an audacious but ultimately unsuccessful effort to shut down traditional spice routes entirely.

Albuquerque's achievements were undeniably impressive, and his methods left no question about his dominance. He ordered massacres of Muslim populations in conquered cities, deployed Portuguese squadrons to patrol the Red Sea and Persian Gulf, targeting any shipping that refused to pay for Portuguese safe passage. His name instilled fear from Mozambique to the Moluccas.

Tristão, however, operated under Albuquerque's shadow, a shadow that has obscured his achievements ever since. Unlike Albuquerque, who inspired epic poems with his conquests, Tristão was known for his administrative efficiency. While history often remembers conquerors, it tends to forget the men who make such conquests sustainable. Yet without individuals like Tristão da Cunha, Albuquerque's victories would have crumbled into chaos.

Someone had to ensure that captured cities were governed, trade networks continued to operate, forts were adequately supplied and guarded, and treaties were negotiated and respected. This responsibility often fell on those not named Albuquerque, whose genius lay in dismantling rather than constructing. Tristão and others like him patiently built the institutional frameworks that enabled Portugal's Asian empire to outlast its founding figures.

THE FINAL YEARS

Tristão returned from his second expedition to India in 1516, now middle-aged and having spent much of the previous decade in tropical climates. The toll on his body must have been immense. Few Europeans who ventured to Asia during this period returned to Europe unscathed. Diseases rampant in the tropics, a subpar diet, the stressors of leadership, and simply surviving multiple Atlantic crossings, a feat in which only one ship in ten returned, accumulated their toll on both body and mind. These effects manifested as unseen scars deeply felt by the men themselves.

Upon returning to his estates in Portugal, Tristão received accolades from Manuel I in gratitude for his service. The king recognized that figures like Tristão were irreplaceable, vital to an empire built upon commanders who melded military prowess with diplomatic acumen and administrative ability. Successors would not be easily found. Hence, those who had served well were rewarded handsomely, with titles, lands, income, and a comfortable retirement.

Tristão da Cunha lived until 1540, passing away at approximately eighty years old, an extraordinary age, especially considering his extended time in the disease-infested tropics of Asia. He witnessed the Portuguese empire in Asia reach its zenith and begin to wane as other European powers ventured into the Indian Ocean, and the expenses outstripped the revenues from maintaining distant strongholds and fleets.

Yet, these future challenges were largely unforeseen in his final years. Looking back at the 1530s, Tristão saw his homeland transform into a pivotal hub within a global trade network. Portuguese vessels sailed to Brazil, Africa, India, and the Spice Islands. Merchants traded goods that had once been exotic luxuries but were now commonplace due to direct access from Asia. With bases fortified and ships heavily armed, Portuguese power dominated the Indian Ocean's trade routes as no other European force had in any Asian region.

His hands had shaped this reality through decades of steadfast service, never a single heroic act but years of diligence. His ability to navigate both the perils of ocean travel and the intricacies of Asian politics was unparalleled. He mastered building systems that would endure beyond him.

THE TESTIMONY OF AN ISLAND

Discovered yet unclaimed, Tristão da Cunha remained uninhabited for centuries. Its volcanic peaks, lack of natural harbors, and extreme isolation discouraged colonizers from settling there. British sailors finally marked it in 1816 by deploying a small garrison to prevent it from being used as a base to liberate Napoleon from his exile on Saint Helena, located several hundred miles away. A modest civilian settlement followed, with no more than a few hundred inhabitants, living off farming, fishing, and the occasional resupply ship that passed through.

Today, Tristan da Cunha (the island's contemporary spelling) maintains its isolation as one of Earth's most remote inhabited places. The nearest populated landmass is Saint Helena, which is accessible by boat in 7 days. South Africa, 1,750 miles to the east, constitutes the nearest continental land. Ships may call once a month. Absent an airstrip, communication with the outside world relies on satellite connections and infrequent passing vessels.

The approximately 250 island inhabitants descend from a small group of nineteenth-century settlers, predominantly British with some American, Dutch, and Italian ancestry. Their English accent is distinct, reflecting a blend of British dialects frozen in time and evolved over their isolated existence. They subsist through fishing, primarily for rock lobster, which they export to Japan and the United States, and by selling postage stamps to collectors who value the island's remoteness and small population. Life on the island is challenging yet stable, characterized by a level of equality and mutual support that isolation fosters.

None bears the surname da Cunha today. This name was bestowed upon the island by Europeans long after Tristão navigated its waters, in honor of a figure known only from charts. The island stands as a monument to discovery, marking history's maps and featured within gazetteers. In this regard, it ironically overshadowed all else that Tristão da Cunha had accomplished.

THE WEIGHT OF NAMES

This is the paradox at the heart of Tristão da Cunha's legacy. He is remembered, but only for what he did not do. He is famous, yet his fame persists as nothing more than a geographical label. The island that bears his name reveals very little about him.

History has this tendency to arbitrarily select and elevate certain achievements while disregarding others. It focuses on the dramatic and forgets what is truly essential. History reduces human beings with all their complexities and struggles into simple stories that neatly fit into textbooks and encyclopedias.

Tristão da Cunha, however, was not a simple story to be told. He dedicated his life to building an empire, navigating both treacherous seas and the political landscape of the Portuguese court. He managed the seemingly impossible logistics of operating thousands of miles away with limited supplies and uncertain support. Tristão negotiated with sultans, rajas, and governors, fought when necessary, and compromised wisely. He constructed forts, established systems, and trained those who would maintain Portuguese power in Asia long after he left.

He was also a man who fell ill, felt the weight of responsibility for his soldiers' lives, witnessed friends perish from disease or violence, and grappled with the frustrations that arose from working with superiors who did not fully comprehend the Asian landscape and subordinates who did not always share his vision.

And yet, on one March morning in 1506, Tristão looked upon an island rising defiantly from the South Atlantic. He chose not to conquer it, nor to establish a foothold there, but merely to record its position before moving forward. Because the mission was paramount, and because glory could wait. The real work of empire-building required patience, discipline, and wisdom to recognize when opportunities were distractions.

WAVES BEYOND

Portuguese influence in Asia reached its zenith in the mid-sixteenth century, only to gradually wane thereafter. Maintaining distant strongholds became costlier than their profits, particularly as competition from Dutch, English, and French traders intensified. The vast distances proved too great, and the supply lines too vulnerable, with populations too numerous to control with limited forces. By the seventeenth century, Portugal had lost most of its Asian bases to the Dutch, retaining only Goa, Macau, and a few minor outposts.

Yet for a century, Portugal dominated trade in the Indian Ocean. For that century, Portuguese ships carried spices, textiles, and precious goods from Asia to Europe, generating wealth that transformed both Portugal and Europe's global relations. This century was built upon the work of men like Tristão da Cunha – men whose names did not echo through history like those of explorers such as

Vasco da Gama or Magellan, yet whose steadfast service enabled these legendary achievements.

The Portuguese empire in Asia was never upheld solely by grand gestures. Its sustainability stemmed from administrators ensuring forts were adequately guarded, diplomats maintaining alliances with local rulers, captains keeping supply lines open, and officials managing the complex trade across vast distances and cultural divides. These men labored behind the scenes, addressing mundane yet essential issues: feeding garrisons when ships faced stormy delays, negotiating with leaders whose political stances were shaky, upholding discipline among soldiers and sailors far from home in a sea of temptations, and making decisions without the support or guidance of superiors.

Tristão da Cunha excelled in this unglamorous yet vital work. His tenures as an Indian commander showcased qualities rarer than bravery or navigational genius: sound judgment, great patience, and the capacity to create systems that functioned even in his absence. Upon leaving India after his second governorship, he bequeathed a stronger Portuguese presence than when he arrived. Forts were better supplied, alliances steadier, and administrative procedures more efficient. He had not conquered kingdoms or discovered trade routes, but made the empire enduring.

This was an achievement that history largely overlooked. It lacked drama, failing to possess the narrative power that immortalizes some historical figures. There were no last stands, brilliant victories, or legendary tales. Instead, there were years of diligent administration punctuated by crises Tristão handled with unwavering professionalism.

THE PATTERN OF FORGETTING

The forgetting of Tristão da Cunha is not an isolated incident. History teems with individuals whose contributions were overshadowed by their more charismatic or dramatic peers. The second wave is consistently overlooked in favor of the initial breakthrough. The administrators are constantly overshadowed by conquerors, and the builders are perennially eclipsed by discoverers.

Part of this reflects the inherent prejudices of historical memory. Human beings gravitate towards tales of individual heroism, striking moments, clear triumphs, and crushing defeats. We remember the first person to scale the summit more than those who established the base camps that facilitated the ascent. We recall the general who secured the decisive victory over the logisticians who ensured his army's supply. We honor the scientist who achieved the breakthrough more than the researchers conducting the myriad experiments that made it possible.

Yet this pattern of amnesia carries consequences. It distorts our comprehension of how significant achievements truly transpire. It makes these accomplishments appear more individual, more dramatic, and more sudden than they actually were. Moreover, it becomes challenging for us to recognize and value the work that most of us will undertake, the unglamorous yet essential task of building and maintaining systems that underpin society's operations. Just as Tristão da Cunha's efforts were vital to the success of the Portuguese empire, so too are modern software engineers, who work behind the scenes to ensure the functionality and reliability of the technology we rely on daily. These contemporary builders, like the unsung figures of the past, contribute significantly to the progress we celebrate, reminding us that the fabric of success is often woven by those who labor quietly, laying the foundation for the milestones that capture headlines.

Tristão da Cunha's story serves as a counterpoint to this distortion. His career illustrates that an empire was not constructed solely by heroes, but by skilled professionals. Success demanded not only brilliant tactics but also steadfast implementation. It showed that the distinction between fleeting conquest and enduring dominion rests with administering the quotidian yet indispensable work of making systems function across time and space.

THE PERSPECTIVE FROM THE ISLAND

Briefly revisit that March morning in 1506. Tristão da Cunha stands on his flagship's deck, observing an island rising from the fog. His pilots are enthusiastic as land in the open Atlantic is scarce, offering potential sources of fresh water, sustenance, and a landmark for future navigators. Some crew members likely expect him to order launches, formally claim the island for Portugal, plant a cross and marker, and emerge as its discoverer.

But Tristão gazes upon the surf crashing onto black rocks and notes the wind whipping clouds across the peak of the isle. He calculates the time required to attempt a landing, the risk of losing vessels and men, the possibility that the island holds no value, and the certainty that his mission lies thousands of miles eastward, with each delay complicating its success.

In that moment, Tristão made the choice that defined his legacy in unforeseen ways. Continuing their voyage, he documented the island's position.

Sketching its outline and recording it in their log, they sailed on without halting. The mission surpassed any monument or immediate triumph.

Within this instance, Tristão da Cunha unveiled the core of his character. He was not one driven by the desire for glory or the need to be foremost, to conquer or claim. Instead, he pursued a subtler and ultimately more valuable impulse: the wish to fulfill the mission successfully, serve his king adeptly, and advance Portuguese interests strategically rather than theatrically.

This trait, this inclination towards substance over spectacle, strategy over showmanship, the long view over immediate gratification from dramatic gestures, rendered him invaluable to Manuel I. It assured his success in the treacherous realm of Portuguese India. Yet paradoxically, it made him forgettable by history, which perpetually favors drama above discipline.

THE CALCULATIONS OF EMPIRE

The Portuguese empire in Asia was built on calculations that many Europeans at the time lacked the capacity to perform. Contemplating how many ships to dispatch so that a sufficient number survive the voyage to complete the mission is one such question. Estimating the size of a garrison, feasible, considering mortality rates from disease, poses another challenge. Determining how much intimidation local rulers can withstand without provoking an uprising that could overwhelm your limited forces presents a further challenge. Elucidating the threshold of violence required to maintain your invincibility and where it becomes counterproductive due to generating enemies faster than they can be defeated is yet another.

These were the queries that dictated success or failure. And they necessitated not just courage or vision but mathematical acumen, the capability to calculate risks, assess probabilities, and make decisions grounded in possibilities rather than certainties.

Tristão da Cunha excelled at these calculations. His decision to bypass the island in 1506 exemplifies this. So too does his careful stewardship of resources in India and measured application of force to establish Portuguese credibility without triggering widespread resistance demonstrate a third. He recognized that Portuguese supremacy in Asia rested on thin margins, with a single misstep potentially unwinding everything. During this era, supply lines were so precarious that only 60% of vessels returned safely from an arduous voyage, and sailor mortality rates from disease often approached 50%, highlighting the fine balance Tristão maintained to ensure the success of Portuguese undertakings.

This comprehension distinguished successful Portuguese commanders from their less fortunate counterparts. Men akin to Albuquerque could seize spectacular victories yet also precipitate disasters through overconfidence or misjudgment. Men similar to Tristão achieved fewer spectacular triumphs but avoided spectacular fiascos. And in the long run, averting calamities often surpassed winning victories in significance.

THE QUESTION OF CHARACTER

To truly understand Tristão da Cunha's legacy, it is crucial to look beyond the dates and events recorded in historical records and seek a glimpse into the man behind them. Unfortunately, information on his personal thoughts and experiences is limited as he left no diaries or extensive correspondence - unlike some of his contemporaries. What remains are official reports, formal letters, and occasional mentions in others' writings, which we must piece together to form an image of him.

Despite this scarcity, fragments can offer valuable insights into a person's character. Tristão's official reports reveal a systematic thinker who could organize complex information and propose potential solutions to problems. His formal letters showcase his skills as a diplomat, adept at managing relationships, finding the right tone for different audiences, and advancing Portuguese interests without unnecessarily antagonizing others with whom Portugal needed to cooperate.

Other accounts portray him in a positive light, expressing respect for Tristão's judgment, reliability, and leadership effectiveness. This kind of practical leadership is vital to an empire's success, even if it doesn't inspire ballads or epic poems. Tristão was the type of leader who could hold the front lines, manage complex operations, make difficult decisions with limited information, and accept the consequences.

Outside of his professional life, he married, had children, maintained estates in Portugal, attended court, and participated in Portuguese noble life rituals - all while bearing unique memories of storms in the South Atlantic, negotiations with Indian rajas, fort construction on hostile shores, and an empire taking shape through distant decisions that understood their role in history.

THE EMPIRE AFTER TRISTÃO

Tristão da Cunha passed away in 1540, but his contributions to the empire endured for over a century. Portuguese ships continued sailing between Lisbon and Goa, Malacca, Macau, and the Spice Islands. Forts controlled key trade routes, merchants profited from access to Asian goods, and Portuguese power remained rooted in fortresses, alliances, administrative systems, and knowledge of winds, currents, and local politics - elements Tristão himself would have been familiar with.

Initially, Portugal's expansion was marked by practical adaptability; however, this approach gradually gave way to a more rigid, religiously aggressive stance. The Inquisition was established in Goa in 1560, introducing religious persecution reminiscent of that in Europe. Portuguese officials became more corrupt and self-serving rather than focused on serving the crown. Moreover, Portugal's vast military power, stretched across significant distances and challenged by Dutch and English rivals with greater resources, started to weaken.

By the early seventeenth century, Portugal was losing its Asian empire piece by piece. The Dutch seized Malacca in 1641, ending Portuguese control over the straits. The English established their own trading posts in India, breaking up Portuguese monopolies. By mid-century, Portugal held only a few bases: Goa (until 1961), Macau (until 1999), and smaller settlements that survived primarily because of geography rather than power.

For a century, however, Portugal had transformed the world by connecting Europe and Asia through regular maritime trade, demonstrating that small European states could project power across oceans, and establishing patterns of trade, conquest, and empire-building adopted by other powers.

Tristão da Cunha lived during Portugal's ascendancy and did not witness its decline or the loss of bases he helped establish. He died believing in his service to his king and the security of Portuguese power in Asia - beliefs that history would later contradict, but not within his lifetime.

THE FINAL CALCULATION

What are we to make of Tristão da Cunha's life? How do we weigh his accomplishments against the broader moral issues that any discussion of an empire inevitably brings up? He contributed to building a system that enriched Portugal by dominating Asian trade routes, using military force to secure cooperation, and giving Portuguese merchants preference over Asian competitors. This system brought prosperity to some while causing hardship for others. It opened opportunities for certain individuals but denied them to others. And it was built on a foundation of violence or at least the credible threat of it, which made cooperation preferable to resistance.

These are truths that are difficult to accept. An empire is rarely comfortable. But these are truths we must acknowledge if we want to understand the world Tristão da Cunha helped create and the one we still inhabit today, a world influenced by the European empires that followed Portugal's example and eventually surpassed Portuguese power.

Tristão da Cunha was not a philosopher. As far as we know, he did not grapple with the moral implications of Portuguese expansion. He was merely a professional, serving his king and country with competence and loyalty, doing what was required of him and doing it well. It would be anachronistic to judge him by standards he never claimed to uphold. However, refusing to recognize the costs of the system he served would be dishonest.

The Portuguese empire in Asia was founded on coercion disguised as trade, military superiority presented as cultural superiority, and the belief that European power gave Europeans the right to reshape Asian commerce according to their interests. Tristão da Cunha accepted these assumptions. He operated within a system that took them for granted. And he worked diligently to ensure the system functioned smoothly.

This is the uncomfortable reality for most historical actors: they are not heroes or villains, but professionals working within systems with moral foundations they rarely question. They make decisions based on the information available to them, the values they have inherited, and the pressures they face. And they leave behind legacies that are always mixed, complex, and open to multiple interpretations depending on whose perspective we consider.

The island named after Tristão da Cunha still exists. Its volcanic peak still rises from the South Atlantic. Its small population lives in isolation, connected to the wider world through satellite technology and the occasional ship that braves the challenging seas to reach the island's limited harbor. The islanders go about their daily lives largely unaware of the man whose name they bear, the man whose decision on a March morning in 1506 led to the island being marked on maps, given a European name, and, however tenuously, incorporated into European consciousness.

The island serves as an appropriate tribute to Tristão da Cunha. Not because it is grand or impressive but because it is overlooked, a footnote in most discussions of exploration, a curiosity rather than a destination, a name that one encounters casually rather than a place someone actively seeks out. The island is forgotten like the man, for similar reasons: it lacks drama. It offers no treasure, strategic importance, or any compelling reason for anyone to care about it beyond the simple fact that it exists and was once noted by someone.

But existence is significant. The island is real. It is inhabited. It has a history, culture, and community that call it home. And the name links that island, however tenuously, to a man who once sailed past it on his journey to help establish an empire, a man whose life was defined not by that moment of sailing past but by decades of diligent service that made the empire possible.

If Tristão da Cunha could return and see the island today, what would he think? Would he be pleased to have his name associated with this remote rock in the South Atlantic? Or would he feel disappointed that rather than his years of service in India, this became his most enduring legacy? Would he find it ironic that he is remembered for something he did not choose to do instead of everything he accomplished?

We cannot know. But perhaps the island, in its seclusion and obscurity as well as its quiet resilience, is not such a bad tribute after all. It endures. It provides shelter for a small community. It acts as a landmark for the few ships that pass it. And it carries a name into the future, reminding anyone who looks at a map of once, long ago, a Portuguese navigator sailed these waters and documented what he found, even if he did not stop to claim it.

That act of documentation, of recording, of adding to the sum of human knowledge without seeking immediate reward was characteristic of Tristão da Cunha. He was a builder rather than a claimer, a maintainer rather than a conqueror, a professional who recognized that the real work of an empire was not in the grand gestures but in the patient accumulation of small achievements that over time amount to something significant.

The island remembers that moment of documentation. And by remembering it, the island conserves something essential about the man: his willingness to prioritize duty above glory, to prefer substance over showmanship, and to choose the mission over the monument. These were the qualities that made Tristão da Cunha invaluable to his king and crucial to his country. However, these are the qualities that history, in its preference for drama over discipline, has largely overlooked.

Yet, the island remains. A mountain emerging from the sea. A name on a map. A tribute to the navigator who found nothing, claimed nothing, and through that choice of restraint, preserved his name forever, connected to the most remote inhabited place on Earth, a fitting emblem for the empire's forgotten builders, standing guard over the empty ocean. In this endurance, the island epitomizes how Tristão da Cunha's patient, disciplined approach enabled the enduring structures of empire, even as history tends to overlook the indispensable work of those who prioritized substance over spectacle.

Dom Francisco de Almeida

March 1, 1510. Table Bay, Cape of Good Hope.

The sun began to wane above Table Bay. Its radiant beams cast long shadows over the glassy waters below. The surface was smooth as a mirror's polish; it perfectly replicated the tranquil sky above.

Three Portuguese vessels, the Garcia, Belém, and Santa Cruz, swayed gently at anchor. Their aged wooden hulls creaked, crusted with salt from endless voyages. Their crews moved listlessly on deck, drained by months spent adrift. Almeida observed them and knew his journey was not just a return home. It was also a trial of what remained of his storied legacy. He had witnessed the rise and fall of empires. Here, at Africa's edge, he confronted the specter of his own mortality. What would history remember him for? The conqueror of oceans or the father who sought solace beyond horizons?

Dom Francisco de Almeida stood upon the shore, sixty years of his life trailing behind him like a shadow. His face was etched with lines from hardship, and his hands trembled not with fear, but with age and suppressed rage. Behind him rose the imposing mountain, unmoved by the turmoil unfolding below.

Initially, the Goringhaiqua Khoikhoi had been affable and offered cattle for glass beads and sheep. Then the mood shifted. Almeida's men acted rashly, seizing cattle that did not belong to them. Now, their hosts had turned hostile and were amassing.

He watched as they crested the ridge, a hundred seventy formidable warriors advancing with their herds like an unyielding wall of muscle and horn. Spears glinted in their hands, bodies adorned for war, faces as hard as stone.

Nearby stood his captains, Pedro and Jorge Barreto, panting from their raid on the village, while one Portuguese sailor lay slain, and their boats were strategically positioned away at a watering point. Master Diogo d'Unhos had gravely miscalculated, leaving them without a means of retreat.

Almeida unsheathed his sword. It caught the late afternoon sun, a familiar gleam that reminded him of battles won, Granada's fall, and North African triumphs that solidified Portuguese dominion over the vast ocean.

He surveyed his men: sixty-four seasoned warriors with eleven captains among them. They were Portugal's nobility elite. They had claimed Kilwa and Mombasa, razed Dabul, and humbled Diu. They instilled fear in kings from Mozambique to this desolate coast. Here, they would face not monarchs, but herders fighting for restitution.

"Hold fast," Almeida commanded, his voice unwavering. "We make our stand here."

The Khoikhoi warriors charged with a tumultuous roar. Their cattle surged ahead in a living bulwark of flesh and horn, with the warriors following close behind. Stones zipped through the air, assegais stabbed forth rapidly, and Portuguese steel clashed against wood in frenzied combat. Despite their proud armor, it was no match for the fury of men fighting to reclaim what had been taken from them.

The battle raged for less than an hour. When it subsided, sixty-five Portuguese lay among the dead, including Dom Francisco de Almeida, the first Viceroy of Portuguese India and conqueror of Diu, whose life slipped away on that shore.

His remains were recovered in the afternoon's waning light; they buried him by the sea with the mountain looming in silent testament. The man who had once devoured a rooster had now been consumed in his turn.

THE MAKING OF A WARRIOR

To comprehend how Dom Francisco de Almeida met his demise on the beach, we must embark on a journey from Table Bay and gaze backward. Our narrative returns to Lisbon circa 1450. This world was still deeply rooted in medieval tradition. Yet it began to dream of horizons beyond its known boundaries.

Francisco de Almeida was born into Portuguese nobility, destined for greatness. He was the seventh son of Lopo de Almeida, the second Count of Abrantes. In this era, such lineage did not guarantee inheritance but demanded service. Francisco's path diverged. He chose to wield a sword, not scripture.

His family boasted distinction even within aristocratic circles. Two of his brothers rose to become bishops, another served as an ambassador to the Holy See, and yet another led Portugal's Knights of Malta. The Almeidas counseled kings, engaged in battles against infidels, and stood as pillars of a nation that had recently secured its independence, now with eyes cast towards ambitious horizons.

Young Francisco swiftly mastered martial arts. He learned equestrian skills. He fought with sword and lance. He developed the ability to command men and take life when necessary. These were not abstract concepts in a peaceful setting. Fifteenth-century Portugal was a fervent crusading nation. The memory of the Reconquista, the centuries-long efforts to drive the Moors from Iberia, lived on. Tales of Santiago Matamoros, Saint James the Moor-Slayer, who allegedly appeared mounted to turn the tide of battle in favor of Christians, were part of every nobleman's upbringing.

Francisco believed in these tales or at least in the power they represented: the divine right of Christian Europe to expand, conquer, and dominate.

In 1476, at approximately twenty-six years old, he participated in the Battle of Toro. It was a brutal conflict, a struggle between Portugal and Castile for control of the Castilian throne. The Portuguese suffered defeat. Yet Almeida learned valuable lessons: that battles could be lost even when one believed God favored one's cause; that politics and warfare were intertwined; and that nobility did not shield one from mortality, as steel found all alike.

He then spent time in Morocco. In 1415, Portugal conquered Ceuta, establishing its first permanent European foothold in Africa since Roman times. Nonetheless, the Moors controlled the coast beyond this toehold. Portuguese nobles ventured into Morocco. Like English nobility later in India, they sought fortune and glory, testing themselves against an enemy deemed inferior but dangerous.

Almeida fought amidst the dusty hills and burning cities of North Africa, commanding a diverse force: knights, infantry, Portuguese allies, alongside Moorish forces, mercenaries, and zealots. He faced harsh realities; logistics proved as crucial as courage, water and bread secured victories as much as swords did. Even the sun became an adversary.

Hatred, too, was forged in this period. The wars in Morocco were characterized by cruelty on both sides. No quarter was given. Cities were ravaged, populations enslaved. This was considered a holy war, justifying atrocities unthinkable within Christian kingdoms. In response, captured Portuguese were sometimes flayed alive or roasted over slow fires, or castrated and sent as gruesome warnings.

It wasn't the quiet of Lisbon's courts or the sanctity of church libraries that shaped Francisco de Almeida, but the blood-soaked landscapes of Morocco that forged his character.

GRANADA: THE LAST CRUSADE

In 1485, King Ferdinand of Aragon and Queen Isabella of Castile, known as the Catholic Monarchs, embarked on the final push of the Reconquista. The Emirate of Granada was the last Moorish stronghold in Iberia, ruling parts of Spain for nearly eight centuries. It was time for Christians to complete their ancestors' work.

It was around this period that Francisco de Almeida visited Granada. Aged approximately thirty-five, he was a seasoned soldier with numerous battle scars and ready to take lives again.

The conflict lasted seven years, characterized by sieges rather than open battles, due to the challenging mountainous landscape of Andalusia. The Moors controlled fortified cities and castles that had to be captured one by one. Almeida played his part in medieval warfare: digging trenches, erecting siege towers, and slowly starving out garrisons.

He beheld Christian forces seizing urban areas. He saw women and children sold into slavery. Mosques were burnt down, libraries were razed. Conquest was no longer cloaked in grandeur; it was merely necessary, so they argued, as God's will unfolded before them.

In January 1492, Granada fell. The Moors capitulated, with Boabdil, the final Nasrid emir, leaving the Alhambra Palace and handing over the keys to Ferdinand and Isabella. Legend recounts that Boabdil's mother chastised him for grieving like a woman when he couldn't defend his land like a man.

Francisco de Almeida was present at this historical moment. He witnessed the ensigns of Castile and Aragon hoisting above the Alhambra. It felt as if Christian destiny had been fulfilled right before his eyes.

Now forty-two, Almeida reflected on a lifetime at war. He had demonstrated bravery, competence, and the capacity for ruthlessness when required. His strategic prowess did not go unnoticed by King John II of Portugal, who valued men who were both warriors and tacticians. Almeida became a trusted advisor to the king, mingling within Portuguese power circles.

Despite his elevated status, an undercurrent of restlessness persisted in him. The conflict had concluded, and Granada was now part of Christendom. What future could there possibly be for someone like him?

It was then that the call of the ocean resonated with a new promise.

THE OPENING OF THE EAST

In July 1497, four ships departed from Lisbon under the command of Vasco da Gama. Their objective was to discover a sea route to India by circumnavigating Africa. Achieving this would allow Portugal to circumvent the Ottoman Turks and Venetian merchants who monopolized the overland spice trade, granting them direct access to the riches of the East.

By May 1498, da Gama had arrived at Calicut on India's Malabar Coast. The local rulers were perplexed by these Europeans with their modest fleet and limited cargo. As the Portuguese marveled at the renowned city before them, a local merchant nonchalantly returned to his merchandise, more preoccupied with the regular clientele than the foreign spectacle. Nevertheless, da Gama's return to Portugal yielded pepper and spices worth sixty times the cost of the expedition.

Portugal erupted in jubilation. This was the prize they had been seeking. Not gold, although there was gold. Not conquest, though there would be conquest. But spices pepper, cinnamon, cloves, and nutmeg were more valuable than silver in European markets. Control of the spice trade meant access to immense wealth. King Manuel I, who succeeded John II, immediately recognized these implications. A new empire was within reach, not in Europe, where every inch was contested, nor in the Mediterranean, dominated by Venice and the Ottomans, but in the Indian Ocean, unexplored by any European power in force.

However, the Indian Ocean was not devoid of activity. It was frequented by Arab dhows, Indian merchant vessels, and Chinese junks. It was controlled by a network of Muslim traders who had dominated maritime commerce for centuries. These merchants maintained connections with the Mamluk Sultanate of Egypt, which in turn had ties to Venice. Together, they formed an extensive economic network that spanned from the Spice Islands to Amsterdam's markets.

Portugal would have to dismantle this network through force. Manuel understood that he required a viceroy, a warrior capable of constructing fortresses, commanding fleets, annihilating adversaries, and, if necessary, establishing Portuguese dominion through acts of violence.

He chose Dom Francisco de Almeida.

THE APPOINTMENT

March 1505. Lisbon.

Almeida, aged fifty-five, was an old man, not just in years but in terms of his era. Yet Manuel sought wisdom and experience over youthful vigor. He yearned for someone who understood power's essence and would remain steadfast against the lure of riches or diplomacy. His desire was clear: a man to prioritize Portugal's interests above personal glory.

The role offered was unprecedented, carrying the title Viceroy of India - not a mere governor or captain but ruler in the king's name over untamed territories and undeveloped peoples, an empire yet to be forged. Almeida's mission was multifaceted:

- Construct four fortresses along southwestern Indian shores: Anjediva Island, Cannanore, Cochin, and Quilon.
- Secure Portuguese dominance over the lucrative spice trade.
- Forge alliances with local rulers where feasible.
- Annihilate hostile Muslim shipping.
- Establish Portugal's supremacy across the Indian Ocean.

His tenure was set at three years, after which he would return to Portugal, making way for another to take his place. Manuel aimed to prevent viceroys from becoming independent rulers, desiring administrators who understood their roles were temporary.

Almeida accepted this endeavor, perhaps viewing it as his final grand adventure or aiming to secure prosperity for his family. The call of battle was too strong to resist.

He assembled a formidable fleet: twenty-two ships comprising fourteen carracks (the backbone of Portugal's expansion), six caravels (swift and agile), two galleys (ideal for close combat), totaling fifteen hundred soldiers and approximately three thousand men in all.

The São Rafael, captained by Fernão Soares, led the fleet. Almeida's son, Lourenço, accompanied him, brave, eager, yet untested as his father had been. The bond between them was strong.

On March 25, 1505, they set sail from Lisbon, a spectacle of crowds and celebrations. Blessings from priests mingled with the cheers of men and the tears of women. As their sails swelled with wind, they departed into the Atlantic's embrace.

None aboard could envision the tumultuous conflicts ahead or the empires destined to crumble beneath their might. They could not foresee reshaping the world itself.

Almeida stood at São Rafael's rail, watching Lisbon disappear on the horizon, propelled towards destiny and unknowingly marching towards his downfall.

THE AFRICAN BLOODBATH

The voyage south proved treacherous, with tempests relentlessly pounding against our fleet. A dire circumstance arose when the Bella sustained damage and was abandoned; her crew and cargo were redistributed among the remaining vessels. Mortality rates soared due to sickness and accidents. The ships bore resemblance to sprawling urban hubs afloat, congested with both humanity and livestock, permeated by the stench of perspiration, waste, and brine. Rats found solace in the depths while weevils infested rations. Even water barrels became tainted. Yet these sailors hailed from Portugal; they'd braved the Cape before, familiar with such trials.

Upon arrival in June, our armada doubled the Cape of Good Hope, though a tempest on the other side caused several to scatter, only to regroup at Sofala and Mozambique by early July.

The true test lay ahead. King Manuel's edicts were explicit: solidify Portugal's reign along Africa's eastern coast. Almeida had to subdue the Swahili city-states that dominated commerce between the interior lands and Indian Ocean cities like Kilwa, Mombasa, Zanzibar, and Barawa. Rich, diverse, and Muslim, these places harbored ill will towards Christians and eschewed Portuguese "commerce," which entailed submission and tribute.

On July 23rd, Almeida's fleet loomed off Kilwa. Kilwa glittered with stone homes with ornate doors, minarets piercing the skyline, a harbor brimming with dhows bearing gold, ivory, and human lives from within. The populace numbered around four thousand souls; the Sultan thereof was obligated to Portugal for tribute, an obligation unmet by his rule.

No heralds were sent; eight ships readied for assault. Five hundred Portuguese soldiers disembarked, overpowering a few hundred defenders armed with tenacity but no match against Portuguese firearms and archery. Kilwa yielded within hours. A fort was commissioned, troops left in garrison, and the Sultan's wealth was seized as Almeida advanced.

Next on the agenda was Mombasa, bigger and stronger. Tens of thousands inhabited it; a powerful Arab sheikh governed with robust defenses. Upon spotting our fleet in mid-August, he marshaled for battle. A grim clash unfolded. Portuguese naval gunfire rained down before troops waded through the surf to engage house-to-house. The sheikh's warriors fought valiantly yet futilely against Portuguese firepower and martial discipline. Mombasa burned; plundered riches were set ablaze by Portuguese hands. Golden wealth, silks, and spices were claimed and hoarded in Almeida's vessels. A local sultan supportive of the invasion was rewarded with preferential trade status, a hallmark of Portuguese favor.

Zanzibar soon followed as a Portuguese caravel seized it for the kingdom. By the time our sails caught the winds bound for India, Portuguese supremacy was enforced through sheer intimidation; resistance meant annihilation, compliance could spare survival. This was Almeida's strategy: Portugal's Estado da Índia was constructed not in commerce or negotiation but on dread. Yet inwardly, Almeida pondered the burden of terror cast across seas. Would such a reign last or crumble under its own savagery? In fleeting contemplations, doubts arose whether he'd be remembered for glory or infamy; these musings were swiftly buried beneath demands of conquest. Was terror a sustainable currency of empire? This question loomed silently as the next battle awaited on the horizon.

COCHIN: THE INDIAN FOOTHOLD

Almeida arrived at Anjediva Island off India's west coast on September 13th, 1505, where he promptly commenced construction on a fort. The Portuguese erected their first permanent structure in Indian territory, stone by stone. Anjediva served as an intermediary station for ship repairs and replenishment.

By October 23rd, with the endorsement of Kolattiri Raja, a local leader favorable towards Portugal, construction began on Fort St. Angelo at Cannanore under Almeida's guidance. He left Lourenço de Brito in charge, accompanied by 150 men and two ships.

Upon arrival in Cochin on October 31st, only eight ships remained. The rest were either engaged in different missions or had been lost at sea.

Cochin was the heart of Portuguese India, where Fort Emmanuel had already been established by Afonso de Albuquerque in 1503. The Raja of Cochin allied with the Portuguese, considering them useful mercenaries against his rivals, particularly the Zamorin of Calicut, the most influential ruler on the Malabar Coast.

However, Almeida received disturbing news about an attack on the Portuguese trading post at Quilon, where some of their traders were killed. This aggression could not stand unchallenged.

In response, he dispatched his son Lourenço with six ships to seek revenge. Upon reaching Quilon, Lourenço encountered twenty-seven ships from Calicut in its harbor and destroyed all of them without mercy or exception. The message was clear: Portuguese blood demanded retribution, and it would be exacted through fire.

Lourenço's success marked his first independent command, impressing Almeida, who noted that his son possessed his own iron resolve despite being no child.

After settling in Cochin, Almeida outlined a strategy to assert Portuguese supremacy in the East. This approach, later referred to as the Blue Water Policy, although not originally coined by Almeida, emphasized controlling sea routes rather than expanding land territories. It was both a simple and revolutionary concept that relied on naval dominance.

Unlike traditional empires, which focused on territorial conquest and control, Portugal lacked the manpower for extensive landholdings in India. Instead, Almeida proposed leveraging naval power to secure trade by dominating the seas. This strategy included building a navy, defeating enemy fleets, blockading hostile ports, compelling maritime commerce through Portuguese-controlled channels, and making it impossible to trade without their approval, a maritime choke point akin to modern-day toll systems at places like the Suez Canal.

While this approach focused on ships over forts, cannons over castles, and admirals rather than governors, later viceroys like Afonso de Albuquerque would diverge from this strategy by pursuing territorial gains such as capturing Goa, Malacca, and Hormuz. However, in 1505, Almeida was convinced that sea control alone sufficed.

His predictions were not unfounded. For over a century, Portuguese naval dominance ruled the Indian Ocean, with no local power matching European ships or weaponry. Their carracks served as floating fortresses, equipped with cannons that easily defeated Arab dhows and Indian merchant vessels.

Nevertheless, maintaining navies required manpower, and sailors needed supplies and repair facilities, necessitating ports and, by extension, forts. Thus, while Almeida built his forts at Anjediva, Cannanore, and Cochin, they served as naval bases rather than territorial strongholds.

The cornerstone of the Blue Water Policy was Portuguese ships patrolling the Indian Ocean and demanding cartazes or naval passes from all vessels. Ships without a Portuguese pass could be seized or sunk, effectively making it protection money. This outraged Arab and Muslim merchants who had long traded in these waters unimpeded. The Zamorin of Calicut, the Sultan of Gujarat, and merchants from Cairo saw their profits under threat.

They would not submit quietly.

LOURENÇO'S TRIUMPH: CANNANORE 1506

The Zamorin of Calicut made preparations for war. He gathered a fleet consisting of 200 ships and dispatched it towards the Portuguese forces. This took place in March 1506, during the final stages of the monsoon season, when weather conditions favored the Indian side.

Nonetheless, the Indians did not highly estimate the Portuguese capabilities. Leading the Portuguese squadron was Lourenço de Almeida, who had around a dozen ships under his command, some historical accounts suggest as few as eight. He encountered the Zamorin's fleet at the entrance to Cannanore harbor.

The events that ensued were nothing short of catastrophic for the Indians. Built primarily with Atlantic storms in mind, the Portuguese vessels boasted thick hulls and reinforced decks equipped with cannons capable of delivering devastating broadside attacks. These European ships' firepower was more than enough to shatter Indian ships' wooden hulls, tear their sails apart, and send their crews plunging into the sea as fragmented remains.

In contrast, the Zamorin's fleet consisted mostly of dhows and smaller craft, which, although fast and agile, were unsuitable for countering European broadsides. The Portuguese warships effortlessly penetrated through them like wolves infiltrating a flock of sheep. Within just hours, dozens of Indian ships either burned or sank. Those who survived fled the scene in terror.

The Battle of Cannanore was ultimately a resounding victory for Portugal and established a crucial lesson: even with their overwhelming numbers, Indian naval forces stood no chance against European warships in open combat. Consequently, the Zamorin had to explore alternative methods in order to resist Portuguese dominance.

For Lourenço, this triumph was deeply personal as he sought to emulate his father's achievements, embodying bravery, aggression, and efficacy. As a reward for his outstanding performance, Almeida appointed him as Captain-Major of the Seas of India, thereby making him the chief naval commander of Portuguese forces operating within the Indian Ocean.

Together with his father, he further solidified Portuguese supremacy by forging a trade agreement with Malacca that paved the way for access to Southeast Asian markets. Their ships explored territories such as the Maldives and Ceylon, mapping coastlines while identifying potential sites for future fortifications.

One of Lourenço's captains, Fernando Soares, may have been among the first Europeans to catch sight of Madagascar. Everything seemed to be unfolding precisely according to plan, with Portugal ascending its dominance. The spice trade flowed almost exclusively through Portuguese hands; Arab merchants submitted themselves to their influence, and even Indian rulers sought their favor.

Then, Egypt entered the geopolitical arena.

THE MAMLUK RESPONSE

The Egyptian Mamluk Sultanate was indeed an influential force within Islamdom. These Mamluks, comprising mainly Circassian and Turkish slaves, had managed to take control of Egypt and Syria during the 13th century, successfully repelling Mongol incursions and Christian crusades alike. By the early sixteenth century, though, their power was dwindling, becoming increasingly tainted by corruption, yet still posing a threat.

The Mamluks held sway over the Red Sea and acted as intermediaries in the lucrative spice trade. Spices would arrive in Arabian ports from India before being transported overland to Cairo, then onward to Alexandria, where Venetian merchants purchased them and shipped them across to Europe. This commerce was highly profitable for both the Mamluks and Venice.

However, Portuguese explorers undermined this system by sailing around Africa to bring spices directly to Lisbon, thereby underpricing their Venetian counterparts in European markets. Consequently, Venice's revenues drastically decreased, while the Mamluks saw their income dry up as both powers felt their positions being threatened.

Venice and Cairo decided to join forces in response. The skilled Venetians contributed technical expertise, including shipwrights, gunners, and naval architects. Meanwhile, the Mamluks provided financial support and manpower. Together, they planned to assemble a fleet capable of tackling the Portuguese in the Indian Ocean.

This was certainly no small undertaking, since Mediterranean-style galleys were not built for oceanic warfare, and the logistics of transporting ships from the

Mediterranean to the Red Sea and onward to India seemed quite daunting. Yet desperation drove the Mamluks forward, as evidenced by their efforts in 1507. They constructed a fleet at Suez, disassembled it, overland transported it to the Red Sea, then reassembled it.

Led by the Kurdish Mamluk admiral Amir Husain Al-Kurdi, this rebuilt fleet sailed down the Red Sea and arrived in Gujarat, India. There, they allied with Malik Ayyaz, Diu's powerful governor, and the Zamorin of Calicut to eliminate the Portuguese presence in Indian waters.

In March 1508, this combined Mamluk-Gujarat armada encountered a Portuguese squadron off Chaul on India's western coast, commanded by none other than Lourenço de Almeida.

CHAUL: THE DEATH OF A SON

Chaul, March 1508.

Lourenço had eight ships: two galleys and five caravels, alongside his personal flagship. He was assigned to escort merchant vessels from allied nations. Upon receiving intelligence that a Mamluk fleet was approaching, Lourenço made a decision that would haunt his father for eternity.

He chose to strike first. The Mamluks possessed approximately thirty ships, along with the advantage of surprise and superior numbers. However, Lourenço firmly believed in Portugal's supremacy, having never tasted defeat before. So why should this time be any different?

The beginning of the battle did not bode well for Lourenço's forces. The Mamluks strategically maneuvered to isolate their flagship while Portuguese captains struggled and were repelled by the might of Mamluk galleys. Moreover, the wind proved unfavorable for the Portuguese, hindering their ability to maneuver effectively.

Lourenço fought amidst the chaos on the deck of his flagship, as arrows rained down upon him. Two of them struck their target, one penetrating his body and another his face, causing blood to pour down his armor. Yet he did not relent in his defiance. His captains pleaded with him to withdraw, but Lourenço refused. Retreat or surrender were not options for him. The battle lasted three days, culminating on the final day when reinforcements from Gujarat, led by Malik Ayyaz, arrived to overwhelm the Portuguese forces. Lourenço's vessel was left crippled beyond repair. The commander met his end upon the deck, with sword in hand and blood streaming down his visage. His remains were never recovered, as it is believed that the Mamluks took them away or discarded them into the sea. Nine Portuguese sailors were captured during this debacle, while the rest of the fleet managed to escape further south towards Cochin.

The survivors carried forth their tale to Dom Francisco de Almeida, the news of which left an indelible mark upon him.

THREE DAYS IN THE DARK

Almeida received news while in Cochin; his captains relayed a sorrowful tale: his son had perished, and his body lay undetected after a crushing loss to the Mamluk fleet. Almeida bore this account silently before sending away his captains.

He then retreated into solitude, locking himself away for three days without interaction or sustenance, as the Portuguese settlement held its breath in muted anticipation of his next move. Some whispered among themselves, contemplating whether he'd unleash an assault or retreat to the ship's safety.

Within his quarters, Almeida was alone with his grief, whether expressed through tears, fury, or prayer, no one would ever truly know, for there were neither letters nor journals to reveal the depths of his despair during this period. On the fourth day, he emerged, a shadow of his former self with eyes swollen from crying, yet maintaining a firm resolve in his voice.

Almeida vowed revenge against the Mamluk fleet, swearing to raze every harbor that had given them shelter and make the Muslims answer for his son's death with floods of retribution. In private, Almeida clutched Lourenço's dagger, a small but potent token of what was taken from him; its edge still keen, a testament to their past triumphs together, now serving as an emblem of his loss and his vengeful quest.

The message was clear: the Mamluks had slain the young bird of prey. Now they would confront the elder and more relentless one, Almeida, no longer the youthful seeker of accolades but a merciless avenger set upon a path of fiery retribution.

THE COMPLICATION: ALBUQUERQUE

The arrival of Afonso de Albuquerque in Cochin introduced a distinct perspective on Portuguese expansion in the East. A formidable figure known for his relentless drive, ingenuity, and unyielding ambition, Albuquerque had made significant strides in North Africa and India, securing victories that solidified Portugal's presence. However, unlike other commanders, he harbored ambitions to build a strong fortress network to control strategic points, thereby diverging from Almeida's maritime-focused strategy.

Almeida, steadfast in his refusal to relinquish authority, found himself at odds not only with Albuquerque but also with the dictates of duty and honor. The death of Lourenço weighed heavily upon him, fueling an insatiable quest for vengeance against those responsible, a personal crusade that overshadowed all else.

Albuquerque, armed with a royal commission granting him the authority to succeed Almeida as viceroy, confronted his rival in hopes of resolving their differences peacefully. But Almeida, consumed by personal vendettas and defying both reason and protocol, chose a path of rebellion. His actions led to Albuquerque's imprisonment, setting the stage for a confrontation that could have dire consequences, including potential execution if the king were to learn of the uprising.

Meanwhile, the Portuguese in India found themselves caught between two competing visions: Albuquerque's authority now superseded Almeida's, yet they faced resistance from those who remained loyal to him. Tensions escalated as preparations for conflict loomed, painting a picture of turmoil and uncertainty amid an already volatile geopolitical landscape.

THE ROAD TO DIU

December 1508: The Portuguese assembled a massive fleet at Cochin, comprising eighteen ships and fifteen hundred soldiers alongside four hundred Indian auxiliaries from various kingdoms - an unprecedented gathering of European military might in Asian waters.

Almeida gathered his captains, informing them of their imminent mission to annihilate the Mamluk fleet. This was not merely for revenge but to establish Portuguese dominance once and for all. Interestingly, Almeida offered any man hesitant about the venture the option to stay behind; however, no one chose this path.

The fleet sailed north along the coast, stopping at various ports. While some interactions involved negotiations, others were marked by aggressive actions - burning Moorish ships, pillaging ports sympathetic to the Mamluks. This display of power was Almeida's way of alerting adversaries, particularly the Mamluks, to his approach and instilling fear.

On February 2nd, 1509, the Portuguese fleet sighted Diu, a highly prosperous port in India situated on an island off Gujarat's coast under the governance of Malik Ayyaz, who had previously fought at Chaul. The Mamluk fleet was anchored there under Amir Husain's command, though their total numbered around two hundred ships, with most being small coastal vessels rather than warships.

Malik Ayyaz, apprehensive about Almeida's arrival, given Portugal's reputation for brutality towards enemies, composed a respectful letter to him. Praising Lourenço's bravery and assuring good treatment of Portuguese prisoners from Chaul, he hinted at the possibility of negotiation.

Almeida responded with cold determination, vowing to reach Diu, destroy the Mamluk fleet, and offering Malik Ayyaz two options: unite all forces in preparation for battle or surrender and plead for mercy.

Choosing a middle path, Malik Ayyaz abandoned Diu's defense to Amir Husain, keeping his own ships near shore for quick retreat if necessary. That night, Portuguese anchored outside Diu's harbor, with city lights illuminating the water within which the Mamluk fleet awaited, warriors prepared weapons, and priests offered prayers. Gunners loaded cannons in anticipation. On February 3rd, 1509, dawn marked the beginning of a battle that would forever define Dom Francisco de Almeida's life.

THE BATTLE OF DIU

Dawn cast its vibrant hues across the seas of Diu, painting them crimson. The water surged with an imposing force as a sea breeze swept across the harbor, setting the stage for what was to unfold. It was a perfect confluence for an impending assault.

Come eleven, Almeida issued his command, and the Portuguese armada commenced its advance. At its vanguard sailed the grand Flor de la Mar, João da Nova's mighty command, leading with unwavering poise. Behind her followed the illustrious ranks: Espírito Santo, Belém, Rei Grande, Taforea Grande - a parade of giants in their own right. And then came the swift and deadly caravels and galleys that trailed in their wake, a total of eighteen vessels, each a formidable floating bastion.

As they neared, the Portuguese unleashed their fire. It was not the uncoordinated volleys from battles past but rather a calculated barrage designed to wreak havoc. Their gunners, honed by years of practice, were adept at striking distant targets and outpacing reloads compared to any other naval force in existence.

Mamluk ships caught ablaze, one after another, as Muslim galleys dared to venture forth for a skirmish but met with relentless fire. Portuguese marksmen manned the decks with crossbows and arquebuses, obliterating anyone bold enough to show themselves above deck.

The melee was a spectacle of confusion, where ships clashed in a desperate dance of combat, Portuguese troops locked in close quarters with Mamluk marines amidst bloodied waters. Cannons roared up close, their projectiles cleaving through hulls, while cries of the fallen melded into screams, and explosions rocked the harbor as flames claimed ships whole.

Amir Husain fought valiantly on his colossal carrack, rallying his fleet's remnants against insurmountable odds. Yet, his captains faltered, outmatched by their Portuguese adversaries. Meanwhile, Malik Ayyaz's Gujarati contingent clung to the coast, hesitating to plunge fully into the fray.

By afternoon's light, the Portuguese had claimed an impressive tally of two Turkish naus, two Gujarati naus, and two Turkish galleys, along with having sent at least four more vessels to the depths. The Muslim armada was in disarray, fleeing or succumbing to the flames.

Almeida, ever the shrewd tactician, saw the winds shift by five o'clock and ordered a strategic withdrawal with their hard-won prizes to avoid entrapment within the harbor's embrace.

One ship remained the formidable flagship of Amir Husain. Its hull fortified against gunfire, it lay too close to shore for the deep-hulled Portuguese ships to engage directly. Almeida called forth an unending bombardment upon this lone holdout, every available gun from his fleet converging on the Muslim vessel for hours.

Just before nightfall, the indomitable ship succumbed and sank beneath the surface.

Thus ended the Battle of Diu.

THE AFTERMATH: BLOOD AND GOLD

The Portuguese secured a decisive triumph. Their losses were limited to thirty-two fatalities and three hundred injured individuals. In stark contrast, the Muslim forces incurred over three thousand deaths and experienced significant numbers taken prisoner. Notably, the entire Mamluk fleet was either decimated or seized, except for a small contingent led by Amir Husain, who managed to escape with twenty-two survivors to Cairo.

Nevertheless, Almeida sought not just victory but retribution. He sought to inform the world of the fate awaiting those hostile to the Portuguese.

The treatment meted out to the prisoners was cruel and merciless. Almeida decreed the execution of most captured Mamluks. Some were subjected to hanging; others suffered the torment of being burned alive, while a few were chained to cannons and blown apart. This reign of terror persisted for several hours, staining the decks with blood and echoing their cries across the water.

Conversely, the Portuguese prisoners from Chaul returned unscathed, demonstrating Malik Ayyaz's adherence to his pledge by treating them humanely. However, Almeida's benevolence did not extend to the Muslim captives.

Malik Ayyaz extended an offer to Almeida, proposing that the city of Diu surrender if the Portuguese would relinquish their claims over it. Almeida declined this proposition since maintaining such a stronghold was deemed costly. Nevertheless, he accepted a substantial payment of three hundred thousand gold xerafins, a remarkable sum that allocated one hundred thousand for the soldiers and ten thousand to the Cochin hospital.

The ruler of Diu, appreciative for sparing his city from destruction, presented daily gifts to the Portuguese fleet, including livestock, produce, rich tapestries, and precious strings of pearls. Almeida declined most personal offerings but sent a luxurious brocade tapestry and an exquisite pearl necklace back to the Queen of Portugal.

On February 12, just nine days post-battle, the Portuguese fleet embarked southward.

Having achieved his objectives, Dom Francisco de Almeida established the Portuguese supremacy over the Indian Ocean and fulfilled his quest for vengeance in honor of his son Lourenço. Yet, instead of exultation, a profound emptiness consumed him, a hollow victory that could never restore his lost loved one.

THE PRICE OF VENGEANCE

The Battle of Diu had a profound impact on global history. The aftermath saw the Mamluk Empire struggle to regain its footing; within ten years, the Ottoman Turks seized Egypt, assimilating the Mamluk Sultanate into their dominion. Consequently, the lucrative Red Sea trade route that had long enriched Cairo dwindled, and Venice's grip on the spice trade was broken. This shift paved the way for European commerce to revolve around Atlantic powers such as Portugal, Spain, and, eventually, England and the Netherlands.

For a century following this pivotal event, Portuguese influence dominated the Indian Ocean. They secured control over strategic locations like Goa, Malacca, Hormuz, and Macao, building an extensive trading network that spanned from Brazil to Japan, and the Estado da Índia, the Portuguese State of India, emerged as a formidable world power.

All these developments were intertwined with the outcome at Diu. Nevertheless, Almeida did not live long enough to witness these transformations.

Following his victory in the battle, he returned to Cochin, where he released Afonso de Albuquerque from imprisonment. Their reunion was described differently across accounts: some depicted a reconciliation, while others portrayed a tense meeting. Despite defying royal orders and imprisoning the designated successor, Almeida's triumph overshadowed his personal motives.

On November 4th, 1509, Almeida officially transferred command of the viceroyalty to Albuquerque in a simple ceremony as he relinquished any desire for power. His priority was to return home to Portugal, seek rest, and reflect on his adventures.

Almeida set sail from Cochin on December 1st, 1509, accompanied by an escort of three ships: Garcia, Belém, and Santa Cruz. Sadly, he never made it back to Portugal.

TABLE BAY: FULL CIRCLE

We return now to where we began. March 1, 1510. Table Bay.

Almeida's fleet stopped by the shores of Table Bay for water replenishment. They had successfully navigated around the Cape and were now far away from Portugal, which meant safety was within their grasp. Eager to put the past dangers behind them, men relaxed, believing they had overcome the greatest obstacles.

The Goringhaiqua clan from the Khoikhoi group approached the beach with cattle for trade. The Portuguese sailors eagerly offered glass beads, iron, and cloth in exchange. Their initial interactions were cordial until arrogance prevailed among the Portuguese, who felt invincible because of their dominance over the Indian Ocean and past victories, such as their defeat of the mighty Muslim fleet.

Some of them ventured into a nearby Khoikhoi village and saw their cattle. Without any hesitation, they stole it. The Khoikhoi people were neither subjects nor feared by Portugal; therefore, the concept of Portuguese power was foreign to them. All they knew was that strangers had wrongfully taken their property.

When these sailors returned to the shore with stolen cattle, the Khoikhoi followed closely behind to demand their return. The Portuguese laughed off their demands, and one sailor attempted to kidnap a Khoikhoi child, either as a slave or hostage. This led to a scuffle resulting in the death of a Portuguese sailor.

The survivors returned to Almeida's ships, blaming the locals for their loss, and pleaded for revenge. Almeida recognized his men's wrongdoings but felt obligated to take action. If they couldn't protect themselves even from such conflicts, what would that mean for their authority?

He assembled 150 soldiers armed with swords, lances, and crossbows to march towards the Khoikhoi village. However, unbeknownst to Almeida, Master Diogo d'Unhos had moved all landing boats to a watering point, stripping them of any means to evacuate from the beach. This left them trapped between the sea and the hostile locals.

Upon reaching the village, the Portuguese captured cattle and children while facing retaliation from Khoikhoi warriors equipped with assegais and clubs, driving their livestock ahead of them. The Portuguese had steel weapons, training, and experience, but they were exhausted, aging, ill, and outnumbered.

The ensuing battle was fierce and short-lived. The Khoikhoi used their cattle herds as shields, pushing them towards the Portuguese lines and causing disarray in their formation. Stones were thrown, assegais struck with deadly force, and clubs crushed skulls.

Almeida himself died in the surf amid these unknown assailants who spoke an alien language, all for the theft of livestock on a distant shore eight thousand miles from his homeland. Sixty-four Portuguese lost their lives alongside him, including eleven captains who had fought in significant battles such as Granada, Kilwa, Cannanore, and Diu. They had conquered oceans and brought kingdoms to heel.

The deceased were buried hastily on the beach, and the survivors departed without their Viceroy. When they reached Portugal with this tale, it moved King

Manuel to declare a day of mourning and to prohibit Portuguese ships from ever docking at Table Bay again. Consequently, the Cape of Good Hope became a haunting memory in the hearts of Portuguese sailors.

THE LEGACY

So what do we make of Dom Francisco de Almeida?

He wielded the sword of war, conquering lands and seas with unyielding force, his name synonymous with victory yet tinged with tragedy as he mourned a son lost to the cruel hand of fate. More than a mere commander, Almeida embodied a transformative vision: he recognized naval supremacy as the foundation of imperial power, anticipating a pivotal shift in global strategy. By formulating and implementing the Blue Water Policy, prioritizing maritime control over territorial expansion, Almeida fundamentally redefined Portugal's imperial ambitions. This approach established a new paradigm in which mastery of the seas determined not only the reach but also the sustainability of empire.

In the tumultuous tapestry of global conquest, his strategic insight echoed through the ages, proving that control of the oceans was the key to unlocking wealth and territory in Asia's distant lands. The Portuguese, once a rising power, saw their hold on India slip away as the ebb and flow of naval supremacy shifted towards the Dutch and English. By the dawn of the 17th century, Portugal's eastern empire crumbled, leaving only remnants of its former glory.

Yet, in 1509, Almeida stood at his zenith, having routed the most formidable Islamic fleet ever assembled and laid a precedent for European imperialism that would echo through centuries - the subjugation of nations through maritime prowess. Diu, where he achieved this monumental victory, sowed the seeds of empires; Britain, France, and the Netherlands each claimed dominion over far-flung lands.

His legacy was not defined by heroism or infamy, but by the consequences of his actions. In death as in life, Almeida's path was paved with conflict, cut short by violence at the hands of the Khoikhoi who fought for their land and dignity, an act devoid of grand design yet resonant with moral justification. The clash between Portuguese steel and Khoikhoi stone tools underscores an irony: the might of advanced technology faltered against the timeless resilience of a people defending their own ground. It leaves us to ponder how strength is measured and whether it resides in armaments or in the unyielding spirit of those who protect their heritage.

History remembers him for Diu, but perhaps it should also honor his struggle against the encroaching European powers in Table Bay, a reminder that empires built on force cannot escape their cycle. The conqueror abroad can find himself conquered at home, and those who spread violence will one day face its reflection.

Underneath an African mountain, Francisco de Almeida's mortal remains were laid to rest, his legacy now dust scattered by the winds of time. Yet the world he helped shape persists in trade patterns and naval tactics, as well as in a conception that underpins much of Western expansion. However, it is important to recognize the limits of Portuguese dominance: their control over these distant territories and seas proved tenuous, repeatedly challenged by Indigenous resistance, rival European powers, and local polities whose agency constrained the reach of empire. The notion of an unquestioned right to dominate lands beyond their shores was frequently contested in practice, revealing the fragility as well as the ambition of early modern imperial ventures.

Though not born from him alone, Almeida's influence accelerated these tenets into reality. We continue to grapple with his legacy, bearing witness to an unending cycle where violence consumes all in its path, and empires rise and fall like the rooster and the chick - a grim feast forever repeating itself.

Afonso de Albuquerque

December 16, 1510 – Goa

Flames ravaged the city as Afonso de Albuquerque stood upon the ramparts of the seized stronghold, observing the ascent of smoke into the monsoon heavens. His armor bore the marks of time at its joints, eroded by salt, which had eaten away the metallic clasps. Despite being fifty-seven years old, he presented a more weathered visage. Concealed within his robust exterior, dysentery gnawed at him, though he would keep this secret for another five years.

In the square below, Portuguese soldiers hauled in the remnants of Muslim defenders. The orders were clear: Each adult male capable of combat was to be killed. Their wives and children would be parceled out among the conquerors. This act was more than cruelty; Albuquerque's ruthless reputation preceded him. It was a calculated incitement to terror, a message written in blood and ash, meant to traverse every sultanate, port, and spice depot from Hormuz to Malacca. A local Goan chronicler later recounted, 'The air itself trembled with the cries of the lost, as if the land itself wept for its children.' This added a new layer of dread and underscored the deliberate nature of Albuquerque's brutal campaign.

The Indian Ocean teetered on the precipice of new rulership.

None in Lisbon comprehended his vision. King Manuel I sought little beyond gold and pepper. The nobility at court derided him as a zealot, consumed by enmity towards Muslims and fixation upon fortresses, while the true riches of the Indies lay in swift raids and cargo holds brimming with plunder. Powers past, like the Venetians, had flourished in analogous trade spheres by monopolizing commerce through alliances and city-states. They focused on economic supremacy rather than territorial claims, relying on mutual understandings and shared interests to sustain their sway. The Mamluks, too, upheld a mercantile dominion without territorial expansion. Yet Albuquerque beheld a different paradigm, an empire unswayed by fleeting triumphs. He foresaw an enduring realm not ships fleeting like night-time thieves, but perpetual conquest. Not the mere manipulation of trade lanes, but sovereignty over vast lands. Not transient merchants, but an imperium.

Smoke still ascended, while beneath it, around six thousand souls lay lifeless. The accounts would later swell to eight, or even ten thousand. Yet the precise tally mattered less than the proclamation: Portugal had arrived with the resolve to persist.

THE MAKING OF AN INSTRUMENT

To truly understand the individual who sought to reshape the Indian Ocean, it is vital to envision Portugal as it existed in the fifteenth century: an underdog nation clinging to the Atlantic's edge with a tenacity akin to a barnacle on a weathered ship.

Afonso de Albuquerque was born sometime around 1453 in Alhandra, just outside Lisbon. The exact year remains unknown due to the rudimentary record-keeping common in provincial Portugal at the time. What is certain is that he entered a world fraught with violence and uncertainty. The Hundred Years' War had recently concluded, Constantinople had fallen to the Ottoman Turks, and the Mediterranean had become increasingly hostile towards Christian merchants.

Portugal itself was impoverished, lacking the grandeur of Florence, the wealth of Venice, the financial dynasties, and the prestigious universities. Yet it harbored a desperate spirit and an expansive coastline that seemed to beckon challenge.

Afonso hailed from the lower nobility, the fidalgos, military landowners who amassed their fortunes by serving the crown rather than engaging in trade. His father, Gonçalo de Albuquerque, was the lord of Vila Verde, while his mother, Leonor de Menezes, had ties to more influential families. These families valued honor, loyalty, and martial prowess above material wealth, considering pride to be their most treasured inheritance.

As the second son, Afonso stood to inherit nothing; his elder brother would receive the family estate. His prospects lay in serving either the church or the king. By the age of fifteen or sixteen, he became a moço de câmara within King Afonso V's royal household, akin to a page. This role entailed dressing nobles, delivering messages, and participating in ceremonies. Although unassuming, it provided him with valuable lessons on power dynamics, the exchange of favors, justifications for war, and how those seeking influence navigated the perils of royal favor.

Access to knowledge was also afforded to him through his position. By the 1470s, Lisbon had emerged as a hub of geographical learning, driven by the legacy of Prince Henry the Navigator, who passed away in 1460 but left an indelible mark on African exploration. Portuguese ships were expanding their reach along the West African coast each year, mapping new territories, establishing trade posts, and bringing back gold dust, enslaved individuals, and knowledge about winds, currents, and stars. Afonso learned navigation, seafaring techniques, and mapmaking, but more importantly, he began to contemplate systems such as trade routes, monsoon patterns, and strategic points of control. Though the Indian Ocean remained a mystery, tools for exploring it were being forged in Lisbon's shipyards and planning rooms.

In 1476, at around twenty-three years old, Albuquerque received his inaugural command. King Afonso V was mobilizing against Morocco, an endeavor that beckoned young nobility to gain their battle scars against the Moors. For Portugal, this was a crusade simmering for generations along North Africa's coast; for the nobility, it served as a proving ground where reputations could be forged and knighthoods won.

During these campaigns, Albuquerque distinguished himself by traits that would define his later career: aggressive strategies, personal bravery, and unwavering dedication to his objectives. Though specific details are scarce, he was wounded at least once, gaining experience in leading under fire, maintaining discipline during sieges, and mastering the management of fear both within himself and amongst his troops.

Yet, Morocco also imparted a harsher lesson, demonstrating that terror could be employed as a weapon, with planned brutality sometimes achieving what courage alone could not. The Portuguese campaigns in North Africa were marred by violence against Muslim populations, including massacres, enslavement, and forced conversions. These actions were justified by commanders citing centuries of religious conflict. The Reconquista on the Iberian Peninsula had concluded only in 1492, rendering memories of Islamic rule quite recent. For figures like Albuquerque, combating Islam was not merely a matter of politics or economy; it was a struggle for survival.

In 1497, the world shifted.

Vasco da Gama navigated the Cape of Good Hope and sailed along the East African coast, accompanied by an Arab pilot. They ventured across the Indian Ocean to reach Calicut on the Malabar Coast in 1499. Upon returning to Lisbon, his fleet carried a variety of valuable goods, including spices, precious stones, and evidence that the Indies could be reached by sea. The implications were profound. For centuries, Europe relied on Arab, Persian, and Venetian intermediaries to access Eastern goods, each taking their share. Now, Portugal had the opportunity to bypass them.

King Manuel I took the throne in 1495. He immediately declared himself "Lord of the Conquest, Navigation and Commerce of Ethiopia, Arabia, Persia and India," although Portugal controlled nothing beyond its shores at the time. This statement displayed his ambitious nature, seeking not only trade but also dominance over these regions.

During this period, Albuquerque, now in his forties, observed these events from the sidelines. He had served in various positions, including master of the king's stables, and was competent and respected but lacked distinction. His moment had yet to arrive.

In 1500, he was appointed to command a ship in a fleet destined for India. Although not a prestigious assignment as he was one of several captains rather than a fleet commander, it marked a turning point. He had the support of his cousin Francisco de Albuquerque, who held the position of fleet commander and likely assisted in securing his role through nepotism.

The voyage followed da Gama's route: down Africa's coast to around the Cape of Good Hope, where the Atlantic meets the Indian Ocean amidst rough currents and storms, then northeast towards India across the open sea. The journey spanned months, during which ships leaked and men succumbed to scurvy, blackened gums, and lost teeth due to poor nutrition. However, this was considered normal at the time; roughly 30% of sailors in India did not survive.

Albuquerque's first encounter with the Indian Ocean must have been overwhelming as he observed a vast web of commerce that made Europe seem relatively small in comparison. In ports like Calicut, Cochin, and Cannanore, Chinese junks, Arab dhows, Indonesian traders, Persian, and African ships crowded the harbors. The abundance of goods available included pepper, cinnamon, cloves, nutmeg, ginger, silk, porcelain, diamonds, and rubies, among others. However, beneath this bustling scene was a complex interplay of languages, beliefs, and ever-changing alliances.

The Portuguese did not blend naturally into this established system, offering goods that the East had little desire for, such as wool, wine, copper, and silver. They assumed trading privileges would be granted automatically rather than negotiated, an expectation that perplexed and sometimes offended local rulers, many of whom operated according to longstanding protocols of careful alliance-building and reciprocal gift exchange. Indigenous authorities and merchants viewed the construction of fortified trading posts without permission and attempts to establish monopolies as direct challenges to their political autonomy and established mercantile norms. These practices disrupted the region's customary frameworks of regulated, multilateral trade, eliciting suspicion and resistance from both rulers and commercial communities. Furthermore, the Portuguese insistence on advancing their strong religious beliefs, often prioritizing conversion or Christian identity in dealings, clashed with the pragmatic, pluralistic attitudes that characterized most Indian Ocean commerce. As a result, many local actors interpreted Portuguese actions not just as breaches of etiquette but as aggressive encroachments on sovereignty and economic independence, thereby fueling a series of conflicts.

In 1503, Albuquerque took part in his first naval battle in Indian waters when the Portuguese fleet clashed with Muslim trading vessels off the Malabar Coast. The Portuguese possessed superior cannon due to advanced European metallurgy and gunnery honed during the endless Mediterranean wars. These weapons were used brutally against Indian Ocean ships that were built for trade rather than warfare, lacking heavy armaments or defenses. However, local shipbuilders soon recognized the imbalance in power and adapted their vessels' designs to better withstand Portuguese attacks. With improved maneuverability and thicker hulls, these evolving designs gradually reduced the technological advantage the Portuguese held, challenging the notion that European naval superiority could be sustained without adaptation from local powers.

Albuquerque swiftly grasped this strategy: superior firepower could compensate for being outnumbered. Just a few Portuguese ships could dominate entire coastlines provided they had secure bases, steady supplies, and were willing to use force when necessary. This idea would guide his actions in the future.

He returned to Lisbon in 1504 with many plans. Over the next two years, he advocated for a more aggressive strategy in the East at court. He argued that Portugal should not only seek trade opportunities but also conquer strategic ports, dominate sea lanes, and dismantle the power of Muslim merchants and rulers. Only through conquest could the spice trade be made truly secure.

While these ideas garnered support from some at court, others preferred a less invasive approach: conducting quick raids, forming trading alliances with friendly Hindu rulers, and maintaining minimal infrastructure. Conquest entailed garrisons, supply lines, administration, expenses, and increased entanglement in foreign affairs. Nevertheless, Albuquerque proved persuasive and held the king's ear. In 1506, Manuel appointed him to co-command a significant fleet bound for India, with instructions to expand Portuguese territories.

THE ARCHITECT OF FEAR

The armada setting sail from Lisbon in April 1506 comprised sixteen vessels and roughly fifteen hundred souls. At the helm of five ships was Albuquerque, while Tristão da Cunha, a nobleman well-acquainted with the Atlantic but unversed in the Indian Ocean's ways, held the reins overall. This leadership division would inevitably lead to complications.

The voyage was fraught with perils. Storms ravaged the fleet near the Cape of Good Hope, damaging several ships and casting some men overboard. Upon reassembling at Mozambique on the eastern coast of Africa, disease began to spread ominously among them. The primary peril in such expeditions was not warfare but illness. Malaria, dysentery, scurvy, and untreated wound infections were commonplace afflictions. The surgeons aboard were often mere barbers with a saw as their main tool; amputations proceeded without anesthesia, while fellow sailors restrained the patient for swifter execution. Many succumbed to shock or later infections.

By late 1506, when they reached the Indian subcontinent's shores, Albuquerque had come to grips with Portugal's tenuous foothold in the East. They lacked a land base, reliable allies, and control over their supply lines. Muslim traders could easily paralyze Portuguese commerce by refusing to load spices onto their ships, whereas Hindu rulers could revoke their privileges at whim. It was Albuquerque's conviction that seizing key maritime chokepoints, including narrow straits and crucial harbors, through which all trade flowed, was the solution. These chokepoints were not solely strategic military objectives; they also held significant economic importance. For example, regulating passage through the Strait of Malacca alone could yield massive revenue from spice duties, estimated at around 150 kilograms of gold annually. This transformation of theoretical strategies into tangible profits justified the venture's risks and highlighted the stakes involved in controlling these critical trade points.

His initial target was Hormuz, a small island strategically positioned near the entrance to the Persian Gulf. Its value was immense because all commerce between Persia, Mesopotamia, and the Indian Ocean transited through it; thus, whoever controlled the island commanded access to some of Asia's wealthiest markets. Furthermore, Hormuz was governed by a weak sultan, reliant on the loyalty of his fickle nobility and susceptible to external influences.

In September 1507, Albuquerque arrived at Hormuz with a modest squadron, estimated as comprising seven ships and five hundred men. He dispatched a herald demanding tribute and trading privileges; the sultan, as anticipated, refused. In swift retribution, Albuquerque ordered Portuguese forces ashore while their vessels bombarded the city. The invaders surged past defenders onto the beach, seized control, and unleashed an organized pillage upon the town. Homes were ransacked, storehouses looted, and resistance met with extreme ruthlessness. Albuquerque himself supervised the construction of a fortress in the heart of the city from the debris of demolished mosques, a symbolic gesture obliterating Islamic rule to erect Christian edifices and Portuguese artillery.

Yet Hormuz resisted tenaciously; the sultan escaped to the mainland, marshaling counterattacks as local nobles rallied forces, fearing the demise of their riches. Surrounded and besieged within his fortress, grossly outnumbered by the defenders, Albuquerque's garrison awaited rescue ships from India that never arrived. Morale frayed amid such adversity; some captains openly questioned his strategic judgment. Why had they assailed a formidable city with so scant forces? Why incense the sultan irrevocably instead of negotiating?

Albuquerque was direct in response: they would endure within the fortress until reinforcements arrived or meet death valiantly. Retreat signified weakness, encouraging adversaries to threaten other Portuguese strongholds. Every fort seized, regardless of cost, solidified Portugal's permanent and indomitable presence.

The siege endured through winter's chill, claiming lives from both disease and combat wounds; provisions dwindled dangerously. Albuquerque too fell ill with a fever, likely malaria, remaining in delirium for weeks while lieutenants upheld the defenses. Upon convalescence, he confronted hopelessness reinforcements abjured arrival, garrison troops depleted utterly. Retreat was inevitable.

In February 1508, the Portuguese evacuated Hormuz in a retreat laden with humiliation; Albuquerque's adversaries exploited it as clear evidence of his rashness and profligacy with lives and resources on a doomed endeavor, unwisely attempting to hold Hormuz in perpetuity. But Albuquerque drew contrasting lessons: he demonstrated that a compact Portuguese contingent could temporarily seize and maintain an Asian metropolis. The fortifications erected stood resilient; the sultan was humbled and capitulated to negotiations; merchants of Hormuz beheld Portuguese cannons firsthand. Next time and plans for next encounters were already in the works. Albuquerque would bolster his forces numerically, maritally, and in terms of armament. He refused repetition of past errors.

APPOINTMENT AND BETRAYAL

In November 1508, King Manuel made the decision that would drastically alter the course of Portuguese expansion. He appointed Afonso de Albuquerque as governor of India, effective upon the death or replacement of the current governor, Francisco de Almeida.

Almeida was a formidable figure in his own right, an experienced military commander who had served in North Africa and been in India since 1505. He had adopted an aggressive naval strategy, bolstering Portuguese sea power and launching devastating attacks on Muslim shipping. In February 1509, he achieved a resounding victory at the Battle of Diu, where a Portuguese fleet vanquished an Egyptian-Gujarati armada. This victory guaranteed Portuguese dominance over the Indian Ocean and made Almeida a celebrated figure in Lisbon.

This success also made him disinclined to relinquish his position.

When Albuquerque arrived in India late in 1509 with royal orders for Almeida's replacement, the incumbent refused compliance. He contended that his term had not yet concluded. That Albuquerque's credentials were improper. That the king's directives must have been misinterpreted. Underlying these legalistic arguments was a simpler truth: Almeida was unwilling to cede power to a competitor.

The dispute between the two leaders paralyzed Portuguese operations in India for several months. Albuquerque, with a fleet and royal authority, opposed Almeida, who controlled the fortresses and enjoyed support from many Portuguese settlers. Neither man would compromise, leading subordinates to choose sides, factions to form, and at one point, even resulted in Almeida having Albuquerque arrested and imprisoned within a fortress at Cochin.

The absurdity of this situation was apparent to all. Two capable and loyal Portuguese commanders were engaged in an unyielding power struggle while enemies observed their weakness with satisfaction and made their own calculations.

The perception of Portuguese invincibility began to diminish.

This stalemate eventually ended in November 1509 when new orders from Lisbon explicitly confirmed Albuquerque's appointment. With his position now indefensible, Almeida stepped down from authority. However, the damage was already done. Valuable time had been lost, resources squandered, and Albuquerque had learned a hard lesson about Portuguese politics: official appointments hold little weight without force to back them up. In the East, far from royal oversight, power belonged to those who could seize and maintain it.

Almeida departed India shortly afterward and never reached Lisbon. He met his end in March 1510 while stopping for water at Table Bay near the Cape of Good Hope; sixty-four men accompanying him were killed in a skirmish with indigenous Khoikhoi people. It was an inconsequential, almost comical end for someone who had once commanded fleets and conquered cities. The news of Almeida's death eventually reached Albuquerque, whose response to his rival's demise remains unrecorded.

GOA ,THE PERMANENT CONQUEST

With undisputed authority finally in hand, Albuquerque swiftly acted upon his strategic vision. His need was for a territorial base, an actual colony, not just fortified factories. A place where the Portuguese could establish residence, cultivate crops, and build and repair ships. He found the ideal location: Goa.

Goa, a modest yet flourishing port on the Malabar Coast under the jurisdiction of Adil Shah of Bijapur, sat on an island, ensuring its defensibility. It boasted a fine harbor that could accommodate large fleets, was centrally located along India's western coast for operational convenience both north and south, and had a Hindu majority restless under Muslim governance, suggesting potential alliances.

In March 1510, Albuquerque gathered twenty-three ships, fifteen hundred soldiers, and diverse Hindu allies to sail for Goa. Surprised, the Adil Shah's garrison offered little resistance; within days, the city was in Portuguese hands. Perhaps too effortless. Yet, Goan elites saw it differently. A contemporary Konkani account titled 'Beyond the Horizon' recounted a local noble expressing disbelief: "The winds of fate have swept away our world, leaving us adrift."

Albuquerque swiftly fortified Goa by reinforcing its walls, installing cannons, and establishing supply depots. He issued decrees to cement the Portuguese presence, encouraging soldiers to marry local women (though conversion to Christianity was expected), offering tax incentives for settlers, and constructing churches and government buildings. It was a colony in earnest.

Adil Shah, however, wasn't finished. In May 1510, only two months post-Portuguese seizure, an enormous Muslim army approached Goa. The exact figures are debated, but suggest upwards of sixty thousand soldiers arriving against Albuquerque's approximately fifteen hundred combat-ready men.

The ensuing siege was dire; Muslims blocked all land access to the island, while Portuguese control over the harbor couldn't break the encirclement or evacuate the garrison. Conditions inside deteriorated rapidly: food shortages, tainted water sources, and disease spread, claiming more lives from illness than combat.

Albuquerque's captains advocated retreat, but Albuquerque refused. Retreat at Hormuz was necessary; retreating from Goa risked proving Portuguese vulnerabilities, alarming India's rulers, and jeopardizing Portugal's entire position in the region.

Instead of retreating, Albuquerque initiated a bold counterattack in late May, choosing to break the siege through offense rather than defense, a calculated gamble exploiting weaknesses in the Adil Shah's massive but unwieldy forces. Under the cloak of darkness, Portuguese forces exited Goa and struck with devastating effect, shattering the siege by dawn.

But this victory was temporary; Muslims regrouped and returned in July with an even larger force, only to be met with resilience again from Albuquerque. This time, however, his strategy included local Hindu allies attacking from behind, causing chaos among Muslim forces and forcing their withdrawal, securing Goa.

The aftermath was brutal: execution of all Muslim men capable of fighting, estimated between four thousand and ten thousand. Among them was a long-resident trading family, with distraught family members helplessly watching. Women and children were enslaved or given to soldiers, and mosques were converted into churches. Non-Portuguese sources bear witness to the traumatic effects: an early sixteenth-century Persian chronicler, Zayn al-Din al-Malibari, recounts that the "entire city was given over to massacre and devastation, its people scattering in terror as the conquerors spared not even the innocent." This trauma, documented as well in oral lamentations from generations later, left Goans of all faiths in shock, witnessing their neighborhoods upended and collective security shattered. Even Hindus who resented Muslim rule expressed horror at the destruction visited upon their neighbors, fearing for their own futures under a regime treating everyone as subjects to reshape.

This dramatic shift made Goa Christian, Portuguese, and permanent, marking not just a change of rulers but an invasion into the city's soul, leaving indelible scars on its residents and shaping collective memory for centuries. The massacres shocked even some of Albuquerque's men, yet he prioritized strategic terror to deter future resistance.

Albuquerque's actions in Goa demonstrate both his strategic acumen and his willingness to use extreme measures to ensure Portuguese dominance.

THE EMPIRE OF FORTS

With Goa secured, Albuquerque focused on constructing a "strategic system," an extensive network of fortified bases stretching across the Indian Ocean to dominate key maritime chokepoints and regulate trade between Europe, the Middle East, India, Southeast Asia, and China.

His approach was highly intricate. He pinpointed three crucial points: Hormuz, controlling access to the Persian Gulf; Goa, regulating the western coast of India; and Malacca, overseeing the straits connecting the Indian Ocean and South China Sea. With fortresses at these key locations, Portugal could manage trade flows across continents while imposing customs duties, requiring merchants to obtain Portuguese licenses, or risking the seizure of their ships.

In 1515, Albuquerque returned to Hormuz with an overwhelming force of twenty-six ships, fifteen hundred soldiers, and valuable insights from past setbacks. Realizing resistance was futile, the sultan yielded without a fight. A treaty was imposed, making Hormuz a de facto Portuguese protectorate, maintaining the sultan for show but granting real authority to the Portuguese fortress captain. Customs revenues were divided, primarily benefiting Portugal, which was now collecting tribute from historically dominant Muslim merchants.

Later that year, Albuquerque set his sights on Malacca, one of Asia's wealthiest port cities, where merchants from China, India, Arabia, and Southeast Asia converged for trade. The Sultan controlled access to the Spice Islands, where cloves and nutmeg were in high demand among Europeans due to their astronomical prices.

In June 1511, Albuquerque approached Malacca with his fleet. Demands for trading rights and the release of Portuguese prisoners were met with refusal from Sultan Mahmud Shah, who possessed a formidable fleet and a well-fortified city, seeing no need to accommodate intruders.

Albuquerque launched an aggressive assault on Malacca, which became one of the most intense battles in early colonial history. Street by street, Portuguese forces battled through, facing fierce resistance from Malay defenders and mercenaries from Java and Sumatra. Combat was brutal with swords, pikes, arquebuses, and fires engulfing buildings, piling bodies in narrow alleys.

Albuquerque himself led assaults, his armor scarred by swords, face darkened by gunpowder, a sixty-year-old suffering from tropical ailments but fighting alongside the youngest soldiers out of deep conviction. He saw his mission as God's work to combat Islam and spread Christianity, valuing little the lure of gold, spices, or glory compared to his spiritual objectives.

After weeks of relentless warfare, Malacca fell, with Sultan Mahmud Shah retreating into the interior. Albuquerque swiftly transformed the city into a Portuguese stronghold, constructing the A Famosa fortress, impervious to cannon fire, expelling or harshly taxing Muslim merchants, fostering trade ties with China, and sending expeditions to the Spice Islands.

Malacca's capture marked a turning point, demonstrating Portugal's power to extend beyond India and Arabia into Southeast Asia. It granted access to markets and products previously controlled by intermediaries while completing Albuquerque's strategic triangle: Hormuz, Goa, Malacca. These three strongholds secured Portuguese dominance in the Indian Ocean for the next century.

THE COST OF EMPIRE

But the empire's toll measured in lives, silver, and the weight of one's conscience continued to grow.

The Portuguese forces in Albuquerque were often understrength due to limited manpower. Each soldier stationed overseas was a precious resource since Portugal itself had only about a million inhabitants who couldn't maintain substantial garrisons abroad. Recruitment was challenging, made worse by high mortality rates in tropical climates that families feared for their sons' survival in Goa or Malacca.

Given the circumstances, Albuquerque found himself heavily reliant on local alliances: Indian Hindus, Southeast Asian Malay Christians, and enslaved Africans from Portugal's Atlantic colonies formed his troop base. These soldiers lacked the reliability of Portuguese forces and required distinct management strategies, including varied payment structures and promises of future rewards. The army remained on the brink of collapse unless held together by Albuquerque's tenacity and the promise of plunder.

Fear was employed to enforce discipline. Mutineers were executed by hanging or burning, and deserters were tracked down and publicly executed. Albuquerque felt that extreme measures were necessary given the operation's fragility and the distance involved. Unlike larger empires such as the Ottomans or Mughals, whose vast manpower allowed for a different approach to military justice, Portugal's limited personnel meant that high levels of brutality were a strategic necessity. Any display of weakness or defiance could unravel everything. He made sure his men feared him, which was part of his strategy.

Financially, Portugal's eastern empire in the Estado da Índia was expected to be self-supportive through customs revenues, trade profits, and tributes from subordinate rulers. However, this ideal never materialized due to rampant corruption among local Portuguese officials who enriched themselves through smuggling and embezzlement. Supply contracts were awarded at inflated prices to favored associates, and underpaid garrison soldiers engaged in contraband trading to supplement their incomes.

In an effort to impose order, Albuquerque introduced administrative reforms, mandated regular audits, and imposed severe penalties for corruption. But he was essentially fighting the system itself, given the vast distances, scarce oversight, and abundant opportunities for abuse. Ships leaving Goa or Malacca carried contraband alongside official cargo, customs houses leaked revenues, and garrison commanders took their own cuts.

Furthermore, the physical environment posed a merciless challenge to Europeans. Tropical diseases like malaria, dysentery, cholera, typhoid, and dengue fever proved deadly in the region's hot and humid climate. What might have been minor illnesses in Europe were lethal in these conditions, and ship surgeons lacked understanding of tropical medicine, offering ineffective treatments that could even be harmful.

Albuquerque himself was gradually succumbing to his ailments. He suffered from recurring fevers, chronic dysentery, and likely advanced syphilis, a disease he may have contracted earlier in Morocco or Portugal. By 1512, he appeared gaunt with yellowed skin and dwindling energy; still, he refused rest. Rest, in his view, equated to surrender.

THE POLITICS OF DISTANCE

In Lisbon, King Manuel's patience waned as he awaited results from his appointment of Albuquerque to secure the spice trade for Portugal. He didn't order territorial conquests or costly fortifications; he only sought trade advantages. The king's letters expressed frustration over the costs of fortifications, provocations, and missed profits.

Albuquerque was unyielding in defense, insisting that only through establishing bases with permanent garrisons could lasting power be achieved in the East. While acknowledging upfront costs, he believed long-term gains would be substantial.

However, geographical distance hindered communication, with letters taking months to travel between Goa and Lisbon. By the time Albuquerque justified his actions, the king's concerns had shifted, and by the time royal directives reached India, conditions on the ground had changed, leading to a persistent disconnect between policy and reality.

Albuquerque faced opposition from envious rivals, nobles who supported other candidates for governorship, and merchants annoyed by monopolistic practices. They portrayed him as reckless, autocratic, and insubordinate, magnifying his setbacks while downplaying successes, hinting at the need for new leadership in India. The communication delays between Lisbon and Goa exacerbated these tensions. A letter sent from India in January would only reach the Portuguese court by October, and by then, circumstances on the ground could have changed drastically, leaving decisions founded on outdated information.

Despite knowing of these plots against him, Albuquerque was limited in his response due to the unavailability of travel to Lisbon; abandoning his post could risk losing control of the Estado da Índia. He relied on letters and allies' interventions, hoping that achievements would speak for themselves.

In 1512, he launched a final major assault, fixated on Aden for as long as it controlled the entrance to the Red Sea. Its significance extended beyond being one of the key ports in the Islamic world to serving as a gateway to Egypt and the Mediterranean. Capturing Aden could eliminate traditional spice routes, compelling all trade through Portuguese-controlled channels.

However, March 1513's attack on Aden with twenty ships and seventeen hundred men, accompanied by heavy artillery, ended disastrously. The city walls were more robust than expected, the garrison was well-prepared, and the barren landscape offered no advantage for attackers or resources for a prolonged siege.

Despite futile bombardment causing casualties and dwindling supplies, withdrawal became necessary.

His most embarrassing defeat occurred at the worst time, fueling speculation by his detractors in Lisbon that Albuquerque had lost effectiveness due to age and infirmity, further intensifying calls for his replacement.

THE FINAL VOYAGE

In 1515, King Manuel made his decision. He appointed Lopo Soares de Albergaria as the new governor of India. The official letter informing Albuquerque of his dismissal was carefully worded, expressing gratitude for his service, citing health concerns, and offering a position at court. But the subtext was clear: his time was over. Albuquerque received this news while at sea, en route from an expedition to the Red Sea back to Goa. He lay weak on a makeshift bed due to dysentery, unable to stand. When his secretary read the letter aloud, Albuquerque remained silent; then, according to chroniclers, he wept. It wasn't the dismissal itself that devastated him; he had always known royal favor to be fleeting. It was the ingratitude. He devoted everything, his health, wealth, and moral convictions, to building Portugal's eastern empire. He secured the spice trade, conquered crucial fortresses, and established Portugal's dominance in Asia. Yet, his reward was humiliation and replacement. He penned a final letter to King Manuel, bitter, rambling, and defensive. It detailed his achievements, justified his expenses, and blamed court enemies for poisoning the king's mind against him. Beneath the bitterness lay hurt; loyalty had not been reciprocated. The letter remained unfinished. On December 16, 1515, Afonso de Albuquerque passed away off the coast of Goa. Age was estimated around sixty-two, but his body bore signs of disease, exhaustion, and years of tropical warfare. Years of dysentery finally defeated his weakened system; he died in his cabin aboard his ship, overlooking the fortress city he constructed, surrounded by a few loyal officers. Chronicler Gaspar Correia described his death with rare gentleness: "The Governor died with much patience, as a good Christian, forgiving all who had wronged him." Yet, even this final act held complexity. Some accounts suggested that in his last moments, Albuquerque instructed on defending fortresses and warned against certain rivals. He departed life as he had lived, preoccupied by strategy, power, and the endurance of his legacy. His body was transported to Goa for burial within the Church of Our Lady of the Serra, a construction built upon a mosque's ruins that he himself founded. The funeral assembly comprised Portuguese officials, soldiers, Hindu allies, and Muslim merchants whose livelihoods had been upended by Albuquerque's actions, a strange blend of Catholic rites in tropical climes, mourning an architect of empire. It took months for news of his death to reach Lisbon. Upon its arrival, King Manuel reportedly expressed genuine sorrow. Despite any frustrations with Albuquerque, the king recognized the magnitude of his achievements. The governor secured an empire for Portugal; the once Muslim and Venetian-controlled spice trade now flowed through Portuguese hands. Fortresses at Hormuz, Goa, and Malacca stood as testaments to their power, laying the foundations of a colonial system that would last for centuries. Yet, the true cost in human suffering defied quantification.

THE ARCHITECTURE OF VIOLENCE

To grasp what Albuquerque built, it's crucial to confront what he demolished.

The Indian Ocean before the Portuguese arrival was not a tranquil place; no commercial network involving such magnitude and rivalry could be without tension, but it operated within certain boundaries. Commerce was cosmopolitan, handled by Muslim, Hindu, Buddhist, Jewish, and Christian traders who, despite their religious differences, shared a common trading language and methods. Metropolises like Calicut, Malacca, and Hormuz were hubs of trade where diverse cultures coexisted, not always in unity, but with practical tolerance.

The Portuguese disrupted this setup. They brought forth a level of systematic aggression unparalleled in the area's commercial history. Earlier powers, such as the Delhi Sultanate, the Vijayanagara Empire, and various Southeast Asian realms, undoubtedly engaged in warfare and conquest. However, none sought to monopolize maritime commerce through permanent fortifications or naval dominion as Portugal did.

Albuquerque's bloodshed at Goa wasn't an isolated incident; it was part of a pattern. Upon conquering Malacca, he permitted the widespread slaughter of Muslim defenders and civilians. In subjugating Hormuz, he obliterated mosques and enslaved citizens. These actions weren't spontaneous outbursts; they were calculated strategies aimed at dismantling Muslim commercial webs to cow potential resistance.

The religious element was central. Albuquerque genuinely believed he was waging a holy war. His dispatches overflowed with references to divine intervention, Christian obligation, and the combat against Islam. He viewed commerce and crusade as synonymous. Every fortress erected signified victory for Christendom. Every Muslim merchant forced out of business struck against the infidel. This religious perspective enabled him to rationalize acts that might otherwise have tormented his conscience.

Yet, economic incentives were equally present. The spice trade was remarkably profitable. A cargo of pepper procured in India and sold in Lisbon could yield profits multiple times over. Cloves and nutmeg from the Moluccas were worth their weight in silver. Commanding this commerce signified dominance over one of the major wealth machines of early modernity.

Albuquerque's system of cartazes, or naval trading permits mandatory for all vessels within Portuguese jurisdiction, was a safeguard mechanism on an enormous scale. Merchants who procured licenses gained protection from Portugal. Those who refused faced the risk of losing their ships and even their lives. The funds generated were used to bolster the Estado da Índia's military forces, which then upheld the licensing scheme. It was straightforward, brutal, and highly successful. To clarify its impact for a broader audience, one might compare the cartaz system to a form of compulsory shipping insurance in the modern world. Under such an insurance model, merchants are required to pay for a policy to legally operate in certain waters; in exchange, they receive basic protection and assurance of recourse in case of attack or loss. Failure to obtain this coverage would result not only in denial of protection, but potentially in serious penalties. In both cases, the arrangement compels adherence to rules and generates revenue for the authority providing this 'security.' Despite its coercive nature, the cartaz enforced compliance among traders, similar to how insurance policies today enforce safety and operational standards in maritime commerce. However, while the Portuguese system raised funds and imposed greater order on turbulent seas, non-European responses were often deeply adverse. Many Muslim and Hindu merchants, who had long savored relative autonomy, saw the cartaz as a tool of alien rule and economic duress. Some attempted to circumvent the system by utilizing alternative ports outside Portuguese control or forging licenses, while others stood up openly, leading to violent clashes on the high seas. For these communities, the cartaz symbolized not just a curtailment of commercial independence but also an encroachment of European authority upon established trade conventions, eliciting both conformity and active defiance.

Nonetheless, the system teetered on the brink of fragility. It relied on sustained naval surveillance, precise intelligence, and readiness to unleash overwhelming force against offenders. Any hint of weakness spurred opposition. This is why Albuquerque reacted with such ferocity to resistance. He understood that his entire structure rested on the perception of Portuguese infallibility. If this perception fractured, the empire would crumble.

LIVES IN THE BALANCE

The tales woven into Albuquerque's conquests are harder to reconstruct. The Portuguese writers who recorded these events were primarily captivated by the feats of their compatriots. They chronicled skirmishes, fortifications, and negotiations. Rarely did they mention the "enemy," often portrayed as indistinct crowds overcome by Portuguese heroism. The narratives seldom included individual stories, especially those of non-Portuguese individuals.

Yet remnants persist. In Goa, Hindu merchants and craftsmen who had endured Muslim rule initially welcomed the Portuguese as deliverers. Some embraced Christianity, intermingled with Portuguese troops, and thrived under their rule. Others discovered that Christian overlords were not so different from Muslim rulers, sometimes worse due to strict religious adherence and heavy taxation. The rich tapestry of pre-conquest Goa was oversimplified into a dichotomy: Christian subjects versus non-Christian outsiders. Muslim communities within Portuguese domains faced dire decisions to convert, flee, or accept diminished status. Many chose flight, settling in cities such as Diu, Surat, and Calicut beyond Portuguese control. In these new bases, displaced Muslims rekindled their mercantile networks, contributing to the emergence of competitive trading centers that challenged Portuguese outposts. Consequently, Portugal's aim to eliminate Muslim commerce backfired: rather than vanishing, these communities adjusted and sustained their influence in regional trade, countering the monopoly Portugal sought. Women in these overrun cities were particularly susceptible. Albuquerque's advocacy for intermarriage stemmed from demographic needs, there being too few Portuguese women to establish lasting colonial populations, but it was enforced through pressure. Women seized during assaults or left unprotected by men were "encouraged" to wed Portuguese soldiers, with conversion to Christianity obligatory. These unions gave rise to the mestiço class, which would occupy a distinct place in Portuguese Asian colonies, neither fully Portuguese nor fully Asian, inhabiting an uncertain social realm. Among Albuquerque's own ranks, reactions varied. Some seized the opportunities for riches and prestige that conquest offered.

A common soldier from Lisbon might ascend to become a landholder in Goa, possessing servants and a household. The allure of upward mobility was potent. Others were simply shattered by their experiences, the violence they witnessed, the diseases claiming comrades, the estrangement of living abroad; these took psychological impacts without contemporary language to articulate. João de Barros, a Portuguese historian of the subsequent generation, interviewed veterans of Albuquerque's campaigns. Some spoke with pride about their service. Others seemed tormented. A soldier, contemplating the bloodshed at Goa, reportedly uttered: "We did what was necessary, but I pray God forgives us." It reveals moral intricacy in accounts usually depicted triumphantly.

THE SYSTEM AFTER THE ARCHITECT

Albuquerque's death led to significant shifts within the Estado da Índia. Though he no longer held power, his successors continued fortification efforts and monopolistic trade practices; however, none possessed Albuquerque's strategic vision or ruthless ambition. The empire became increasingly bureaucratic, routine, and corrupt.

Lopo Soares de Albergaria, who succeeded Albuquerque, excelled at administrative duties but lacked the drive for expansion. He focused on consolidating existing territories rather than venturing into new conquests, a decision likely wise given Portugal's stretched resources. This signified a shift from growth to preservation, marking a change in priorities.

Albuquerque's fortresses demonstrated remarkable endurance. Goa served as Portugal's Asian capital for over four centuries before being annexed by India in 1961. Malacca was held until 1641, when it fell to the Dutch, and Hormuz lasted till 1622. Even after the Portuguese decline and the rise of the Dutch and English empires, Albuquerque's strategic acumen proved valid: control over maritime choke points equaled control over trade.

However, the moral and cultural impact was complex. Portuguese colonialism in Asia mirrored the patterns of future European empires: employing superior military technology against larger populations, imposing monopolistic trade, imposing European norms, and forming hybrid societies. Albuquerque did not invent these methods but systematized them, setting a precedent for subsequent colonizers.

Posthumously, his reputation in Portugal grew. Chroniclers like João de

Barros and Damião de Góis lauded him as one of Portugal's greatest heroes, likening him to Caesar and Alexander the Great. His supposed final words, "I have served my king," were adopted as a national symbol of loyalty and sacrifice, leading to statues and street names in his honor. He became a legendary figure, simplified into an idealized hero.

In territories he conquered, narratives diverged significantly. In India, Goa's capture marked the onset of centuries of colonial suppression, remembering massacres, forced conversions, and the destruction of cultural institutions. In Malaysia, Albuquerque remains a villainous figure, seen as an oppressor who dismantled Malacca's independent sultanate. Here, he is cast not as a hero but as an invader driven by greed and religious zeal.

Both perspectives hold truth: Albuquerque was a strategic genius and brutal conqueror, a man of sincere beliefs capable of calculated violence. His establishments endured while cultures were destroyed. He cannot be merely seen as heroic or villainous. An embodiment of early modern imperialism's logic, he surpassed contemporary norms in the scale and organization of his violence.

THE PERSONAL MAN

What kind of individual was Afonso de Albuquerque beyond his roles as conqueror and administrator? The sources available are frustratingly brief, offering only glimpses into his character.

We're provided with official correspondence letters to the king, orders to subordinates, and diplomatic exchanges with Asian leaders. These reveal a strategic thinker, skilled in organization, and possessing a prose style that was straightforward, occasionally elegant, and often defensive. He reacted defensively to criticism, felt compelled to explain his choices at length, and then swiftly justified them.

We know he had some education; it was uncommon for military men of his time. His references spanned classical authors, Christian theology, geography, and natural philosophy. After his death, his library was found to include works on navigation, military engineering, history, and religious texts.

He remained unmarried but had at least one illegitimate son named Brás de Albuquerque, who later became a chronicler and even wrote a biography of his father. This biography is both valuable and questionable. It alone provides personal details but leans heavily towards hagiography, aiming to enhance Afonso's reputation and perhaps benefit his descendants.

Brás describes him as tall and broad-shouldered, with an imposing figure, and notes that he suffered health issues from middle age but rarely spoke of them. His lifestyle was simple, with plain meals and limited wine consumption. He showed personal courage, often leading charges into danger, which unsettled his subordinates. He had a volatile temper but could be unexpectedly gentle, particularly around children and animals.

Verifying the accuracy of these details is impossible, yet it's evident that Albuquerque inspired fervent loyalty in some followers while stirring intense animosity in others. He did not inspire neutrality. His inflexible principles, intolerance for dissent, and unwavering conviction in his cause made him effective but also isolated him. Subordinates and enemies; he had many of both, few friends.

His relationship with religion was central yet complicated. Daily Mass, regular confession, and churches founded in every city: his language and letters are steeped in religious references and divine providence. Yet he showed an adaptable practicality rare among zealous believers. In Goa, he allied with Hindus against Muslims; in Malacca, he protected Chinese merchants for their economic value; he forced conversions but allowed non-Christian populations to exist when it was impractical to do otherwise.

This suggests that Albuquerque's religious belief, while sincere, was secondary to strategic goals. God was on Portugal's side, and the nation's interests coincided with divine will. Not out of deceit, but out of true belief, it enabled policies that some more rigidly religious might have rejected.

THE TECHNOLOGY OF CONQUEST

Albuquerque's successes were grounded in technological advantages that Portuguese chroniclers touted as evidence of European supremacy, but this tale is more complex than it initially seems. Portuguese ships, particularly the nau and carrack models, excelled in combat in the Indian Ocean. Their robust construction enabled them to deploy numerous cannons effectively and navigate challenging waters safely. The high sides offered a significant advantage against boarding since Portuguese crews were often outmatched numerically. European metallurgy produced more dependable and powerful cannon than most Asian counterparts.

However, technology alone was not the clincher for success. Portuguese vessels were slower and less agile compared to dhows, junks, and other opponents they faced. These ships proved vulnerable in calm seas or shallow waters. Extensive upkeep, supplies, and repair facilities were needed; thus, Albuquerque prioritized securing safe harbors as part of his strategy. Without a strong base, the fleet was susceptible to defeat.

Moreover, Asian forces quickly adjusted to these challenges. By the 1520s, Ottoman fleets in the Red Sea began deploying cannon-equipped ships reminiscent of European-style warfare. Local leaders employed Portuguese turncoats and European mercenaries to train their troops in Western tactics. This led to a steady narrowing of the technological gap throughout the sixteenth century.

What Albuquerque leveraged was not a permanent technological edge but rather a fleeting window of opportunity. For approximately two or three decades, Portuguese naval supremacy was sufficient to secure key positions, which were then maintained through fortifications and stationed troops, even as Asian nations began to close the technical gap.

The constructed fortresses represented a unique European skill set; Portuguese military engineers versed in Italian theories of fortification created structures able to withstand cannon fire. Thick walls with sloped bastions, overlapping fields of fire, and protected entrances composed intricate defense mechanisms requiring specialized knowledge for both construction and assault.

Yet these fortifications were economic liabilities, as they required garrisons, resources, upkeep, and periodic rebuilding. As the Portuguese empire expanded, maintenance costs outpaced the revenues generated from these fortresses. By the late sixteenth century, the Estado da Índia was perennially bankrupt, heavily reliant on subsidies from Lisbon, and threatened by Dutch and English adversaries.

Albuquerque recognized these financial burdens yet believed them essential for securing bases without which Portuguese influence in Asia would vanish. He was correct in his assessment, but he also established a system that could endure only for a specific period when Portugal maintained naval dominance with no significant European competitors in Asian waters. With a shift in conditions, as Portugal lost its maritime superiority and faced increased rivalry, Albuquerque's creation began to crumble.

THE QUESTIONS THAT REMAIN

What would have transpired if Albuquerque had lived an extended life? What difference might it have made had he not been relieved from duty? How differently would his health have influenced events? These are questions that elude us yet fascinate us. He harbored ambitions for further campaigns. Aden was a certainty, and perhaps Jeddah and the sacred cities of Islam were within reach as well. Some of his correspondence suggests visions of even attacking Mecca, hitting at the core of the Islamic world. Whether such intentions were genuine or merely grandiloquent gestures to influence royalty is unclear. Had he endeavored in these pursuits, it's almost certain he would have faced defeat. The Ottoman Empire was fortifying its dominion over the Red Sea and possessed resources far beyond Portugal's capacity to defend itself. An assault on Aden had already been thwarted back in 1513. Advancing on Jeddah or Mecca would have led to disaster. Yet, failure had never discouraged Albuquerque previously. He might have still attempted it, convinced that divine intervention would secure his victory. On the other hand, he could have oriented towards reinforcement, enhancing existing fortifications, broadening trade networks, and forging bonds with Eastern potentates. This course of action would have been more prudent but less in keeping with his nature. Albuquerque was a conqueror, not an administrator; his brilliance lay in seizing chances and enforcing his dominance through brute force. The meticulous toil of maintaining an empire did not align with his temperament. It is likely that his ousting and demise coincided with a pivotal juncture. The Estado da Índia was at the zenith of its territorial growth, but had not established stable governance frameworks. Successive governors lacked his sway and foresight. Consequently, the empire meandered, decayed in integrity, and lost its martial spirit. One may argue that Albuquerque's approach was bound for failure; no individual could sustain such a fever pitch indefinitely. Nonetheless, his singular momentum is sorely missed.

THE MIRROR OF

Afonso de Albuquerque forces us to confront uneasy questions about empire, violence, and the judgments we impose on history.

He achieved remarkable feats, but at a great cost. The fortresses he constructed, the trade routes he secured, and the strategic system he established all left a lasting impact on history. They enriched Portugal, connected distant lands, and set up global trade patterns that persist in some form today. But these accomplishments were built upon mass murder, enslavement, cultural destruction, and systematic exploitation. What legacy would you accept at such a price?

Can we acknowledge both realities simultaneously? Can historical figures be both brilliant and monstrous, effective and evil, visionary and barbaric?

Modern sensibilities are repelled by the celebration of conquistadors and colonizers. Contemporary scholarship has sparked intense debate over how to interpret and represent the legacies of European imperial expansion, with historians critiquing triumphalist narratives that overlook or downplay the severe human costs involved. Scholars argue increasingly that the victims of empire, the slain defenders of Goa, the enslaved women, the decimated communities, deserve recognition beyond brief mentions in chronicles written by their conquerors, prompting reevaluations of sources and incorporation of subaltern perspectives in historical analysis.

Yet simplistic condemnation is also insufficient. Albuquerque was not uniquely cruel compared to his contemporaries. Asian rulers at the time engaged in conquest, massacre, and exploitation as well. The Mughal expansion into India, Ottoman wars in Europe and the Middle East, and conflicts among Southeast Asian kingdoms all involved comparable levels of violence. The early modern world was brutal in ways our age struggles to comprehend.

What set Portuguese imperialism apart was its global reach and organized approach. Albuquerque developed methods that later European empires improved upon: using superior technology to control larger populations, extracting wealth through trade monopolies, creating permanent colonies, and combining religious objectives with commercial exploitation. These practices became the model for centuries of European imperialism.

Understanding Albuquerque requires embracing multiple perspectives simultaneously. He was a man of genuine conviction, certain of his righteousness, operating according to the logic of his time. He was also responsible for atrocities that caused immense suffering. He built structures that lasted centuries and destroyed societies that had thrived for generations. He was both a product of his time and an exceptional individual who shaped historical outcomes through force of will.

History does not offer clear moral lessons. It presents complexity, contradiction, and the unsettling recognition that human beings possess equal capacity to create and demolish with comparable vigor.

SMOKE OVER THE HARBOR

December 16, 1510. The smoke over Goa starts to disperse.

Afonso de Albuquerque stands on the ramparts overlooking the harbor where Portuguese ships are moored. His victory is complete; he has conquered the city. The fortification walls rise as the first stones of a colonial capital are laid.

He senses, though cannot yet express it, that this moment marks a turning point in history. The Indian Ocean will never revert to its previous state. Long-established trade networks and Muslim commercial dominance are violently overthrown. A new order emerges: European, Christian, and militaristic.

Five more years for him, years filled with conquest, governance, struggles against rivals, deteriorating health, and eventual ousting. Yet in this moment on the ramparts, triumph is his. He has achieved the unthinkable by permanently implanting Portuguese power into Asian soil.

Bodies are interred, pyres extinguished, and the surviving populace cautiously resumes their lives at home, attempting to grasp their altered reality. Mass is held within a hastily converted mosque chapel, while weary Portuguese soldiers find whatever refuge they can, and others plan how to divide spoils. Some merely strive to survive another day in this foreign land fraught with disease.

Albuquerque observes as the dissipating smoke reveals the harbor, sea beyond, and the vast Indian Ocean that he sought mastery over for years. He has succeeded partially. Nonetheless, the ocean remains indifferent to empires it was present before Portugal and will remain long after its Asian dominion perishes.

For this transient instant, it is his. Smoke wafts away, sunset descends, casting elongated shadows from the fortress walls. Amidst the narrow streets of subjugated Goa, a wife mourns her absent husband, a child shelters in ruins, a merchant tallies losses, and a soldier quells memories with drink. An empire burgeons amidst bloodshed and flames.

History may immortalize the conqueror, but it will remember the conquered with less clarity. Their narratives persist, scattered and partly obliterated in fortress stones, altered city demographics, hybrid cultures arising from coerced interactions, and the vast shadow cast by empires.

The Indian Ocean retains its memory. The walls do too. And as the smoke finally clears, a tale of empire and defiance, exploitation and resilience, and the frightful yet glorious intricacy of human ambition commences, a saga sprawling across centuries.

Afonso de Albuquerque descends from the ramparts. There is still much to accomplish. The empire cannot erect itself.

Jorge Álvares

Tamão Island, Pearl River Delta, China, July 8, 1521

The air within the cabin clung heavily with scents of decayed salt and perspiration, an intense reminder of the ocean's unyielding dominance.

Jorge Álvares lay in his final moments, embraced by Duarte Coelho, a friend of many shared years. Humidity passed through the slats of bamboo shutters. Below, water gently lapped against the wooden pilings. Outside the ramshackle huts and makeshift warehouses, Chinese war junks glided across the Pearl River estuary. Their hulls, adorned with dragons, sliced through the morning fog.

Coelho's hands remained steady, even as his jaw trembled with profound sorrow.

"Inter my remains here," Álvares pleaded weakly. His Portuguese was thick with inertia. "Beside the boy."

Next to the padrão, the stone monument is inscribed with both Christ's cross and King Manuel I's arms. Beside a small grave, where his son had been laid seven years prior, when this expedition was young and the world seemed full of promise.

Álvares tried to speak again but coughed, blood staining his lips. No official record marked the cause of death; it might have been fever, or the slow poison of exhaustion. For eight years, he endured on the world's fringe, caught in a place where no common language bridged two civilizations. He wove merchant tales into Chinese silk fabric. He trod a narrow path between the Portuguese drive and Ming suspicion. He tried to build something fragile yet audacious in these southern waters, where foreign presence was little more than an outline.

Out in the open, cries rang through the settlement. Men scrambled as it became known that the Chinese navy was drawing nearer.

Álvares closed his eyes, not knowing if he was fifty or younger; no records confirmed. Already, history's currents were beginning to erase him from memory. His breaths grew shallow, and Coelho clutched him tighter. And in that very moment, as life slipped away from the first European to reach China by sea during the Age of Exploration, neither Coelho nor the fleeing Portuguese merchants nor the advancing Chinese officials comprehended what would be lost. None foresaw how completely this man's accomplishments would be overshadowed by events grander still, or how his name would shrink into a mere footnote in the annals of those who followed.

Yet the veil of oblivion had not yet descended. Why do some names persist while others fade? It begins with the journey itself.

THE EASTERN HUNGER

Lisbon, Kingdom of Portugal, 1505

To truly appreciate Jorge Álvares, one must first immerse oneself in the fervor that enveloped Portugal at the dawn of the sixteenth century. It was an era brimming with anticipation, a fever pitch of distance explored, opportunities seized, and spices coveted for their wealth beyond anyone's wildest dreams.

For nearly a hundred years, Portuguese explorers tentatively navigated along the African coastline. Their caravels ventured further southward, driven by the quest to circumvent the vast continent that barred direct access to the Indies. In 1498, Vasco da Gama succeeded in this monumental feat by rounding the Cape of Good Hope and traversing the Indian Ocean to reach Calicut. This breakthrough opened a new spice route.

This revelation sent shockwaves through the kingdom of Portugal. King Manuel I, known as "The Fortunate," saw his destiny unfolding before him like an unrolling scroll. For centuries, the Venetians, in partnership with their Muslim counterparts, had monopolized the lucrative spice trade via overland routes from Asia. Now, Portugal could bypass them entirely by addressing the source directly.

Yet the source was not a single entity but an extensive network, a vast tapestry of ports and kingdoms stretching from India to the Moluccas, from Ceylon to the enigmatic Cathay. In every bustling port they encountered, the Portuguese discovered treasures that Lisbon had never before laid eyes on, markets brimming with goods highly coveted in Europe.

Emboldened by this realization, the crown embarked on a ruthless, ambitious strategy, capturing key ports, erecting fortresses, and exerting control over pivotal points of Asian commerce. By 1505, Cochin, India, fell into Portuguese hands; Goa followed suit by 1510. In 1511, under Afonso de Albuquerque's command, an armada laid siege to Malacca.

Malacca, the very name resonated profoundly within the global trading community. Situated at a narrow strait between Sumatra and the Malay Peninsula, it served as Southeast Asia's foremost entrepôt. Ships from all corners of the world, Arabs, Persians, Indians, Chinese, Japanese, and Javanese alike, would stopover in Malacca. The sultan's warehouses overflowed with an array of exotic goods: silks and porcelain from China, cloves and nutmeg from the Moluccas, sandalwood from Timor, and gold from Sumatra.

Albuquerque understood the strategic value of Malacca; he shared his vision with his troops before the onslaught: "The King of Portugal has often commanded me to go to the Straits, because... this was the best place to intercept the trade which the Muslims carry on in these parts. By taking Malacca, we would close the Straits so that never again would the Muslims be able to bring their spices by this route."

The ensuing battle was swift and merciless, as Portuguese naval might, comprising nineteen ships and approximately twelve hundred soldiers, clashed with a city garrisoned by twenty thousand defenders. Albuquerque's arsenal included modern European artillery capable of breaching fortifications. He also enjoyed the support of Chinese merchants discontented with the sultanate and willing to supply their junks for this campaign.

On August 24, 1511, Malacca succumbed to Portuguese dominance. The city endured brutal pillage; its Muslim populace faced massacre, mosques were torched, and a citadel known as A Famosa, "The Famous One," was erected from the ashes. With their hold on Malacca secure, the Portuguese cast their gaze eastward.

What secrets lie beyond the horizon? The Moluccas, those fabled Spice Islands, are now tantalizingly within reach. Yet, whispers began to circulate about something even grander: a vast and potent empire known as China. The salty wind whispered tales of distant lands, carrying the scent of exotic spices mingling with the sea's salt. The rhythmic creaking of the ship as it sailed onward, combined with the hushed murmurs of the crew, all contributed to the mounting tension. It was as if the very air was alive with anticipation, hinting at the profound mystery that awaited across the horizon.

MALACCA 1512

Jorge Álvares landed in Asia around this time, though the records are vague.

Born in the small town of Freixo de Espada à Cinta near Portugal's Spanish border, he likely set sail east with an armada, perhaps alongside Albuquerque or later fleets. Without noble titles or landownership, Álvares was most likely a merchant adventurer, a trader driven by ambition and seeing Portugal's Asian empire as a fresh start.

By 1512, he established himself in Malacca, possessing a junk, a Chinese-style ship with a flat bottom and battened sails, ideal for coastal waters. He traded wherever possible: Burma, Siam, and possibly the Moluccas. Gaining language skills, especially Malay and potentially some Chinese from local merchants, he amassed resources and built his reputation.

Beneath Álvares's cautious mercantile exterior lay a yearning to unlock China's secrets, not solely for profit, but for deeper understanding. At times, unnamed fears crept in, doubts that all efforts might be undone by uncontrollable forces. In Malacca's warehouses and taverns, amid Chinese traders selling porcelain and silk, conversations frequently turned to China. Such discussions fueled Álvares's ambition while revealing the deep asymmetries between European and Chinese civilizations, heightening his awareness of both cultural opportunity and risk. Significantly, these cultural contrasts, which lay at the heart of Álvares's journey, framed not only the practical challenges he faced but also shaped his entire experience as the first European to engage China by sea. Portuguese mercantile motives, centered on individual profit and the exploitation of new markets, starkly contrasted with Confucian commercial ethics, which emphasized harmony and moral duty over profit. This divergence underscored how each side understood commerce differently, profoundly influencing every encounter and ensuring that both perceived the same trades in vastly different lights. Portugal seemed minuscule compared to China, whose ancient and advanced civilization amazed even educated Portuguese who read Latin and knew Aristotle.

This awe represented more than mere curiosity; for Álvares and peers, China's vastness and sophistication challenged European worldviews and necessitated a rethinking of cross-cultural navigation. Private Chinese merchants in Malacca operated outside official channels despite Ming law, which made their trade technically illicit. The strict tribute system mandated that foreign trade occur through official missions, yet private merchants defied these rules for decades, sailing south with goods, returning with others, and amassing wealth. They sometimes spoke of the Pearl River Delta, Guangzhou, and opportunities in Chinese waters. For Álvares, these merchant tales revealed not just paths to profit but also contested spaces where imperial regulation, illicit exchange, and cultural interaction converged; his interest extended beyond trade to understanding how such networks might shift as power balances changed.

The Portuguese yearned for a foothold. China was no small sultanate to be conquered with a few hundred men and cannons. The Ming Empire was among the most powerful in the world, with Emperor Zhengde ruling over about 150 million people, more than all of Europe combined. Chinese cities dwarfed anything Portugal had; their infrastructure, canals, roads, and granaries operated with impressive efficiency. For centuries, Chinese shipbuilders sent massive treasure fleets across the Indian Ocean, eclipsing any European endeavor.

Conquering China was beyond the reach of the Portuguese. Trading with it was the only viable path. So Jorge Álvares mapped his strategy.

THE VOYAGE INTO THE UNKNOWN

Pegu (Burma) , May 1513

The six junks convened at the Irrawaddy River's mouth during May's initial days.

Rui de Brito Patalim oversaw the venture as a Portuguese captain stationed at Malacca, designated by Albuquerque himself. However, Patalim did not physically participate; instead, he facilitated and organized the journey and obtained official approval. The actual sailing would be handled by crew members such as Jorge Álvares.

Three Portuguese sailors embarked on this voyage: Álvares, along with two companions whose names have been lost to history. They were aboard Chinese-style junks, manned by Malays and possibly some Chinese. Their plan was to follow the coastline north, staying close to the shore while relying on landmarks and local knowledge for navigation. Essentially, they blended in as passengers on an Asian trading expedition without revealing their Portuguese identity or intentions of conquest.

This approach proved prudent, given that the Portuguese reputation in Asian waters had become problematic. They were associated with Christianity, making them suspicious to Muslims; foreigners that posed a threat to local rulers; and known for their violent methods, particularly under Albuquerque's rule in capturing Goa and Malacca.

Álvares carefully assimilated among merchants, observing, learning, and establishing new connections without arousing attention as a Portuguese conqueror.

That May, monsoon winds favored their journey. The junks hoisted their distinctively battened sails - large canvas panels reinforced with bamboo that could be quickly adjusted during storms. These vessels displayed remarkable seaworthiness, honed through centuries of Asian maritime expertise. They sat low in the water, partitioned into watertight compartments to ensure buoyancy even if damaged. Below deck, they carried an assortment of trade goods: Portuguese wine, spices from the Moluccas, and weapons, as the Portuguese always traveled prepared for conflict.

Their path led northward along Burma's coast, passing through the mouths of the Salween River and into Siam's waters. The sea transformed its hues like a painter's palette - deep blue transitioning to jade green, reflecting the lush greenery of the mountainous coastline. As they stopped at fishing villages to replenish water and supplies, local inhabitants gazed in wonderment at the three Portuguese: unusual men with fair skin and beards, clad in European doublets despite the tropical climate.

Álvares meticulously observed their surroundings - local boat construction methods, agricultural practices, and village organization. Over the next eight years, he would accumulate extensive knowledge of southern China and its surrounding regions. This information was later shared by Álvares with King Manuel I, as recorded by João de Barros, detailing Chinese culture, religion, financial matters, and military capabilities (Barros, Décadas da Ásia). Portugal sought this intelligence to foster relations with China.

But that objective lay in the future. First, they had to reach there.

The junk advanced further northwards, entering the expansive South China Sea dotted with islands, reefs, and sandbanks. They navigated past the Gulf of Tonkin and approached the Pearl River Delta by late May.

And then, through the shimmering haze, an unmistakable sight emerged on the horizon.

China.

TAMÃO ISLAND, PEARL RIVER DELTA , MAY 1513

The island emerged faintly from the horizon, a verdant mark upon the slate of water. Yet which island this was remained uncertain to the scholars of our time, a question posed for centuries. It was called Tamão by the Portuguese, who believed it lay off Guangzhou, the famed port known as Canton. They described its location at the confluence of the Pearl River and the South China Sea. Still, they did not pin down an exact spot, their instruments lacking in 1513. Various researchers propose Lantau Island or others that no longer exist, such as small islets swallowed up by land reclamation over time. The landscape of the Pearl River Delta shifts continuously: rivers carry sediments, landforms emerge and vanish, leaving a coastline today unlike that seen in 1513. Yet Álvares knew his destination. Likely, he consulted the Chinese merchants aboard their junks. As the junks drifted closer to the mysterious shore, the aroma of smoked juniper wood spiraled through the air. Amongst the busy island life, fishermen busied themselves, mending nets with nimble fingers on the sand, while junks gently nudged against bamboo poles securing them to the shore. The Chinese merchants informed him of restrictions imposed by the Ming Emperor on foreigners entering the mainland. However, this island served as a buffer zone for trade. The junk moored steadily. Ávares and his two companions boarded a smaller craft and rowed to shore. In that moment when they set foot on the beach, Jorge Álvares became the first European to touch Chinese soil during the Age of Exploration, an occasion unmarked by fanfare or a grand welcome from local officials. Only fishermen would have gazed at these peculiar strangers with their light skin and unusual attire. Ávares adhered to Portuguese tradition by planting a padrão: a stone column crowned with a Christian cross. Such monuments were erected to signify Portugal's reach, found along the African coast, where they endure despite harsh elements. Bartolomeu Dias planted one at the Cape of Good Hope, while Vasco da Gama set his in India. Now Ávares erected one in China, its design featuring the arms of King Manuel I and a Christian cross. Positioned near the shore for all passing ships to see, it was both an assertion of Portuguese sovereignty and a testament to their beliefs. Yet, beyond its immediate declarative purpose, the padrão epitomized the symbolic intersection of two civilizations: for the Portuguese, it was an emblem of faith and imperial ambition, while for the local Chinese, it was an unfamiliar signifier whose meaning invited contemplation and uncertainty. In this liminal space, the stone symbolized both the promise of cross-cultural encounter and its underlying tensions. The local Chinese watched silently, contemplating the purpose and significance of this foreign relic. They wouldn't have long to ponder. Ávares came with trade in mind.

THE MIDDLE KINGDOM AND THE SEA BARBARIANS

To comprehend China's reaction to Jorge Álvares, one should first understand the perspective the Middle Kingdom held of itself and the rest of the world.

In 1513, China was seen as the epicenter of its own universe, not an issue of pride or arrogance. Instead, it reflected a worldview shaped by three millennia of continuous civilization. The nation identified itself as Zhongguo, or the Central Kingdom, with lands around it considered either tributary states or barbarian territories. Korea, Vietnam, Siam, and others paid homage to the Chinese emperor, acknowledging his sovereignty by sending missions to Beijing with offerings and receiving, in return, imperial favor, trade rights, and, when necessary, military protection.

The tribute system served as the blueprint for China's engagement with foreign lands. Such interactions were never seen as equal among nations but rather as exchanges between superior entities - civilization and barbarism, the Son of Heaven and rulers who recognized his cosmic supremacy.

In Ming law, engaging in private foreign trade was outlawed. The Hongwu Emperor, who established the Ming dynasty in 1368, enacted the haijin or sea ban, which mandated that all maritime commerce must be conducted through official tribute missions. Punishment for violating this decree could range from death to exile for the merchant and their family, while shipyards along the coast were dismantled and ports blocked with obstacles to prevent any ships from docking.

The motivations behind these restrictions included a desire to disrupt the networks and supply lines of Japanese pirates - the wokou - who had been raiding China's coasts. Additionally, political factors played a role, as the Hongwu Emperor fought a civil war to solidify his dynasty. Some adversaries fled to sea, becoming pirates or smugglers, prompting the implementation of a sea ban to cut off their economic support.

However, enforcing the ban completely across thousands of miles was impractical. Private merchants continued their trade through unofficial channels in port cities' gray markets. They would travel to Southeast Asia, the Philippines, and Japan, all outside official trading routes. Local officials often turned a blind eye to these activities, sometimes accepting bribes or acknowledging the limitations in enforcing the ban comprehensively.

By 1513, things had become even more intricate with Emperor Zhengde on the Dragon Throne. He deviated from typical Confucian norms because he was young, energetic, yet impulsive, and more interested in military pursuits and personal pleasures than in administrative tasks. The court during his reign was opulent, with incense filling the air and sandalwood scent wafting from ornate burners. Early in his rule, he allowed eunuchs to gain considerable influence over policy, causing officials, clad in silk robes that rustled with each move, to lament his behavior. The empire's finances were strained by Mongol raids from the north, which required resources and attention.

Despite this, Zhengde was more open to foreign contact than some of his predecessors. He recognized the potential benefits trade could bring, especially after the maritime expeditions led by Admiral Zheng He showcased China's capacity to project influence across maritime Asia. Unlike Emperor Hongwu, he wasn't as strict in enforcing the sea ban.

This created a window of opportunity – a space where cautious foreign merchants might operate if they respected Chinese customs and did not provoke the mandarins overseeing the southern provinces.

It was within this context that Jorge Álvares arrived at Tamão in May 1513.

TAMÃO 1513-1514

The Chinese referred to the Franks as Fo-lang-ki, an old term derived from "Frankish" and used broadly to describe Western foreigners appearing in Chinese waters.

Initially, the local officials, mandarins, or itau kept their distance from Álvares but closely monitored his actions, sending reports up the chain of command and assessing any potential threat.

Álvares managed to establish contact with private Chinese traders who saw an opportunity in the Portuguese goods. These included wine, European textiles, and possibly firearms, despite concerns over selling weapons to outsiders. The Chinese, on their part, had desirable items for the Portuguese, such as silk, porcelain, tea, precious stones, and medicinal rhubarb.

Trade commenced tentatively, involving a few chests of goods and silver coinage, with negotiations conducted through interpreters, such as Malays who could speak both languages. The exchange rates were not favorable to the Portuguese because of their status as foreigners, but Álvares made the payments necessary to build relationships.

Word of the trade reached Guangzhou, where Chinese officials imposed restrictions on Ávares' activities while allowing controlled trade. He was only allowed to operate from an island and required permission to travel to the mainland or visit Guangzhou, with meetings between him and Chinese merchants taking place under official supervision.

This approach by the mandarins was a typical Chinese policy of containing barbarians, limiting their movements, and regulating trade interactions that minimized contact with the general population while keeping them at a distance.

Álvares accepted these limitations, understanding his vulnerability as one of only three Portuguese in a nation of millions, without military force or threats to challenge the Middle Kingdom. His presence was by sufferance, and he knew it.

Thus began his endeavor to build a small settlement on Tamão, including a warehouse, living huts, and a shelter for merchants, Portuguese, Malays, and possibly some Chinese supporters who had gathered for this trading opportunity. It may have been basic and temporary, but it served its purpose well as a place where East met West for trade and relationship-building.

Adjacent to his padrão, near the stone cross, Jorge Álvares buried his son. The chronicles provide scant information: no age, name, cause of death, birthplace, or details about his mother. However, this burial signifies that Álvares was not merely a passing merchant; he brought his family to this remote island and sought to establish roots despite the limitations.

He was effectively planting roots at China's edge.

MALACCA , LATE 1513

News took time to travel across maritime Asia, with months passing before a letter from Lisbon reached Malacca and weeks before information from the Pearl River Delta reached Portuguese headquarters. However, by late 1513, Afonso de Albuquerque had learned of Jorge Álvares' success.

Elated, Albuquerque recognized China's significance, not only in terms of wealth but also in legitimacy. The Portuguese needed the Ming Empire to regard them as more than just pirates and to establish official relations through treaty ports for structured trade that would ensure their safe operation and profitability.

Albuquerque sent Rafael Perestrello, a captain from Malacca, on a ship crewed by Malays to follow up on Álvares's initial contact. Late in 1513, he arrived at Guangdong Province's southern shores, becoming the first Portuguese to set foot on the mainland – something Álvares had been denied.

Perestrello, a cousin of Christopher Columbus and involved in Portuguese Atlantic exploration for years, carried a letter from King Manuel I to the Ming Emperor requesting official trade relations. This formal, respectful document, composed in accordance with European diplomatic standards, addressed Zhengde as an equal sovereign, offering friendship and mutual benefit.

The Chinese officials who received it must have been perplexed, as addressing the Son of Heaven as an equal clearly violated the fundamental principles of the tribute system. Foreign rulers were expected to send tribute missions and humbly request trade privileges while acknowledging China's superiority.

However, Perestrello's mission did not immediately fail. Guangzhou mandarins forwarded the letter to Beijing and reported on the Fo-lang-ki, their ships, weapons, Malacca's conquest, and their unusual requests. These foreigners were not a minor barbarian tribe that could be ignored.

The slow-moving Ming bureaucracy processed reports and held meetings, debating whether to permit trade with these foreigners, assess them as a threat, or consider their usefulness. No immediate decision was made, but the Portuguese were not expelled. Álvares continued operations at Tamão as the settlement grew, attracting more junk from Malacca carrying Portuguese goods and increasing the number of Chinese merchants.

Trade flourished.

THE SETTLEMENT YEARS

Tamão 1514-1520

For seven years, Jorge Álvares existed in a state of suspension. Neither fully embraced nor entirely repudiated by the Chinese authorities, he navigated through the ebb and flow of political tides.

The settlement at Tamão flourished. Approximately two dozen Portuguese inhabitants coexisted with Malay sailors, Chinese merchants, and other opportunistic traders seeking to benefit from commerce. The settlement was rustic, constructed from bamboo, thatch, and procured or purchased materials. Unlike Malacca's A Famosa – a stone fortress, the Portuguese outpost on a Chinese island persisted only by the grace of Ming officials' tolerance.

The Portuguese traded wine, spices, and European textiles, which were less favored by Chinese buyers, in favor of their own silk. Coral was highly prized in China, and firearms were increasingly in demand; Portuguese matchlock muskets and small cannons attracted interest as they were perceived as more advanced. Despite official policy prohibiting the sale of weapons to foreigners, demand remained strong.

Álvares treaded carefully, cultivating amicable relations with Chinese merchants who profited from Portuguese goods. He forged connections with lower-ranking officials who could be bribed or benefited indirectly through trade. He assimilated into Chinese customs, etiquette, gift-giving, and concepts of face and hierarchy, while adjusting to expectations, while maintaining his Portuguese identity.

Throughout, he amassed intelligence. João de Barros chronicled that Álvares "acquired knowledge about China's culture, religion, finances, and military." This information was systematically gathered; he studied the Ming administration's structure and the tribute system, inquired about Chinese military strength, fortifications, and weapons, and understood trade networks for Portuguese advantage.

Back in Lisbon, his findings were likely dispatched via letters aboard the annual Portuguese fleets sailing between India and Portugal. King Manuel I received these reports with keen interest, piecing together a comprehensive view of China to inform decades-long policy decisions.

Yet, security was always an issue; Chinese politics shifted continuously. Emperor Zhengde's rule was erratic, often disregarding Confucian advisors in favor of military pursuits against Mongol invaders. In the court, opinions diverged: some advocated expulsion, others proposed limited trade as beneficial to the empire, while concerns arose about Portuguese potential alliances with external threats.

In 1517, a larger Portuguese embassy arrived under Fernão Pires de Andrade's command, equipped with eight ships and numerous men. This official mission aimed to establish treaty relations and to present gifts to the Emperor, including European curiosities and luxury items, along with letters from King Manuel I, surpassing earlier correspondence.

Pires de Andrade met Emperor Zhengde in May 1520 in Nanjing; the encounter was formal but inconclusive, as Zhengde focused on leisure rather than diplomacy. He accepted gifts but made no commitments or agreements.

Meanwhile, tensions escalated in Guangdong Province. The Portuguese displayed military might with cannons and fired salutes considered aggressive by the Chinese, leading to brawls between sailors and locals and incidents involving women. Some Portuguese behavior fueled resentment among the local population.

Things deteriorated further as Simão de Andrade, brother of Fernão, engaged in transgressions, including the kidnapping of children for slavery and constructing a fortified outpost on an island, asserting colonial dominion. These atrocities outraged Chinese officials, portraying the Portuguese as barbaric, disrespectful, and contemptuous of Ming authority, prompting calls for action.

Then, in 1521, Emperor Zhengde died.

Beijing , April 20, 1521

The demise of Emperor Zhengde left no successor to inherit his throne. His life ended at merely thirty-one years old under uncertain circumstances; it might have been an illness contracted by accidentally falling into a lake during a moment of intoxication, though there could be other reasons behind it. The official records remain silent on this matter.

Zhengde's passing triggered a succession crisis for the throne. Grand Secretary Yang Tinghe took swift action to ensure that Zhu Houcong, Zhengde's cousin, ascended as Emperor Jiajing, despite his tender age of fifteen and lack of experience in governance or foreign policy matters. The court faced internal divisions, especially concerning the presence of Portuguese traders, known as Fo-lang-ki.

The faction opposed to Portuguese trade in China saw this as an opportunity to adopt a new stance. With Zhengde gone, they aimed to overturn his tacit approval of the Fo-lang-ki and to expel them entirely from Chinese waters. Beijing issued a strict prohibition on any dealings with these foreigners.

In Guangdong Province, an immediate halt to all trade relations with Portugal was enforced; they were required to leave or face military force. Defiance of this edict would result in the harshest punishment – execution by death. This policy was not only severe but also absolute in its enforcement.

The remnants of Pires de Andrade's mission, located at the Portuguese embassy in Beijing, faced immediate arrest. Some members were imprisoned while others met a more gruesome fate: execution. Amongst those who suffered this fate was Rafael Perestrello, who had followed up on Álvares' journey. As a result, the hopes for official diplomatic relations between Portugal and the Ming Empire were utterly shattered.

Tamão , Summer 1521

News trickled to Tamão, carried by Chinese traders whose contacts extended upriver in Guangzhou. The Portuguese outpost faced peril as the new emperor turned against them. Preparations for war were underway; an attack was imminent.

Jorge Álvares must have understood the gravity of this situation. After eight years of cultivating relationships and trade, he had worked tirelessly to prove that the Portuguese could be reliable neighbors. But incidents orchestrated by individuals such as Simão de Andrade sullied those efforts. The Chinese deemed the Fo-lang-ki as pirates rather than merchants, a threat to be vanquished rather than a partner for commerce.

Some Portuguese seized their opportunity and fled, setting sail for Malacca before the Chinese naval forces arrived on the scene. Others chose to remain, either out of stubborn resolve or due to significant investments in Tamão that they could not abandon so easily. Álvares remained as well. Perhaps ill health prevented him from leaving; perhaps he clung to hope of negotiations or salvage from this unfolding disaster. Or maybe he had nowhere else to go. This island outpost, which had consumed nearly a decade of his life's work, had become an integral part of who he was.

The Chinese fleet anchored off Tamão in early July, deploying war junks larger than the usual merchant vessels that navigated these waters. These ships carried soldiers armed with swords and spears, along with possibly some artillery. The Chinese naval tactics were advanced; they conducted boarding operations adeptly, employed fire ships, and executed coordinated assaults. In contrast, the Portuguese had only about a dozen men capable of combat and a few firearms at their disposal. However, they faced overwhelming odds without a fortress or any defensive stronghold to call upon. Tamão was fundamentally a trading post, not designed for military defense.

The assault began rapidly and ruthlessly. Chinese forces burned down the settlement, dismantling the warehouses and huts constructed by the Portuguese over eight years. The trade network that had flourished was completely destroyed. It remains unclear whether they clashed with the Portuguese defenders or drove them out at sword's point, forcing them to retreat to their ships. The accounts of this event are sparse on specifics.

The Portuguese took flight. They hastily boarded their junks and set sail southward, abandoning Tamão to the mercy of the Chinese navy. The settlement was erased from existence, reduced to a watery grave, its remnants left as charred stumps and ash. Yet before the final evacuation unfolded on July 8, 1521, Jorge Álvares drew his last breath.

THE DEATH AND THE FORGETTING

Duarte Coelho interred his companion beside a stone cross, standing firm against odds, be it indifference on the part of the Chinese or physical resilience. It may also have been regarded as an anomaly unworthy of demolition. Yet in that moment, the cross transcended its role as a symbol of Portuguese dominion to represent the convergence and limitations of two cultures. So Coelho buried Álvares by this silent witness, aligning him with his son under an emblem now signifying both individual loss and the delicate, enduring encounter between East and West on foreign soil.

João de Barros, decades later, chronicled Coelho's sentiments: "That land may consume his body, yet for the honor of his Fatherland at Earth's ends, having set this padrão of his discoveries, he will not see his memory eaten, as long as our deed endures." However, despite the poignancy of this conviction, collective remembrance diminished swiftly. Memories faded; Álvares's tomb disappeared, Tamão's location became uncertain, and even the island's name became a puzzle for later generations of scholars. Within a single generation, Álvares was forgotten, exemplifying how historical memory can rapidly recede when individual achievements are overshadowed by subsequent events. This erasure underscores the central argument: that the significance of pioneering figures like Álvares in shaping cross-cultural encounters often vanishes from historical consciousness, highlighting the fragility of historical memory and the processes by which some names endure while others are lost.

Why? Because his narrative was overshadowed by grander events. The Portuguese withdrawal in 1521 seemed like a failure at the time. Unbeknownst to them, this inaugural contact, even though it ended badly, laid the groundwork for Macau's future. Insights gleaned at Tamão, including maneuvering with Chinese officials and trading approaches lessons that would resurface in 1557, proved invaluable when the Portuguese attempted anew thirty years hence.

By 1557, Portuguese traders secured permission to establish a permanent settlement in Macau, a small peninsula near Álvares's arrival point. Context had shifted: The Portuguese exhibited greater prudence and respect for local customs, paying dues and acknowledging Ming authority. Their intent was purely commercial, eschewing conquest or conversion.

It worked marvelously. Macau thrived as the conduit between China and beyond. For four centuries, it remained under Portuguese administration, a singular amalgamation of Chinese and European cultures. Jesuit missionaries traversed through Macau en route to Beijing; Chinese silk flowed out via Macau into European markets; Japanese silver changed hands in Macau's warehouses for Chinese goods. The city flourished immensely, a gem of the Portuguese empire.

None of this would have been possible without Tamão or Jorge Álvares's pioneering voyage. Those eight years spent in China, immersing, networking, and learning from mistakes, informed future Portuguese strategies.

Yet by then, Ávares himself had slipped into obscurity.

THE STONE THAT REMAINS

Duarte Coelho's survival during the evacuation from Tamão enabled him to return to Portugal, where he bought a captaincy in Brazil with his wealth from Asia. As one of the most successful administrators in Portuguese America, he established thriving settlements in Pernambuco through sugar cultivation and slave labor. His achievements in Brazil far surpassed those of his Asian endeavors.

There are no known instances of Duarte Coelho speaking publicly about his friend Jorge Álvares. The Portuguese attempted multiple times to reestablish relations with China after 1521, but all of their efforts either failed or achieved only limited success until they finally secured the Macau settlement. By this time, the Age of Discovery was coming to an end, and a new era had begun, one characterized by ongoing colonialism, missionary work, and global trade networks that would change both Europe and Asia.

Jorge Álvares was reduced to a mere footnote in history, according to one historian. Although his name appeared briefly in the works of João de Barros and Damião de Góis, Portuguese historians who documented early Asian exploration, even their accounts were vague and uncertain about dates and details. Later historians almost completely ignored him. The story of Portuguese-Chinese relations primarily focused on Macau, the later arrival of Jesuit missionaries, and broader political and economic developments.

For four hundred years, Jorge Álvares remained obscure until a historian named José Maria Braga attempted to revive his memory in the 1950s. Braga, a scholar of both Portuguese and Chinese descent working in Macau, authored "China Landfall 1513: Jorge Álvares' Voyage to China," which aimed to reconstruct what was known about the explorer. By thoroughly examining the chronicles and piecing together dates and locations, Braga argued for the historical significance of Álvares's accomplishments.

In 1954, a bronze statue of Álvares was installed in Macau. The sculptor Euclides Vaz created a figure depicting a middle-aged man wearing sixteenth-century clothing, pointing upwards towards the sky and standing before a stone cross. The statue was placed in Jorge Álvares Plaza, a small public garden, not far from where he might have first seen the Chinese coast.

Even with a statue and plaza bearing his name, Álvares remains largely unknown. Visitors to Macau pass by his monument without stopping, while locals barely acknowledge his presence. The story of the first European to reach China by sea has been overshadowed by subsequent history, the Macau colony, the Opium

Wars, and Macau's return to Chinese sovereignty in 1999.

The exact location of Tamão remains a mystery, with scholars suggesting Lantau Island, Lintin Island, or smaller islands within the Pearl River Delta. However, its precise location is still unknown. Additionally, the padrão that Álvares erected has never been discovered, and Duarte Coelho's friend's grave has vanished, lost to time, tides, and the ever-changing geography of the Pearl River Delta.

Yet, his achievements endure. Jorge Álvares, a merchant adventurer from a small Portuguese town, was the first European to reach China by sea during the Age of Discovery. He planted a cross on Chinese soil and established the first European trading settlement in southern China. He gathered information that would influence Portuguese policy for generations. For eight years, he resided between two civilizations, helping them begin to understand one another.

His death on July 8, 1521, in Duarte Coelho's arms marked the end of an era. The Portuguese initial attempt to engage with China had failed, and the settlement at Tamão was destroyed. However, the door he opened could never be fully closed. Eventually, the Portuguese would return, learning from their failures at Tamão and approaching China with greater respect and understanding of Chinese customs and power. This eventually led to their success in establishing Macau and initiating a long, complex, and often troubled relationship between China and the West that shaped the modern world.

Jorge Álvares did not live to see this future; he died in obscurity, forgotten even before his burial. Nonetheless, his journey to Tamão in May 1513, planting the stone cross, founding the first settlement, and dedicating years to trade, learning, and striving to bridge the gap between civilizations, significantly shaped the course of Portuguese-Chinese relations and the broader encounter between East and West. Today, the monument in Macau endures, overlooking a city that embodies the very cross-cultural legacy Álvares helped initiate. The bronze statue has withstood tropical rains, and the stone cross continues to mark a horizon now dominated by skyscrapers and jet trails. In this quiet garden, Álvares stands memorialized as tangible evidence that, even when immediate recognition is absent, endeavors to connect disparate worlds can leave enduring marks on history, affirming the essential thesis that the forgotten journeys often underpin transformative moments across civilizations.

That land of idolatry may consume his body. Yet his accomplishments endure.

Ferdinand Magellan

September 20, 1519, Sanlúcar de Barrameda, Spain

The water smells foul with the stench of rotting matter and a pungent hint of human refuse.

Five ships are anchored in the pre-dawn darkness. Their cargo holds overflow with supplies that seem insufficient. On the Trinidad's quarterdeck, Ferdinand Magellan oversees the final loading of barrels. The stench of rotting matter from the surrounding waters seems to seep into the spirit of those onboard, an invisible but potent omen of the voyage's deeper expectant tensions. At thirty-nine, his body tells the story of many campaigns fought for empires. His right leg drags behind him, a memento from a Portuguese blade in Morocco. This injury is a private torment, turned into a public jest by his adversaries and a silent signal to his crew.

A morning fog blankets the Guadalquivir River delta. Below deck, Magellan listens as five languages collide in chaos: Spanish, Portuguese, and Italian are predominant, with the occasional murmur of French and Basque. Most of his men cannot communicate with each other. None truly comprehends the risky deal they've made. Two hundred sixty-five souls prepare to pursue the setting sun, hopeful it will guide them eastward. They follow a path that exists solely in Magellan's mind. He navigates by a tapestry woven from pilfered secrets, ethereal maps, and what others call madness.

He is not deaf to the rumors; Portuguese captains, cloaked by their noble pride, resent taking orders from someone branded a traitor. They whisper about his obsession being driven by old wounds rather than reason. Perhaps they're right, but they haven't witnessed what he has nor understood what he knows; they lack the vision of the immense prize: a western route to the Spice Islands. Whoever claims this passage seizes control of the world's wealth. Cloves, nutmeg, and mace, these dried-up treasures are worth more than gold. Men will sacrifice their lives for them; empires will rise on their scent.

The ships are small and old. Their timbers have weakened from rot. Spanish officials spared only what they deemed disposable. These are dilapidated vessels captained by men marked for failure and rations barely enough to keep them afloat. Spain would rejoice in victory but lose nothing should he fail. The battered ships and the life of a foreigner are losses the kingdom can afford.

Magellan's coat conceals a letter, a royal charter from King Charles I, barely out of boyhood, risking his fragile kingdom on this quest. This document grants Magellan sweeping authority: dominion over new lands, a share of profits, and noble titles for his kin. Yet, all these promises are as insubstantial as paper unless he returns successfully. The strait he vows to discover and the secret passage through the Americas may be no more than a myth.

The sun rises above the horizon. Magellan gives the order. Anchors stir up, patched sails mend and unfurl against the morning breeze. Five ships, Trinidad, San Antonio, Concepción, Victoria, and Santiago, float down the river towards the vast sea that awaits them. There are no cheers; most men cannot fathom what lies ahead. Those who can only feel apprehension.

Ahead lies the longest voyage ever dared, a journey destined to reshape the world and consume many of those aboard. Magellan will not witness its conclusion. But on this morning as his ships slip from Sanlúcar into the Atlantic Ocean, none of that is apparent. There is only an endless, silent horizon and one man's steadfast belief in taking a leap of faith where others see impossibility.

The world is much bigger than anyone can imagine, and someone must be the first to measure it.

THE MAKING OF AN OBSESSION

To truly comprehend why Ferdinand Magellan embarked on such a perilous journey, we must first understand what he had already lost during his lifetime.

Born around 1480 in northern Portugal's Trás-os-Montes region, Magellan hailed from a family of minor nobility distinguished by lineage rather than wealth. His parents passed away when he was still young, and at twelve, he moved to Lisbon to serve as a page under Queen Leonor, the wife of King John II. This arrangement wasn't altruistic; it was customary for Portugal to immerse the sons of lower noble families in court culture, teaching them navigation, astronomy, languages, and political intrigue to prepare them for roles within the empire.

Portugal, at the close of the fifteenth century, was a beacon of global expansion. Amidst European kingdoms that often engaged in internal conflict, Portugal turned its ambitions outward. The Portuguese expeditions along Africa's coast sought a sea route to Asia that would bypass Muslim-controlled overland trade. In 1488, while Magellan was still a child, Bartolomeu Dias navigated the southern tip of Africa, proving that the Atlantic and Indian Oceans were connected. Vasco da Gama finalized this journey in 1498 by reaching India and returning with spices that yielded immense wealth for his investors.

These voyages propelled Portugal from its position on Europe's periphery to a dominant maritime empire. The young Magellan absorbed all aspects of this world, including the use of the quadrant for celestial navigation and the precision required to calculate latitude from stellar tables, a foundation for future exploration that was tangible and practical.

By his early twenties, Magellan had left the court and enlisted in expeditions. In 1505, he accompanied Francisco de Almeida on a fleet of twenty-two ships and fifteen hundred men as they sailed around Africa. The journey took months, marked by the tragic loss of life due to scurvy. Upon arriving in India, Magellan encountered an empire of complex Sultanates, merchant networks, and intense trade rivalries.

Magellan fought in battles that cemented Portuguese rule along the Indian coast, including the siege of Kilwa and the bombardment of Mombasa, experiences that taught him that empires are not built solely on exploration but also on the capacity to wield overwhelming force to control trade routes. The Portuguese did not come as equal traders; they sought to monopolize, dominate, and extract wealth from others.

In 1509, Magellan participated in the Battle of Diu, where a Portuguese fleet defeated Egyptian, Ottoman, and Indian forces, securing dominance over the Indian Ocean trade. He gained invaluable knowledge about the empire's geography, prevailing winds, monsoon seasons, and distances between islands, knowledge that he understood was power itself, closely guarded by the Portuguese crown.

Then came the conquest of Malacca in 1511, a strategic port city controlling access to the Spice Islands. Whoever held Malacca commanded the world's most coveted spices. The assault was brutal; thousands perished as Portugal secured this final piece of their eastern trade route. It was here that Magellan encountered Enrique, a slave and interpreter fluent in Malay who accompanied him on his historical voyage around the globe.

In the sleepless nights leading up to this decision, Magellan reflected on his years of service to Portugal, which were met with disdain rather than recognition. He rationalized that this betrayal was not only necessary for survival but also a step towards a future he envisioned, one in which knowledge of Portugal's eastern trade routes and the tantalizing possibility of a western route could be leveraged for greater gains.

Magellan's career in Portugal abruptly ended when he returned to Morocco to fight against Muslim forces. He suffered a knee injury during combat and was accused of trading with the enemy, charges dismissed by King Manuel I coldly and contemptuously. With no future in Portugal, Magellan made the choice deemed treasonous by his former countrymen: to offer his skills to Spain, its greatest rival.

THE ARGUMENT

In October 1517, Ferdinand Magellan arrived in Seville, Spain, accompanied by Enrique, his servant, and a proposition that could either immortalize him as a legendary explorer or brand him as delusional.

The concept of reaching Asia by sailing westward was not novel; Christopher Columbus had been convinced he had reached Asia's borders but had actually stumbled upon a new continent. By 1517, geographers recognized that the Americas acted as an obstacle, a massive land obstructing the western path to spices. Yet Magellan believed there must be a passage - perhaps through a strait in the southern part of the Americas - that would allow ships to traverse into the ocean beyond.

His evidence was sparse and somewhat speculative; tales from previous Spanish explorers described a significant bay or river mouth on the southern Atlantic coast of South America, while rumors circulated about secret Portuguese expeditions venturing farther south than documented. Magellan even drew upon logic: if Africa boasted a point that could be circumnavigated, then why not South America? In his vision, the world was symmetrical.

What truly set Magellan apart from other seekers of the western passage was not just his maps but his experience. He knew the enormity of sailing to the Spice Islands via the Portuguese route around Africa and could accurately estimate provisions, calculate navigation time, and discuss the winds and currents of the Indian Ocean. In presenting his case to Spanish officials, Magellan drew directly on this seafaring knowledge, affirming in Antonio Pigafetta's later account: "The Captain-General, being an experienced mariner and very circumspect, did not neglect anything which seemed necessary." Thus, it was not a naive dreamer speaking, but a weathered, wounded sailor with two decades of practical expertise.

The political landscape also favored him. The Treaty of Tordesillas, signed in 1494 by Spain and Portugal, divided the non-European world between them along a meridian in the Atlantic Ocean. Everything eastward belonged to Portugal; everything westward belonged to Spain. However, this treaty remained ambiguous about where that line fell on the other side of the globe. The Spice Islands, located near the land division, were particularly coveted if they could be deemed Spanish and thereby break Portugal's monopoly.

This was more than a trade issue; it concerned the balance of power between kingdoms that were once united but now vied for global supremacy. King Charles I of Spain, who had also been elected as Holy Roman Emperor Charles V, saw in Magellan's plan an opportunity to challenge Portuguese dominance without provoking direct conflict. If successful, Spain would gain access to the spice trade and could claim the Moluccas legally as Spanish territory. Conversely, failure would mean a loss for Charles but no significant damage to his kingdom.

Magellan gained a crucial supporter in Juan de Aranda, a member of the Casa de Contratación, the royal agency overseeing Spanish trade and exploration.

Aranda foresaw substantial profits and helped Magellan navigate the bureaucratic maze of the Spanish court. Together, they crafted a proposal blending elements of exploration and enterprise. In exchange for five ships, two years' worth of supplies, and a crew of 250 men, Magellan promised to find the western route to the Spice Islands, assert new lands in Spain's name, and return with riches that would far exceed the initial investment.

The negotiations stretched over months. Spanish officials harbored skepticism while Portuguese diplomats sought to undermine the endeavor by disseminating warnings that Magellan was untrustworthy or even a potential spy whose expedition would assuredly fail. However, Magellan had honed his skills in Lisbon in presenting persuasive arguments with conviction, leveraging his injuries and scars as evidence of his resolve, and framing his proposal not as speculation but as an inevitability.

On March 10, 1518, King Charles endorsed the capitulación that authorized Magellan's expedition. The terms were exceedingly generous, particularly for a foreigner with no proven allegiance to Spain. Magellan and his partner, cosmographer Rui Faleiro, gained governorship over any lands they discovered, one-fifth of the profits from the first voyage, and one-twentieth of subsequent ventures. They would be ennobled as knights of the Order of Santiago, with their titles passing down hereditarily.

Magellan had been denied such opportunities in Portugal, but now found them offered by a Spanish king young enough to romanticize exploration and ambitious enough to challenge established norms. However, the contract included a stipulation that would prove troublesome: Magellan was obligated to employ Spanish captains as subordinates commanding four out of five vessels. These were not his choices but political appointees who held noble birth, served Spanish interests, and watched Magellan vigilantly for any sign of disloyalty.

Preparations took over a year, during which ships needed to be purchased, refurbished, and supplied with provisions. Crew members had to be recruited from Seville's docks: a mix of seasoned sailors, young adventurers seeking excitement, criminals offered the choice between imprisonment and seafaring, and foreign nationals from various backgrounds signing up out of desperation, curiosity, or a lack of better prospects.

The five vessels were not particularly impressive; the largest, Trinidad, Magellan's flagship, was only 110 tons, around 60 feet long, and a wooden ship prone to creaking even in calm waters. The San Antonio was slightly smaller, followed by Concepción and Victoria, while Santiago barely qualified as an offshore vessel. These were not grandiose ships of future eras but cramped, leaky platforms infested with rats that could only survive the open sea through good fortune and skilled handling.

The Spanish captains designated to lead the fleet included men of status but questionable loyalty. Juan de Cartagena, in charge of the San Antonio and overseeing the entire flotilla as inspector-general, was the most senior and openly dismissive of Magellan's abilities. Gaspar de Quesada led Concepción, Luis de Mendoza commanded Victoria, and Juan Serrano - one of the few individuals Magellan trusted - captained Santiago. Their selection emphasized connections to the Spanish court over sailing prowess, carrying secret orders sealed until specific conditions were met that Magellan would not learn about until it was too late.

As departure neared, tension escalated with Portuguese agents trying to sabotage the mission by spreading rumors that Magellan intended to guide the fleet into Portuguese waters and betray Spain. Spanish authorities withheld final payments, causing supply shortages. The crew grew agitated by inadequate accommodations while in port and began deserting; Magellan had to post guards to prevent nighttime absconders.

On August 10, 1519, the fleet finally departed Seville and proceeded downriver towards the coast, but did not set sail for another month, hindered by last-minute preparations and a supply stopover at Sanlúcar de Barrameda. During this delay, more crew members abandoned ship. The Portuguese ambassador communicated warnings that Magellan's voyage was doomed to fail, claiming he led men to their deaths and that the promised strait did not exist.

Magellan remained silent in his defense; through years of rejection and betrayal, he had learned that words were worthless unless backed by tangible results. Only the actual discovery of the Strait, passage through the Americas, arrival in the Spice Islands, and return to Spain laden with cloves could silence his detractors.

By September 20, 1519, the fleet finally ventured forth. The weather was favorable, ships formed a line, and they advanced into the vast Atlantic. Behind them, the Spanish coast receded into the distance until it disappeared completely. Before them lay three years of the unknown; none could yet comprehend just how immense that unknown would turn out to be.

THE FIRST STORM

The Atlantic crossing to South America, which Magellan embarked upon, was not supposed to be a simple task. The fleet maintained a southwesterly course, avoiding Portuguese-controlled waters near the African coast. During daylight hours, ships stayed in sight of each other and communicated through signal flags, while at night they used lanterns. Life at sea in the sixteenth century was marked by harsh discipline; men were whipped for drunkenness, theft, or failing to carry out orders promptly. Food rations consisted of hardtack biscuits infested with weevils, salted meat that had to be soaked before eating, dried fish, beans, onions, and a daily ration of wine – the only liquid without an overpowering stench of rot. The smell of tar-soaked hemp permeated the air, serving as a constant reminder of the ship's continuous movement, creaking timbers, and the ever-present threat of storms. Below deck, sailors were greeted by the rhythmic slosh of bilge water, a haunting cadence that underscored the vastness and isolation of their journey.

However, the real issues were not the challenging conditions. They were the men themselves.

Captain Juan de Cartagena of the San Antonio began to challenge Magellan's authority almost immediately after setting sail. During evening conferences when captains gathered on the flagship to receive orders for the following day, Cartagena refused to show proper respect. He referred to Magellan as "captain" rather than "captain-general," a slight meant to remind everyone that their commander was a foreigner commanding Spanish ships. When Magellan issued orders or outlined the route, Cartagena would question them and suggest alternatives. This was a slow and calculated campaign of insubordination designed to undermine authority without crossing into open mutiny.

Magellan demonstrated patience over several weeks, surprising those familiar with his temper. However, for him, this patience was not a sign of weakness; it was preparation. He awaited the moment when Cartagena would overstep, providing a clear justification for action.

This tipping point came off the coast of West Africa during one of the evening conferences, where Cartagena openly refused to acknowledge Magellan's command and stated that he would no longer follow orders from a Portuguese traitor. The other captains observed in silence, eager to see how Magellan would react. If he accepted the insubordination, his authority would collapse, leading to fragmentation of the fleet and failure before even embarking on their true journey.

Magellan reacted with surprising speed. He seized Cartagena by the collar, declaring him under arrest for mutiny, and called for the guards. Struggling against being taken into custody, Cartagena argued that Magellan had no authority over a Spanish nobleman. However, the guards, following their captain-general's orders, dragged Cartagena away and placed him in chains on board the Victoria under the watch of Captain Luis de Mendoza.

The other captains remained frozen, as Magellan stared them down, daring them to object. No one uttered a word. He dismissed them and returned to plotting their course. The message was clear: he would tolerate no challenges - not from noblemen or anyone else. They would follow his orders, or he would break the fleet apart by his own hands if necessary.

Cartagena remained in chains with Antonio de Coca, the fleet's accountant, now taking command of the San Antonio. This drastic breach of protocol replaced a nobleman with a bureaucrat, but Magellan had the contract from King Charles and the authority it granted him. For the time being, this was enough.

The fleet continued south, crossing the equator in late November. The heat was unbearable. Below deck, the air was so thick and foul that men could hardly breathe. Water supplies stored in wooden casks grew stagnant and slimy, while biscuits became infested with weevils to the extent that sailors learned to eat them in the dark to avoid seeing what they consumed. Scurvy had not yet appeared, but conditions were laying the groundwork for its onset through gradual nutrient depletion, leading to the betrayal of men's bodies.

On November 29, the fleet sighted the coast of Brazil near present-day Rio de Janeiro. Their encounter with the indigenous Tupí people was peaceful and almost idyllic, in contrast to the tensions on board the ships. They traded food, fresh water, and wood for knives, bells, and playing cards. Men went ashore, bathed in fresh water for the first time in months, ate fruit and fresh meat, and attempted to regain some health before returning to sea.

Magellan allowed the fleet to remain for two weeks to replenish supplies and make repairs. This break provided respite but also deepened divisions within the fleet. Spanish captains, including Gaspar de Quesada and Luis de Mendoza, met privately to consider whether to continue under Magellan or seize control of the fleet themselves. While not daring to open mutiny yet, they prepared for it by sounding out which crew members might support them and identifying officers who could be turned.

In mid-December, the fleet left Brazil and continued south along South America's coast. The weather grew colder; days shortened until January 1520, when they were sailing through unmapped waters, searching for the strait Magellan had promised existed. They explored every bay, every inlet in hopes of finding passage. Each time, they discovered only rivers or dead ends. The estuary of Río de la Plata, enormous and promising, was found to be fresh water, not a strait. They continued south.

By March, southern autumn gave way to winter as seas grew violent, with the wind blowing against their course, forcing ships to tack endlessly and making minimal progress. The coast was barren, with cliffs, rocks, and occasional beaches where giant birds wandered. Men began muttering that there might be no Strait and that Magellan had led them into a wasteland. They feared all dying here for one man's obsession. In the midst of this despair, Gabriel, a young sailor, lifted spirits despite hopelessness. Clinging to rigging each morning with steady hands, mending and adjusting quietly, occasionally he would call out a jest or two, his voice cutting through the wind, drawing chuckles from shipmates. This small act of resilience hinted at the crew's hidden strength, tempering gloom with an echo of hope.

On March 31, 1520, Magellan ordered the fleet into a sheltered bay to weather the winter. The place was bleak, windswept, and almost treeless, offering no respite except the storms. He named it Puerto San Julián. This would be remembered as a place where the expedition nearly fell apart.

THE MUTINY

In Puerto San Julián, it snowed on the first night. The crew members were inside their boat, tightly wrapped up with whatever fabric they could find, listening to the cold wind howling over the vast, empty bay. They were thousands of miles from any European settlement and found themselves trapped in a location where winter was just beginning, and they would have to remain for months until the weather allowed them to move forward. The food rations were cut again. Magellan announced that they would stay here during winter, no further exploration until springtime, and the men had to endure. The Spanish captains saw an opportunity for themselves.

On April 1st, 1520, while Magellan slept aboard the Trinidad, a mutiny took place in three of the five ships. The plan was well-organized and carried out quickly. Gaspar de Quesada, who was still captain of the Concepción, led a boarding party to take over the San Antonio, which was the largest ship in their fleet. They managed to overpower the loyal officers, killed the master Álvaro de Mesquita - Magellan's cousin - and took control of the ship. Luis de Mendoza, captain of the Victoria, joined this conspiracy. Juan de Cartagena, who was technically still under arrest, was released and declared a leader of this rebellion. By dawn, the mutineers had control over three ships: San Antonio, Concepción, and Victoria. Magellan held only Trinidad and Santiago. The number of mutineers was higher than that of loyalists. They sent a message to Magellan demanding that he submit to their authority, suggesting that they should return to Spain, accept the failure, and stop leading them towards death in pursuit of a nonexistent strait. Magellan remained silent. He observed the ships' positions in the bay. The mutineers made one crucial error: they didn't precisely coordinate their positions. The Victoria was anchored slightly away from the other two ships, close enough to the entrance of the bay so it could be accessed easily.

Magellan prepared a boarding party. He selected fifteen men, armed them with swords and crossbows, and gave them a simple task: board the Victoria, eliminate Captain Mendoza if necessary, and take control of the ship. To transport these men, he used a longboat as if they were delivering a message to Mendoza, inviting him to a meeting. When their boat approached Victoria, the men boarded peacefully. Unaware of the danger lurking within, Mendoza came forward to receive the letter. Upon opening it, one of Magellan's soldiers, named Gonzalo Gómez de Espinosa, stepped forward and stabbed Mendoza in his throat. The captain collapsed as blood flowed across the deck. The boarding party then drew weapons and quickly overpowered the crew before they could react. Within minutes, Victoria was back under Magellan's control. Now the balance had shifted; Magellan controlled three ships while mutineers held two. However, more importantly, he controlled the entrance of the bay, trapping the mutineers. The San Antonio attempted to escape into the open sea in hopes of returning to Spain or continuing their rebellion elsewhere. Magellan positioned Trinidad and Victoria to block the exit path. When San Antonio tried to force its way through, Magellan's ships fired cannon shots - not to sink but to intimidate and demonstrate that he was willing to destroy the ship if necessary. The crew of San Antonio, facing being blown apart by their fleet, lost their nerve and surrendered. The mutiny lasted less than twenty-four hours.

Immediate and brutal punishment followed. Magellan ordered the arrest and trial of the ringleaders. Gaspar de Quesada and Luis de Mendoza were found guilty of mutiny - a crime carrying the death penalty. Mendoza was already dead from the boarding, so his body was beheaded and quartered; these pieces were displayed on the beach as a warning. On the same day, Quesada was executed by his own servant, who was promised clemency for carrying out this sentence. Juan de Cartagena and Pedro Sánchez de Reina, a priest supporting the mutiny, were left marooned. When the fleet finally left Puerto San Julián months later, they abandoned these men on a barren coastline with minimal supplies. No one ever knew their fate; they were never seen again.

Forty other men involved in the mutiny were sentenced to death. However, Magellan couldn't afford to kill them as he needed every able-bodied person to man the ships. Instead of execution, they were condemned to hard labor, chained up, and forced to work for the entire voyage duration. It was a calculated mercy - sparing their lives but ensuring that each day in pain of shackles would remind them of their rebellion. The fleet settled into uneasy peace. Mutiny had been crushed, yet trust between Magellan and his crew was shattered. Now knowing that a significant portion of his crew wanted to turn back, believed the voyage was doomed, and could rebel again if they got the chance, he increased security measures by placing more guards on the ships, moved loyalists into key positions, and waited for winter to pass - understanding that every day's delay increased chances for another uprising. In May, Magellan sent Santiago - the smallest ship - on a scouting mission south along the coastline in search of a strait. The Santiago sailed for weeks exploring the coast and was then caught in late May by a storm; it crashed onto a desolate shore. The crew survived but had to travel overland back to Puerto San Julián, arriving emaciated and half-frozen. Losing the ship was disastrous. Now they had only four vessels, and the men's confidence in Magellan's leadership continued to erode. As winter deepened, the cold became unbearable. Men spotted strange creatures on shore: gigantic flightless birds called "strange geese" (which we now know as penguins) and tall, odd humans dressed in animal skin who left massive footprints in the sand. Magellan named these people Patagones - meaning 'big feet.' This region was then known as Patagonia. These encounters illustrated European tendencies to mythologize unfamiliar peoples and landscapes, fueling legends of giants at the empire's edge. Contemporary scholarship recognizes that so-called Patagones were likely Tehuelche, an indigenous group whose physical stature and cultural practices were exaggerated in European accounts. The interactions between these two groups were brief and tense. Indigenous people wanted nothing to do with Europeans, while weakened by cold and scarce rations, Europeans lacked the energy for exploration. In August, as the southern spring approached, Magellan ordered the fleet prepared for sailing. As much as possible, ships were repaired, supplies redistributed among the remaining four vessels, and on August 24th, almost five months after arrival in Puerto San Julián, they resumed their search for the Strait. Although they lost one ship, executed two captains, marooned others, and nearly tore the expedition apart, Magellan had survived. He crushed mutiny with a combination of ruthlessness and strategic brilliance. As his ships sailed south into the unknown territory, he knew that his authority was absolute; even though men hated him, they feared him more, and in the end, fear proved to be enough.

THE PASSAGE

October 21, 1520.

Amid a tempest raging towards cliffs and crags, Trinidad faces off with northeast gales, a cacophony of destruction imminent. Magellan commands sails be drawn tight, a feeble bid to forestall calamity; control eludes them still. The crew braces for the inevitable impact, cognizant that upon striking rocks, disintegration will ensue in moments.

Miraculously, the crags part. A slender channel manifests, not much wider than the vessels, flanked by colossal rock formations. Magellan steers his fleet into this gamblesome passage: a trap if it's an inlet or river's maw, salvation if a strait.

The vessels navigate through the watercourse. Its depth is profound, current strong and west-bound. The men sample the liquid; its essence is saline. Not a river nor a bay, they've found something else.

Their quest reveals the strait Magellan boasted of, scorned as folly by detractors who cost him a ship, lives lost, and five months imprisoned in ice. It's real and tangible now that the passage through America exists. Yet, this is merely the beginning; the strait is not just a straightforward channel but an intricate maze with dead ends and branching fjords that veer westward or loop back into walls of ice and rock. Every mile navigated here feels as if it stretches into ten, amplifying the dread and heightening the stakes. The weather is tempestuous, with winds shifting capriciously and rain unceasing. Four ships probe the pathways with care, deploying longboats to sound depths and scout for the western passage.

Navigating this labyrinth takes more than a month; cautious and slow, they're ever unsure whether the path ahead lies in the right direction or a snare. By nightfall, they moor in tight coves walled by cliffs that blot out stars. To the south, fires smolder; their glow belongs to native folk dwelling here. Magellan christens this stretch Tierra del Fuego, Land of Fire, a name that holds.

In a northern offshoot, the San Antonio vanishes; cannons and expeditions seek it in vain, fruitlessly. Unbeknownst to Magellan or ever to learn, mutiny has seized the largest of his surviving ships. The pilot, Estêvão Gomes, overpowers Captain Álvaro de Mesquita, a relative, and reverses course. San Antonio retraces the strait's path, crosses back into the Atlantic, bound for Spain, bearing away most of the remaining rations.

This setback is devastating; only three vessels, Trinidad, Concepción, and

Victoria, now teeter on the edge of survival, with meager victuals to sustain them. The commitment to the strait is unwavering; retreat is now impossible. They press forward into western seas.

On November 28, 1520, after thirty-eight days spent threading the watercourse's convolutions, they burst through its terminus into open waters. Beyond lies an ocean placid after the turmoil within the straits, a serene, cloudless sky above, radiant sun below. Magellan dubs this new sea Pacifica, the Peaceful Sea.

Unbeknownst to him, it's not diminutive; Europe buzzed with the notion of a narrow band connecting America and Asia. Yet, Waldseemüller's conception would be eclipsed by the reality of its vastness; a third of Earth's expanse lies within the Pacific's domain. And facing this immensity are three ships short on provisions and manpower, about to confront their journey's most arduous phase. In this moment, Magellan permits himself a rare release of joy; tears stream down as he acknowledges the truth of his discovery. No matter what follows, a western passage to the East is demonstrated to be feasible; he has redrawn the world map.

The crew does not share his jubilation. They tally their diminishing supplies and gauge the ocean's enormity before them, recognizing their path now irreversible; either they'll claim the Spice Islands or succumb in pursuit. Magellan commands northwestward, veering into the open Pacific, venturing into an abyss that will test endurance beyond limits previously faced.

THE OCEAN OF HUNGER

The voyage across the Pacific initially fueled the sailors' hopes of reaching land swiftly. Geographers had estimated a distance of approximately six hundred leagues between the Americas and Asia, equating to around two thousand miles - a mere few weeks' journey with favorable conditions. The loss of San Antonio notwithstanding, provisions were deemed sufficient. Calm weather accompanied them, with steady winds propelling their ships northwest into warm waters beneath clear skies.

However, by the second week, concern crept in as land remained elusive. An endless sea surrounded them, devoid of any sign of dry ground. They continued northwest, driven by Magellan's conviction that the Spice Islands lay ahead, but discrepancies in calculated distances and expectations began to surface, unveiling the vastness of the world previously unimagined.

Food rations were tightened by the third week, with the number of biscuits per person drastically reduced. Water, once stored neatly in wooden casks, became undrinkable, teeming with algae and rat carcasses. Desperation led them to consume it mixed with wine, while salted meat and beans vanished. They resorted to eating biscuits infested with weevils and mold, the only remaining option.

Scurvy claimed its place by the fourth week, manifesting through exhaustion and swollen gums, leading to tooth loss. The afflicted faced constant pain, delirium, and hallucinations as their bodies suffered from vitamin C deficiency. Despite these dire conditions, the ships pressed westward, leaving no room for turning back. Insufficient supplies made return impossible; forward movement became their only hope, relying on the belief that land would eventually emerge to defy the seemingly infinite ocean.

By the sixth week, sustenance dwindled further as they resorted to eating leather softened in seawater and roasted sawdust from cask remnants. Even rats, once scavengers of their own food scarcity, became a vital source of protein when hunted and traded among sailors for a ducat each - a price that could have afforded a chicken in Spain.

Italian scholar Antonio Pigafetta documented the crew's grim existence: consuming rotting biscuit powder teeming with worms, foul water rife with rat urine, ox hides, sawdust, and rats sold at exorbitant rates. Death became a slow affliction as men succumbed to scurvy, their bodies unable to maintain health without vitamin C.

On December 18, after seventy-eight days in the Pacific, an uninhabited island appeared on the horizon. Named Isla Desierta for its barren nature, landing was impossible despite the crew's desperate yearning. A week later, they encountered another lifeless rock dubbed Isla de los Pájaros, where they briefly set foot and hunted birds for sustenance but to no avail against scurvy's relentless grasp.

The relentless ocean beckoned them westward, their suffering now routine, as men operated like machines, fulfilling duties mechanically. On March 6, ninety-eight days post-strait departure, land was finally spotted - lush and inhabited. The crew, overjoyed, met the Chamorro people of the Marianas on Guam and Rota.

An initial peaceful encounter offered coconuts and fish, but tensions arose from cultural misunderstandings regarding trade practices, leading Magellan to authorize an attack, resulting in burned villages and deaths. Despite the brutality, the plunder provided temporary sustenance, enabling recovery through a long process. Westward sailing resumed with renewed hope, and on March 16, they discovered the archipelago of the Philippines, named after Saint Lazarus.

The expedition had conquered the Pacific, enduring the longest ocean voyage in history without resupply across over seven thousand miles, suffering severe losses to starvation and scurvy, but arriving at a new chapter filled with unforeseen consequences.

THE BEACH

April 27, 1521, Mactan Island, Philippines, just before dawn.

Ferdinand Magellan wades through shallow water towards a beach defended by warriors with no intention of submitting to a foreign king. He is clad in full armor, including a steel breastplate and helmet, and he is wielding both a sword and a musket. In his wake, forty-eight men from the fleet advance in disarray, similarly armed; the water up to their thighs hampers them, leaving them exposed.

On the shore, fifteen hundred Filipino warriors stand ready under Lapu-Lapu's command, rejecting Magellan's demands for submission and conversion. They arm themselves with spears, swords, and fire-hardened bamboo stakes, and they have an advantage in numbers, terrain, and knowledge of shallow water combat.

This encounter should have been averted; Magellan should've been negotiating instead - using interpreters, offering trade goods, and fostering alliances, as he'd done successfully before. But on Mactan, something has fractured within him; driven by the need to demonstrate Spanish dominance, he aims to crush resistance.

He's been cautioned about his hubris; Rajah Humabon of Cebu, having allied with the Spanish, offered a thousand warriors for support in this endeavor, but Magellan dismissed them, claiming that forty-eight Europeans would topple a thousand Filipinos. This arrogance proves his downfall.

The skirmish erupts into gunfire, with Spanish musketeers and crossbowmen firing upon the beach's amassed warriors, yet the range is too great, and the water slows the bullets; they do not falter. The warriors surge forward, shouting, spear-charging, and launching fire arrows aimed at ships offshore.

Magellan orders his men to burn the village to intimidate the defenders, but the strategy backfires. Enraged by smoke and flames, the warriors swarm aggressively. The Spanish line shatters; retreat begins as they realize their grave underestimation.

Magellan fights valiantly for a retreat despite his limp; he's skilled with the sword, fending off attackers. Yet the warriors identify him as the leader and focus their attacks on him. Pierced by spears, he continues to hack at foes around him until overwhelmed; a bamboo spear lacerates his arm, disarming him.

Antonio Pigafetta, wounded and watching from the boats, recounts the moment: "Many turned upon him, knocking off his helmet twice... An Indian hurled a bamboo spear into the captain's face, but he immediately killed the assailant with his lance. However, he could only draw his sword halfway due to his wounded arm. The heat of the sand seeped through his armored feet, grounding him in the present danger. Witnessing this, all rushed upon him; one struck him on the left leg with a large cutlass... and he fell; they then pounced upon him until our mirror, light, comfort, and true guide was killed."

Ferdinand Magellan dies in the surf at Mactan, felled by those defending their land against invaders. Aged forty, he had navigated farther than any European before him - discovered the strait named for him, crossed the Pacific, proving the world's interconnectedness.

Survivors flee to boats, desperately rowing towards ships. Seven other

Europeans perish in combat. The expedition is devastated; Magellan has died. His slave Enrique, who may have circumnavigated the globe first, will vanish shortly after without a record of his fate. Traumatized and leaderless, remaining captains grapple with control.

Rajah Humabon, suspecting the Europeans of the disaster on Mactan, turns against them. Inviting Spanish officers to a supposed reconciliation banquet, he betrays them; twenty-nine men enter, including João Serrão and Duarte Barbosa. Warriors ambush and massacre during the feast; only one man, João Carvalho, escapes.

The expedition teeters on collapse. Four ships remain, but lack the manpower to operate all of them. The Concepción is abandoned, stripped of supplies, burned. Only Trinidad and Victoria can sail on - barely a hundred men alive between both. They head westward, still hunting for the Spice Islands, now a survival ordeal.

By November 1521, eight months post-Magellan's death, these two ships finally arrived at the Moluccas (true Spice Islands). Anchoring at Tidore, they trade cloves with locals, seeking Portuguese opposition allies; their cargo would justify this perilous voyage financially if returned to Spain.

Trinidad leaks so severely that it can't be repaired here and must remain while Victoria sails west, retracing the route around Africa. Trinidad never makes it back - months of repairs send it east across the Pacific, where a Portuguese fleet captures most on board.

Victoria, under Juan Sebastián Elcano's command (one of Puerto San Julián's mutineers but a proven navigator), embarks westward with eighteen men from the original crew plus transfers from other ships, overloaded with cloves and barely seaworthy. They endure another Atlantic crossing rife with death due to scurvy, starvation, and exhaustion.

On September 6, 1522, three years post-departure from Spain, Victoria docks in Sanlúcar de Barrameda, having survived with eighteen souls out of two hundred sixty-five; they've circumnavigated the globe.

Loaded with cloves worth more than expedition costs, their return justifies the voyage despite catastrophic loss. They bring new knowledge, reshaping European worldviews - maps, journals documenting an Earth once thought flat and now known to be round, united by water.

Yet Ferdinand Magellan is not among them; buried unmarked in the Philippines or lost in Mactan's surf. History will remember this journey as his, name the strait after him, and call it the Magellan expedition, but he never lived to see completion, nor claim glory, wealth, or vindication sought.

THE WEIGHT OF DISTANCE

In the immediate aftermath of their journey, the people of Europe found it difficult to understand exactly what had transpired.

Eighteen individuals emerged victorious and were both lauded and questioned extensively. King Charles granted Juan Sebastián Elcano a coat of arms depicting a globe and the Latin phrase "Primus circumdedisti me", meaning "You were the first to encircle me". These survivors received financial support, although many succumbed within a short period due to injuries sustained throughout their voyage.

The geographical ramifications of their findings were quite remarkable. The Pacific Ocean was not merely a narrow body of water but an extensive expanse bigger than all land masses on Earth combined. The distance from Europe to Asia by sea in a westerly direction proved impractical for trade due to its vastness. Though the route discovered by Magellan, which led through the strait at the southern tip of the Americas, was dangerous and time-consuming, it ruled out Spain's ability to exploit this path for regular commerce.

The Treaty of Tordesillas, which had previously divided the world between Spain and Portugal, became even more contentious. The exact location of the dividing line on the opposite side of the globe remained uncertain. Both kingdoms claimed control over the nearby Spice Islands. Eventually, in 1529, Spain sold its claims to the Moluccas to Portugal for 350,000 ducats, relinquishing dominance over the spice trade to their rival. It was proven that the western route was possible, but not economically viable.

Despite these setbacks, the voyage forever altered the course of human knowledge and understanding. It empirically confirmed what theorists had long argued: the Earth is a sphere with finite and measurable dimensions. The distances were far greater than anyone could have imagined, yet they became known quantities. Furthermore, the oceans became interconnected, enabling ships to set sail from Europe, journey westward, and return from the eastern shores of the globe. This newfound understanding brought about a sense of unity in the world.

Magellan himself became a contentious figure - some remembered him as a traitor who abandoned his homeland for Spanish allegiance, while others honored and yet overshadowed him due to Elcano's completion of the circumnavigation. In the Philippines, Lapu-Lapu was celebrated as a national hero and symbol of resistance against colonialism.

The historical narrative is heavily influenced by European sources from that era, which often viewed indigenous populations merely as obstacles or curiosities for European interests. This skewed perspective overlooked the legitimacy and agency of the encountered societies, such as those of the Chamorro people, Filipinos, and Patagonians. Thus, our understanding of Magellan's expedition remains heavily reliant on the invaders' perspective rather than the perspective of the invaded.

It can be concluded that Magellan was an individual of remarkable determination, strategic acumen, and ruthless discipline. He also demonstrated a capacity for extreme violence in executing mutineers, massacring villagers, and enlisting people for his goals. His motivation stemmed from ambition, wounded pride, and genuine curiosity about the world's shape. Believing fervently in his vision, he persevered through years of rejection to risk everything on an uncertain route.

Although Magellan did not live to achieve his immediate objectives, his voyage proved that global circumnavigation was indeed possible. The Earth's true size could be measured accurately, and oceans were demonstrated as highways rather than barriers.

The Strait of Magellan remains a treacherous waterway today, but it played a pivotal role in the sixteenth-century era of exploration as the only known passage between the Atlantic and Pacific Oceans. This geographical keyhole facilitated the dissemination of knowledge about the Earth's true scale.

Magellan's legacy is complex, reflecting both his accomplishments and atrocities, as well as his visionary capabilities coupled with violent tendencies. He broadened human understanding while participating in colonialism and exploitation that would devastate indigenous populations worldwide. He was an exceptional navigator but also a brutal commander capable of inspiring loyalty or fear equally.

The most striking aspect of Magellan's story is the enormous loss of life involved. Out of 265 men who embarked on this journey, only 18 returned. Although casualty rates were common during early modern exploration, such a high percentage was still shocking. The crew understood these risks and yet chose to pursue their goals driven by desperation, curiosity, poverty, or the promise of wealth.

The world they discovered was far larger and stranger than they had anticipated. It was characterized by complex societies with their own technologies, beliefs, and political systems. The distances traversed were nearly incomprehensible, involving oceans that took months to cross and coastlines that stretched thousands of miles.

At the heart of this discovery stood an individual who refused to accept impossibilities simply because they had not been accomplished before. Magellan was neither the first nor the most educated geographer or experienced sailor; however, he was willing to risk everything - his reputation, life, and the lives of his crew - on the belief that a western route to the East did indeed exist.

He succeeded in finding this route but ultimately lost his life in the process. This paradox is emblematic of exploration: often, discoveries with transformative impact are made by individuals who do not live long enough to witness their consequences. Magellan demonstrated that global circumnavigation was achievable, yet he himself died before completing the journey. His death on a Philippine beach due to a convergence of personal pride, strategic misjudgments, and cross-cultural violence continues to represent the expedition's complex legacy.

Historians have debated whether Magellan should be celebrated primarily as a visionary navigator or criticized for being an agent of imperial violence and cultural misunderstandings. This duality shapes our interpretations of his achievements and highlights the tension between expanding geographical knowledge and the destructive impact of European colonial ventures.

The Victoria returned to Spain bearing cloves, survivors, and crucial knowledge about the Earth's measurable nature. The cost at which this knowledge was obtained - devastatingly high - would influence centuries to follow, enabling global trade networks, colonial empires, and eventually mapping every coastline and island. The Age of Exploration that Magellan helped inaugurate connected the world and wrought both wealth and suffering beyond previous imagination.

However, on that morning in September 1519, as they set sail from Sanlúcar de Barrameda into the Atlantic Ocean, none could have foreseen these outcomes. At the time, all that existed was the horizon - vast and unknown - along with the conviction of a scarred and injured man who believed it could be conquered. History would prove him correct. It would also prove his mortality.

The world was round, as Magellan had proven. And this proof came at the cost of everything he held dear.

Fernão Mendes Pinto

September 23, 1543. Off the coast of Tanegashima Island, Japan.

The storm screamed.

Rain pummeled the deck. The Chinese pirate junk creaked and swayed, its wooden frame obscured by the wind's fury. Clinging desperately to the mast, Fernão Mendes Pinto felt salt sting his eyes, his knuckles tightening around the rope. Reflecting on thirteen wrecks over six years, he almost chuckled, a grim humor surfacing as the vessel lurched violently. He instinctively knew the stories he might one day tell stories that would blur the lines between reality and legend, casting him in the dubious light of what some would call a 'Prince of Liars.' As he caught sight of land, ragged cliffs and pine trees bending under the gale, these shores were utterly unfamiliar. The Chinese captain bellowed orders Pinto barely understood, but the message was clear: the sea would soon claim them all.

The crash sounded like the world itself shattering. Planks shattered. Men soared into the air. The ocean swallowed everything: silk, silver, weapons, every dream of profit. Pinto plunged into darkness. His lungs screamed as he twirled in the current, adrift in a realm without up or down. His hands grazed coral; he kicked frantically, hoping the light above was genuine.

He breached the surface, choking on salt and air. Debris bobbed around him. A few survivors flailed nearby. The shore beckoned, one hundred yards away. The current pulled him back, but he fought it valiantly. His muscles screamed, his mouth seared with brine, yet he pressed onward. Thirteen wrecks had not been the death of him by chance, destiny, or sheer will; he defied death.

Finally, his feet touched sand. He dragged himself ashore and collapsed, wracked by shivers and vomiting seawater. As he lifted his head, he saw a ragtag group emerging from the surf: two Portuguese, the interpreter, and a few Chinese sailors. Many others had vanished with the ship, leaving behind only splinters on the rocks.

Perched high on the cliff, warriors stood against the storm-tossed sky. Their armor gleamed with raindrops. Armed with bows and swords, they watched silently, making no move to descend.

Pinto staggered to his knees, then struggled to stand. Twenty-eight years old, he had endured slavery in Arabia, imprisonment in China, torture in Burma, and seventeen transfers from one master to another. He spoke five languages, none perfectly. Now, all he possessed were drenched rags and the battered arquebus that had survived with him through the shipwreck.

He raised his arms and approached the waiting warriors. Behind him, a Portuguese voice whispered a prayer. The warriors gazed down, unmoved. In that moment, Fernão Mendes Pinto set foot on Japanese soil for the first time just hours after surviving his thirteenth wreck. Whatever lay ahead, his fate hung in the balance.

He knew not that centuries later, in Portugal, his name would become a jest: "Fernão, Mentes? Minto!" , "Fernão, do you lie? I do!" At this moment, as he endeavored to survive along unfamiliar shores, little could he suspect that his tales would become the subject of credibility debates for generations. Nor could he foresee that scholars would debate his veracity for generations or that his book would become one of the most renowned and disputed travel tales ever written.

All Pinto was aware of was this: he was alive impossibly, defiantly alive. Somewhere on this alien shore, perhaps, lay a way back to the Portugal he had left six years prior as a penniless dreamer. But first, survival in the next hour was imperative.

The warriors descended from the cliff.

THE BOY WHO FLED

To truly comprehend how one might find themselves stranded on the fringes of the world, it is essential to examine their roots. Born circa 1509 in Montemor-o-Velho, a quaint hamlet nestled in central Portugal near Coimbra, Fernão Mendes Pinto's exact birth year remains uncertain, whether 1509 or 1510. Given that records of those in dire straits seldom endure, we find little documentation of Pinto's lineage, despite claims to nobility which brought them no material gain.

Portugal in 1509 was a realm reborn. Just over a decade prior, Vasco da

Gama had navigated the vastness of Africa and reached the shores of India, thereby unleashing the floodgates for the spice trade. Portuguese marauders now charted courses through seemingly infinite waters. Lisbon, their capital, swelled with exotic wares: Malabar pepper, Ceylon cinnamon, and silks from distant lands. The kingdom was in the throes of establishing its dominion, each port conquered, each treaty forged, each victory stained by blood.

Yet, in Montemor-o-Velho, destitution reigned supreme. Bereft of a father at a young age and with his mother wrestling against penury, Pinto's family seemed to bear the spectral mark of Jewish marranos, Jews converted under duress yet burdened by their ancestry. Though connected to the wealthy Mendes clan, who controlled the pepper trade, this lineage afforded him naught but knowledge of cousins living in opulence he could only imagine.

Around age ten or twelve, precisely when lost to time, an uncle mercifully extracted Pinto from his squalid existence and ushered him to Lisbon. The city must have seemed an alien realm. The docks teemed with sailors conversing in Arabic, Gujarati, and Swahili tongues. Exotic creatures arrived in enclosures. Rua Nova dos Mercadores dazzled with the affluence of merchants. And looming over all was the royal palace.

Settling young Pinto into the home of a prosperous noblewoman, perhaps tied to the court, he served as an attendant for eighteen months. There he gleaned the subtleties of courtly life, manners of deference, silence when it counted, discerning the sentiments of potentates. He learned his alphabet, though never with eloquence. His status was evident: a charity case amidst silk grandeur.

Then came the inciting incident. Pinto's recounting is shrouded in vagueness: "something happened that placed me in such great jeopardy that I was forced to leave the house at a moment’s notice and flee for my life." What triggered his flight? Did accusations of theft fly? An untoward proposition, resisted or acquiesced? A breached trust? Or merely a glimpse of something that should have remained unseen? The mystery persists. Even in his memoirs, penned decades later, with all key players silenced by the sands of time.

What is certain is his departure. He hastened to Alfama's labyrinthine quarters, where prying questions were seldom entertained. Arriving at the docks as a desperate soul, he secured a berth on a cargo vessel bound for Setúbal, sixty kilometers south. Their destination remained unattained. French corsairs seized the ship mid-journey just fifteen miles shy of its goal.

It was here that Pinto encountered the vast world's harsh realities: seizure, captivity, and abrupt upheaval. The pirates were methodical in their plunder. They stripped away all valuables before jettisoning the crew upon Spain’s shores. A lone thirteen-year-old, he navigated back to Setúbal, begging or laboring for sustenance.

In Setúbal, Pinto secured work with a noble of the Order of Santiago and later with Jorge de Lencastre, a bastard son of King John II. These roles, though menial, were stable and secure, ensuring a life unremarkable but steady, possibly affording small comforts like a house or marriage.

Yet within him stirred a revulsion against the mundane. Mayhap it was the gnawing edge of childhood poverty. Perhaps it was knowledge of the Mendes family's fortune that eluded his grasp. Or simply an insatiable yearning for horizons unknown, an inability to settle for what is, when wonders beckoned from beyond.

At twenty-eight in 1537, Fernão Mendes Pinto abandoned secure positions and joined the Portuguese India Armada, setting sail on this epic quest. He possessed no wealth, influence, or special talents. His only asset was an unyielding ambition to seek his fortune in the East, where riches were won overnight, spices grew abundant on trees, and the impossible beckoned.

On March 11, 1537, he embarked upon a Lisbon-bound vessel destined for the Cape of Good Hope, up Africa's eastern coast, and across the vast Indian Ocean. The voyage spanned six perilous months. Many did not survive its trials. As Pinto's homeland receded into the horizon, Lisbon fading in the ship's wake, he was unaware that twenty-one years would pass before his feet trod Portuguese soil again. Upon return, he would be unrecognizable, a man remade by adventure, harboring tales so fantastic that even his countrymen might dispute their veracity.

The winds caught the sails. Portugal dissolved into mist. Fernão Mendes Pinto sailed forth towards an ordained destiny.

The journey to India was akin to traversing the depths of water and wood. The Carreira da Índia, the India Run, featured massive carracks capable of holding hundreds of men and vast amounts of cargo, yet these same vessels became coffins for many due to disease outbreaks such as scurvy, dysentery, and typhus. The water aboard grew brackish and thick while weevils infested the hardtack, leading to the demise of numerous men who were buried in canvas and submerged into the Atlantic's embrace. The Cape of Good Hope, named aptly as the Cape of Storms, unleashed waves reaching heights of forty feet, threatening to engulf ships entirely, prompting sailors to cling desperately to masts to avoid being swept overboard. The ship groaned under these torments, its timbers creaking and seams leaking profusely, with constant bilge pumping needed to clear the bilges, while the stench of stagnant water, human waste, and decaying food was almost insufferable.

Yet Pinto endured. His innate talent for survival would later serve him well on numerous occasions in the years that followed.

On September 5th, 1537, after a nearly six-month voyage at sea, the ship arrived at Diu, a fortified Portuguese island located off the northwestern coast of India. The fortress bristled with cannon while ships crowded the harbor, filling it with the scent of spices and smoke. Beyond the walls lay the army of Suleiman the Magnificent, Sultan of the Ottoman Empire, as they besieged the city.

Pinto had arrived in India amidst a war.

The Portuguese empire in Asia was not truly an empire but rather a collection of fortified trading posts connected by ships and violence. Portugal lacked the population to conquer Asia; instead, it dominated sea lanes with naval technology. The strategy was to control chokepoints: Hormuz in the Persian Gulf, Goa on the Indian coast, and Malacca in the Malacca Strait. From these fortresses, Portuguese ships enforced a cartaz system functioning essentially as a protection racket. Asian merchants were required to pay for licenses or risk having their ships seized. Penalties for non-compliance were severe, effectively compelling compliance.

The Ottomans contested this control, attempting to push seaward from their land-based empire. Diu was a prize worth taking, but the siege would ultimately fail; Portuguese cannon and European tactics prevailed. Young Fernão Mendes Pinto caught his first glimpse of what it truly meant for Europe to dominate Asia: constant warfare, shifting alliances, the smell of gunpowder and burning flesh.

Pinto did not linger in Diu for long. Within months, he joined an expedition to the Red Sea and Ethiopia under the guise of an embassy. Portugal sought out allies against the Ottomans and intelligence on Turkish naval strength. The mission sailed up the Red Sea towards Massawa in Eritrea, where they encountered Turkish galleys; sleek, fast warships propelled by slaves' synchronized oar strokes. These were pitted against Portuguese ships, which were designed for cargo and survival in storms. When three Turkish galleys intercepted them, it became clear that the Portuguese would not prevail.

The ensuing battle was swift and brutal, with Turkish janissaries boarding with curved swords and composite bows, while Portuguese arquebusiers fired their guns before engaging in hand-to-hand combat. Despite this effort, numbers and speed ultimately led to the Portuguese ships being seized. Those who survived were chained and transported to Mocha in Yemen.

The slave market reeked of human bodies and fear as men were stripped, examined like livestock, with teeth checked and muscles prodded. Pinto witnessed his companions being sold one by one. He learned a crucial lesson that would resonate throughout his life: a man's status could change instantly from soldier to slave, from free to owned, and what mattered most was survival until circumstances changed.

A Greek convert to Islam purchased him with plans to resell, beating him for slowness and starving him for disobedience. Pinto considered suicide but later wrote that he threatened it openly, not out of despair but as a calculated move; a dead slave held no value, potentially leading to better treatment from his captor.

This tactic worked, and he was sold to a Jewish merchant for approximately thirty ducats' worth of dates. The merchant treated him somewhat better, possibly recognizing in Pinto someone reminiscent of himself, possibly with Jewish ancestry. They traveled overland to Hormuz, the great trading city at the mouth of the Persian Gulf, traversing burning deserts where men paid for water by its weight.

Hormuz was a jewel within the Portuguese empire, dominated by a fortress that controlled the strait. Ships from Persia, Arabia, India, and East Africa converged here, selling everything imaginable: rubies from Burma, frankincense from Arabia, horses from Persia, and slaves from Africa. The city shimmered with heat, wealth, and the brutal logic of commerce.

The Jewish merchant had connections among Portuguese factors who relayed word to the Portuguese crown that maintained a fund for ransoming Christian captives from Muslim hands. It was both good politics and faith. Three hundred ducats were paid, and Fernão Mendes Pinto was set free.

Pinto spent months in Hormuz learning about trade mechanisms and power dynamics. He observed how fortunes could rise and fall with the turn of a monsoon. Portuguese soldiers supplemented their meager pay through private trade or outright piracy. The brutality that maintained the empire was evident in the torture of prisoners, massacres of rebellious towns, and the casual violence of colonial rule.

At twenty-nine years old, Pinto had experienced enslavement and freedom, witnessed battles and negotiations, and learned that Asia was not a treasure trove waiting to be plundered but rather a complex web of powers where the Portuguese played one part among many, militarily strong yet vastly outnumbered with disaster always lurking just around the corner.

In 1539, he arrived at Goa, the Portuguese headquarters in India. The city astonished him; situated on an island off the Indian coast, it served as the capital of Estado da Índia or the State of India. The Viceroy ruled from a palace modeled on Lisbon's architecture, while churches lined the waterfront, and the Inquisition made its ominous arrival, setting up machinery for religious enforcement.

Pinto entered service under the newly appointed captain of Malacca, the fortress controlling the Straits of Malacca and gateway to the Spice Islands and China. He sailed eastward past Ceylon and around the Malay Peninsula, delving into waters few Europeans had ever navigated before. The world grew stranger, more complex, and dangerous with each passing journey.

He was about to embark on the final stage of his education: the years in Southeast Asia and China that would transform him from an unremarkable Portuguese individual into a man whose stories were deemed lies yet preserved details of a vanished world like no other European could.

THE THIRTEEN SHIPWRECKS, THE SEVENTEEN SLAVERIES

Memory shifts when waves of trauma crash upon each other relentlessly. Fernão Mendes Pinto later claimed he had been shipwrecked a dozen times over and enslaved countless others during his two decades in Asia. Scholars have long debated these numbers, questioning their literalness or symbolic weight. What is significant is how Pinto's existence morphed into an intricate tapestry of disasters, teaching him resilience against all odds and driving him further from the world he once knew. Thirteen shipwrecks, which can be compared to the experience of the average sailor who might face only one or two in their career, stand as testimonials to his turbulent journey, each wreck imprinted on his soul rather than just a footnote in history. Seventeen times enslaved, a relentless cycle of bondage and liberation, intertwining his life with invisible chains.

Malacca, where he arrived in 1539, was a bustling crossroads, a hub for traders from Malay, Chinese, Arab, Indian, and Portuguese communities who converged in markets saturated with goods such as sandalwood, tin, silk, and ceramics. The Portuguese fortress dominated the port, its cannons controlling the Strait. Pinto worked as an emissary to Sumatra's small kingdoms and the Malay Peninsula, establishing connections, negotiating treaties, and gathering intelligence on rivals.

He mastered Malay. He grasped the intricacies of Southeast Asian courts, where hierarchy was paramount and a slip of the tongue could lead to death or imprisonment. He also learned that Europeans were not always revered; they were seen as useful but barbaric, distinguished by their technological prowess and cultural backwardness, and by their peculiar habit of eating with utensils.

On the eastern side of the Malay Peninsula, in Patani, Pinto set up a successful business trading with Siam, the land later known as Thailand. He was accumulating wealth; his ultimate goal was capital that could secure status back in Portugal.

Yet pirates struck, seizing his goods, ship, and fortune. Rather than succumbing to ruin, Pinto and his partners turned pirates themselves, blurring the line between legitimate trade and outright raiding, a common practice among Portuguese settlers who combined legal commerce with illicit attacks. The Portuguese crown officially disapproved, yet profited indirectly from the chaos.

Pinto raided Indochina's coasts, plundering Chinese coastal towns. In one such raid, his world crumbled temporarily.

The specifics are hazy, filtered through time and Pinto's narrative choices. Yet, it is clear he and his companions targeted a Chinese imperial tomb seeking treasure, only to be apprehended by authorities. The punishment was swift and gruesome: thumbs severed at the joint, leaving deep scars, both physical and psychological.

He faced a year of hard labor on China's Great Wall, a series of fortifications rather than a single entity, continuously rebuilt and expanded. The work was grueling labor under the lash in extreme temperatures, resulting in regular deaths. Pinto survived, showcasing his resilience, that innate ability to endure against all odds.

He did not complete his sentence. Before the year ended, Tartar raiders attacked, capturing him. His status shifted again from convict to slave, then to something akin to an agent. The Tartars utilized his linguistic skills and regional knowledge, making him invaluable.

With the Tartars, he traveled to Cochinchina, now part of Vietnam, witnessing warfare utilizing elephants, Buddhist monasteries with thousands of monks, and agricultural systems supporting denser populations than in Europe. Pinto admired Asian civilizations in a way few Portuguese did, most dismissing them as inferior, pagan, suitable only for conquest or conversion. Pinto observed and absorbed.

The chronology blurs thereafter; he escaped Tartar captivity and sought passage back to Portuguese-controlled territory. The vessel available was a Chinese pirate junk bound for southern waters. He boarded it.

A storm ravaged the coast of China, propelling them eastward and north by typhoon winds, far off course into unfamiliar waters. When the tempest subsided and land emerged, no one recognized the island. It was Tanegashima, an island south of Kyushu, Japan's southernmost main island. The date was September 23, 1543, or possibly 1542, contingent on calendar variations and accounts trusted.

The Portuguese hailed it as the discovery of Japan. The Japanese referred to them as Southern Barbarians. Fernão Mendes Pinto claimed for life that he was among the first Europeans in Japan, introducing firearms, allegedly reshaping Japanese history.

Truth is likely more intricate than claimed.

THE ISLAND OF SILVER AND SWORDS

The master of Tanegashima was Tokitaka, a local ruler holding authority over a small island on the outer fringes of Japan's influence. His interest in foreigners who washed ashore was particularly piqued by their weaponry.

Pinto and his group carried arquebuses, firearms that used a matchlock mechanism, which, although primitive by later standards, were revolutionary at the time. These weapons consisted of a metal barrel mounted on a wooden stock, with powder and ball loaded from the muzzle end. A slow-burning match was applied to a priming pan, igniting the main charge. While cumbersome to reload, inaccurate beyond fifty paces, and susceptible to wet conditions, they could penetrate armor and required less skill to operate than archery.

In 1543, Japan found itself in turmoil, with the Ashikaga shogunate crumbling and power shifting to local warlords, the daimyo, who engaged in a chaotic cycle of alliances and conflicts. Warfare was highly ritualized yet deadly, with samurai employing swords, bows, and spears, using techniques passed down over centuries. The introduction of firearms promised to radically alter this established order.

Pinto stated that Lord Tokitaka was intrigued by the Portuguese arquebuses he showcased, and it is asserted that several weapons were sold or gifted to the lord, leading to local replication of the design that would spur a revolution in samurai warfare through the spread of firearms.

While Pinto's account may have embellishments, historical records support that Portuguese firearms did arrive at Tanegashima during this period. The island became known for firearm production, with "Tanegashima" becoming synonymous with matchlock technology. Contemporary accounts from the era describe the astonishment of local warriors upon seeing these foreign weapons, described as 'bringing thunder and flame in one's hand.' Within a few decades, thousands of arquebusiers were employed by Japanese daimyo, undermining traditional pike formations and enhancing the strategic importance of castles, leading to the eventual unification under central authorities equipped with firearms.

However, Pinto's personal involvement is likely exaggerated. While Japanese accounts mention Portuguese traders arriving, they do not specify him by name, and Portuguese records suggest earlier arrivals such as António Mota and Francisco Zeimoto may have preceded his visit. The exact details are obscured by the complexity of competing narratives, national pride, and the remote nature of these events from centralized record-keeping.

It seems more plausible that Pinto was present during one of the early Portuguese incursions into Japan, observed the interest in European arms, and witnessed the beginning stages of a technological and cultural exchange with profound implications. While not the central figure he claimed to be, he did bear witness to significant historical events.

This pattern repeated throughout his life. Pinto possessed an extraordinary knack for finding himself at pivotal moments in history. He was also a gifted storyteller who recognized that witnessing an event is less captivating than catalyzing it, and that audiences prefer tales of protagonists to those of bystanders.

He later returned to China, carrying stories of Japan's wealth, including pure, abundant silver. This attracted Chinese merchants willing to pay premium prices for the metal. Recognizing the potential, Portuguese traders set sail for Japan, buying silver with goods from China, selling it in China for silk and porcelain, then trading these for spices in Malacca or India before returning to Lisbon for European gold, a trade known as Nanban, or Southern Barbarian trade that made fortunes for those daring enough to engage.

Pinto attempted the voyage again, possibly around 1544-1545, but ended shipwrecked this time on the Ryukyu Islands. The survivors were adrift after their vessel broke apart on coral reefs, and there are accounts of women pleading from the shore for their men to save the foreign castaways. Whether this account is factual or embellished for narrative effect, it highlights the compassion shown by strangers.

He continued his journey, reaching Malacca, where he was commissioned for missions in Burma and the Irrawaddy Delta. Captured by Burmese forces, Pinto traveled through unknown territories to Luang Prabang, Laos, observing Buddhist temples adorned with gold leaf and a variety of religious practices that both fascinated and repulsed him.

Eventually freed or managing an escape, he returned to Goa only to be promptly sent on another mission to Java to secure pepper for the China trade. The mission concluded like many before it, in disaster; their ship was attacked by Japanese pirates and wrecked along Thailand's coast. In desperation, the survivors built a raft to drift towards Java.

The events that followed were so dire that Pinto could barely recount them. Starvation and dehydration claimed many, driving others mad, until they resorted to cannibalism in an attempt to sustain their lives. Upon reaching land, barely alive, they sold themselves into slavery just for the basic necessities of life.

Pinto's fortunes shifted when he was ransomed or freed, allowing him to borrow funds from Portuguese contacts and start anew. His travels took him to Siam, where he became embroiled in Thai-Burmese conflicts, leaving behind a written account that would become one of the first European descriptions of Burmese political dynamics and history. He amassed wealth but lost it again, continuing his cycle of sailing, shipwreck, capture, and freedom.

By his early forties, Pinto had spent over a decade away from Portugal with little to show for it, no estate, family, or security. However, he had experienced phenomena unknown to most Europeans, endured perils that would have ended lesser men, and mastered languages barely known in Europe.

Then, around 1547 or 1548, his life took a dramatic turn as he met Francis Xavier, a Jesuit priest later sanctified as the Apostle of the Indies. This meeting would alter Pinto's path forever.

THE SAINT AND THE SINNER

Francis Xavier burned with a holy fervor. A noble from the Basque region, educated at the University of Paris, he co-founded the Society of Jesus and ventured to Asia on a mission: to convert pagans, establish the cross in new lands, and save countless souls. His tirelessness, charm, and conviction that he was doing God's work were remarkable.

Fernão Mendes Pinto, however, had grown cynical about such endeavors by now. He had witnessed firsthand what Portuguese Christianity entailed: massacres justified under the guise of the cross, slavery masquerading as soul-saving, torture deemed necessary to eradicate heresy. He saw priests blessing ships that raided and burned coastal villages and the Inquisition establishing itself in Goa, turning it into a place where conversion was enforced by sword or fire.

Yet Xavier was distinct from others. Perhaps his sincere faith, willingness to live among the impoverished, fearlessness in peril, or simply the timing, Pinto nearing forty, exhausted, questioning his struggles, set him apart.

They encountered one another in Malacca. Xavier desired to reach Japan after learning from a Japanese convert named Anjirō about its refined civilization, unfamiliar with the Christian message. He believed that if he could clearly present Christianity, the Japanese would convert en masse, given their sophistication, and recognize the truth upon hearing it.

In need of someone who spoke Chinese, knew trade routes, and could serve as an interpreter and guide, Pinto inexplicably agreed to assist him.

In 1549, they embarked for Japan with a small party. They arrived in Kagoshima on Kyushu's southern island. Xavier zealously engaged in evangelization, using interpreters, debating Buddhist monks, and demonstrating Christian rituals.

The outcomes were varied. Some Japanese converted, often politically motivated rather than spiritually driven, aligning with the Portuguese could be advantageous. Debates between Xavier and Buddhist monks left both frustrated as they spoke past each other due to non-translatable concepts. The Japanese language posed significant challenges; early Portuguese missionaries used Buddhist terms to describe Christian ideas, leading to confusion when they likened God to Dainichi, the cosmic Buddha. This implied that God belonged within Buddhist cosmology rather than being its supreme creator.

Pinto observed these encounters with a mix of sympathy and skepticism. He likely fabricated his intellectual ineptitude in his later accounts, or, out of caution, it was safer to claim ignorance in Counter-Reformation Portugal, where theological errors could be dangerous, especially siding with Buddhism during debates.

Pinto did appreciate Japanese culture, though. He admired their craftsmanship, elaborate etiquette, and complex social hierarchies and saw the samurai as warrior-class equals to European knights. He recognized Japan's sophistication rather than its primitive nature.

One anecdote encapsulated his experience: when Lord Ōtomo Yoshishige greeted Pinto with such refined courtesy that he found himself speechless. Claiming he was silent for a hundred thousand years because he stood before such a great king, the tale is likely embellished, but it reveals Pinto's awareness of being an outsider, a Portuguese adventurer who survived by cunning and luck, standing before a refined protocol he barely comprehended.

Pinto accompanied Xavier on his missionary journeys, witnessing the formation of early Christian communities and observing his frustration as only a few converts emerged after months of preaching. Christianity was exotic in Japan, yet it didn't appear superior to established Buddhist beliefs. Why abandon Buddhism for a foreign faith preached by barbarians who behaved poorly?

Xavier fixated on converting China, believing it would lead to the conversion of all East Asia. Japan was important, but China was essential. In 1552, Xavier embarked toward China's coast but never reached the mainland. Francis Xavier died of fever on December 3, 1552, within sight of China, leaving his great mission unfulfilled.

His remains were buried on Shangchuan Island, exhumed and transported to Malacca, then enshrined in Goa's Bom Jesus Church, where miracles were reported. His incorrupt body became a pilgrimage site, boosting his reputation. Saint Francis Xavier, the Apostle of the Indies and patron saint of missionaries, was canonized within decades.

Pinto was in Goa during the display of Xavier's remains. Simultaneously, he received a letter from Lord Ōtomo Yoshishige, offering to convert to Christianity and requesting Pinto's return to Japan, which seemed to him either divine providence or mockery.

In 1554, waiting for a ship back to Portugal with accumulated wealth to return home as a wealthy man, Fernão Mendes Pinto made an extraordinary decision. He joined the Society of Jesus, donated much of his fortune to fund Japan's Jesuit mission, and became a lay brother, dedicating his remaining years to God's service.

What drove this? Genuine religious conversion or guilt from years of piracy and violence? Xavier's influence or simply exhaustion seeking meaning in suffering? The reasons remain unclear. Pinto left no explanation, and his future actions suggest ambivalence at best.

He returned to Japan with Xavier's successor, Father Baltasar Gago. He served as the Viceroy of Portuguese India and as ambassador to Lord Ōtomo, funded the construction of Japan's first Christian church, and worked tirelessly for the mission.

Then in 1557, he left the Jesuits. The reasons remain unclear; officially, he requested release from his vows due to "want of vocation," implying a loss of faith or unsuitability for religious life. Alternatively, some sources suggest he was expelled because authorities discovered his marrano ancestry, Jewish heritage, and the Society of Jesus still enforced limpieza de sangre (purity of blood) statutes excluding those with Jewish or Muslim lineage.

Regardless, Pinto left the order. He arrived in Martaban, Burma, amidst a siege, betrayed by Portuguese mercenaries, imprisoned by Burmese forces, taken inland to Calaminham (now Luang Prabang), escaped, returned to Goa, sent to Java for pepper trading, shipwrecked, resorted to cannibalism, enslaved, and eventually ransomed.

His life became a loop of disasters repeating with variations. Nearing fifty, most men in the Indies died or barely survived twenty years. Pinto had been there for twenty-one, exhausted, scarred both physically and mentally, carrying wounds that would never fully heal.

In 1558, he decided to return home.

THE RETURN OF THE WANDERER

On the 22nd day of September, in the year 1558, Fernão Mendes Pinto disembarked upon the shores of Lisbon's harbor, his age marking forty-nine years. He had left Portugal, an ambitious yet unknown figure, twenty-one years prior. His return to Lisbon was not marked by the wealth he had donated or lost at sea, nor fame, as few in Lisbon were aware of his exploits beyond their borders. Nor did he possess power; no office, no men to command, and no estates to call his own.

He was simply a survivor, but survival alone could not suffice for life's necessities.

Lisbon had undergone transformations; King John III was approaching death, while Portugal's eastern domains were becoming increasingly costly in terms of resources and lives. The spice trade, once believed to be the epitome of boundless wealth, now yielded diminishing returns due to excessive numbers of men, ships, and overhead costs. Dutch and English traders began challenging Portuguese dominance, marking the end of the discovery's golden age and ushering in an era where maintaining the empire came at a great cost.

Pinto made his way to court, petitioning for a pension to compensate him for his twenty-one years of service to the crown. He recounted his missions, diplomatic connections, and his role in initiating trade with Japan and China. However, the court officials were polite yet unmoved, as numerous men arrived each week with similar claims. They questioned how they could verify Pinto's accounts from half a world away, where Portuguese influence was often more fiction than fact. To add insult to injury, he observed that even lesser nobles with minimal service and achievements received land grants through their connections to court advisors. This deepened his frustration, exposing the systemic inequality that overshadowed merit and perseverance.

He received nothing, not a pension, nor a land grant, nor any appointment. For four years, from 1558 to 1562, he persistently sought recognition at court amidst numerous returning adventurers with similar tales. He was, in the most profound sense, a failure.

In 1562, he abandoned his pursuit of recognition. Instead, he purchased a small farm in Pragal near Almada, across the Tagus River from Lisbon. He married a woman whose name was not recorded in history and had children whose names were documented only by legal witnesses. He sought obscurity as a middle-aged veteran of wars that no longer concerned others and of experiences deemed unbelievable.

Around 1569, he began to write.

The manuscript that would become Peregrinação, Pilgrimage, took years, perhaps even a decade, to complete. Pinto wrote in Portuguese but not the polished literary language of the court; his style was conversational, digressive, and brimming with nautical terms and foreign words. Classically educated, he did not know Latin or Greek, nor had he read fashionable authors. He wrote as he spoke directly and vividly with the authority of lived experience.

He claimed to write for his children so that they would understand what he had done, where he had been, and what he had survived. He wrote to explain life to himself because memories burned within him like a fever, demanding release.

The book he produced defied anything in Portuguese literature. With 226 chapters, it spanned twenty-one tumultuous years, thousands of miles, and dozens of kingdoms. It was an adventure narrative and ethnography, an autobiography and critique, a picaresque novel and a spiritual confession, all rolled into one.

Moreover, Pinto's work was deeply subversive for its time. He wrote critically about Portuguese colonialism, describing massacres and atrocities committed by his countrymen. He questioned whether forcibly converting pagans aligned with Christian ideals. His depictions of Asian civilizations, China, Japan, and Siam, were marked by respect and admiration, acknowledging their sophistication while challenging European superiority. This subversion lies in his writing, which embodies a hybrid of history and literature, offering not just an account of events but a narrative that critiques the era's prevailing ideologies.

Yet, Pinto's accounts raised doubts among contemporaries. The sheer number of disasters, including thirteen shipwrecks and seventeen instances of enslavement, seemed implausible. His presence at pivotal historical moments, the first Portuguese landing in Japan, the introduction of firearms, and meetings with key Asian rulers, appeared too convenient. His detailed descriptions of places and events beyond his direct witness seemed fabrications.

Was he lying? Exaggerating? Blurring fact and fiction through literary techniques? Likely all of these. Yet, he also conveyed truths unrecorded by others. His descriptions of Asian societies aligned remarkably with independent sources; accounts of Portuguese violence corroborated other testimonies; and his depictions of trade networks, political structures, and cultural practices preserved details that would have otherwise been lost.

He completed the manuscript in the late 1570s but did not publish it. Perhaps he could not afford publication, or maybe he feared censorship, given the Inquisition's power; his critical views posed a threat. Alternatively, he may have doubted anyone's belief in his narrative.

In 1580, Portugal lost its independence as King Sebastian perished in a disastrous campaign in Morocco, prompting Philip II of Spain to claim the Portuguese throne and merge their crowns. This bureaucratic union led to an award from now-Philip I of Portugal in January 1583, a small pension for Pinto's services in the Indies.

This acknowledgment arrived forty-five years after sailing for India and twenty-five years following his return. Though meager, it recognized that his service was genuine.

Six months later, on July 8th, 1583, Fernão Mendes Pinto passed away at his farm in Pragal, aged seventy-four. Buried in a local church without grand memorials marking his grave, his manuscript remained unpublished and collected dust among his belongings.

His daughter, fulfilling his wishes, eventually donated the manuscript to Casa Pia, a charitable institution for repentant women in Lisbon. Recognizing its value, the administrators facilitated publication. Thirty-one years following Pinto's death, in 1614, Peter Crasbeeck's presses in Lisbon finally released Peregrinação.

Its full title was expansive: Pilgrimage of Fernam Mendez Pinto in which is told the many and very strange things he saw and heard in the kingdom of China, in the one of Tartary, in the one of Sornau, usually called Siam, in the one of Calaminhan, in the one of Pegù, in the one of Martauão, and in many other kingdoms and lordships of the Oriental parts, and that in our Occident there are few or no accounts. Also, the account of many particular affairs that occurred both to him and to many other people. And at the end of it, briefly regards some things, & the death of the Holy Priest Francis Xavier, sole light and brightness of those parts of the Orient, & universal ruler of the Society of Jesus in those parts.

It was a sensation yet immediately controversial.

THE LIAR'S PARADOX

The book's descriptions were so detailed and its adventures so extraordinary that readers began to doubt its authenticity. Within decades, his name became a joke - "Fernão, Mentes? Minto!" or "Fernão, do you lie? I do!", as the pun seemed irresistible. The Portuguese language itself appeared to condemn him.

He was labeled as the Prince of Liars, and his book was classified as fiction. It was entertaining, but not reliable in any way. Historians dismissed it, while travel writers mocked it. His vividness made the story compelling yet suspicious.

However, when later scholars crosschecked Pinto's accounts with independent sources, they found out that many of his descriptions were indeed accurate. His account of Siam matched Thai chronicles, Japanese daimyo politics aligned with Japanese records, and even his account of Chinese trade networks proved correct upon careful examination. Pinto did not lie about everything, but he certainly shaped his narrative to place himself at the center of events, where he might otherwise have been peripheral. To differentiate between 'lie' and 'story,' it is useful to consider two types of truth: 'factual truth,' which concerns the precise, verifiable facts of an event, and 'dramatic truth,' which reflects the emotional and experiential essence of a story. He reported conversations that he could never have heard and described places that he may not have visited personally. He also compressed timelines, conflated episodes, and employed literary techniques that served dramatic truth rather than historical accuracy. So was this lying or storytelling? It all depended on the perspective, as Pinto was not bound by academic standards of evidence and citation like modern historians are. He was a memoirist writing decades after the events, relying on memory, notes, and stories told by his companions. He wanted to make sense of his life, give meaning to suffering, and preserve experiences that would otherwise vanish.

Modern scholarship has partly vindicated him. Rebecca Catz's 1989 critical edition of Peregrinação includes extensive notes that corroborate many of Pinto's descriptions. His account of the first Portuguese missions to Japan matched Jesuit correspondence, while his descriptions of Southeast Asian kingdoms aligned with other European and Asian sources. He even critiqued Portuguese colonialism in a way that was ahead of its time.

Pinto was not exactly lying; he was a man who wanted to make his life cohere, find patterns in chaos, and extract meaning from suffering. All storytellers do what he did - selecting details, shaping narrative, emphasizing certain elements while downplaying others.

The tragedy is that Pinto's own writing style worked against him. He wrote too well, described vividly, and made his adventures too dramatic. If only he had written in a more cautious manner with qualifications and scholarly hedging, then perhaps he would have been believed. But he wrote with passion and immediacy, and that was what was considered suspect.

There is a deeper irony here. Pinto's book preserves details of Asian civilizations at a crucial moment – the mid-16th century, when sustained contact between Europe and East Asia was first established. His descriptions of Japanese society before unification, Chinese coastal communities before the Qing dynasty, and Southeast Asian kingdoms before European colonization are invaluable historical sources.

However, because he was dismissed as a liar, historians ignored him for centuries, treating his testimony as fiction rather than history. It wasn't until the late 20th century that scholars realized what had been lost, and now Peregrinação is considered one of the most important sources for understanding the Portuguese presence in Asia and the Asian response to European arrival.

The book was translated into multiple languages and went through numerous editions, influencing how Europeans imagined Asia - as exotic, dangerous, sophisticated, and strange. It also influenced literature, with some scholars seeing it as an early picaresque novel, while others noted its influence on travel writing, observation intertwined with adventure. Voltaire may have been influenced by this work, along with Portuguese literature, which saw a tradition of critical reflection on empire.

Throughout it all, Pinto remained a paradox - too accurate to be entirely dismissed and yet too dramatic to be entirely believed. He was too vivid to ever truly forget.

THE STORM THAT NEVER ENDED

Return now to Tanegashima's beach on September 23rd, 1543, after the storm has passed. Pinto is alive; warriors descend from the cliff. It's not a moment of discovery or triumph, but one filled with confusion and misunderstanding, the beginning of something unclear to both sides.

The Portuguese called what was happening 'entrada', 'entry', believing they were claiming a new world through their presence and bringing civilization and Christianity to the darkness. The Japanese saw it differently: 'kokka', the state; 'sakoku', closure to the outside; 'joi', the expulsion of barbarians. They viewed foreigners as tools or threats that needed to be managed.

On the beach, Pinto, waterlogged and exhausted with his arquebus still strapped to his back, represented neither civilization nor barbarity but desperation.

He had survived against all odds through luck, stubbornness, and quick situational assessment.

The Warriors took the Portuguese to their lord; negotiations occurred, gifts were exchanged, and demonstrations were held. The arquebus fired, its loud crack echoing across the island, smoke, fire, a ball punching through a wooden target. The Japanese were impressed; the Portuguese left freely. Trade began.

Pinto carried this story for decades, polishing it until the storm became legend, the landing history, and the encounter a tale of European prowess and Asian wonder.

Was he wrong to shape his narrative? History is always told from a perspective; the teller's story shapes it. We don't have an objective record of that day on Tanegashima; only Pinto's account, Japanese chronicles that may not be about the same event, and the fact that Portuguese firearms arrived in Japan around this time, transforming warfare.

It doesn't matter if Pinto was the hero of his own story. He was there or somewhere like it, in encounters repeated dozens of times across two decades. Present at the moment when worlds collided, Asian and European civilizations were engaging in a complex process that would reshape both.

He was a witness and wrote it down, leaving us his testimony. Flawed and exaggerated, yet vivid, detailed, and irreplaceable.

WHAT THE LIAR TOLD

Fernão Mendes Pinto occasionally fabricated facts, yet his narrative held profound truths about human suffering and violence.

He authentically portrayed physical pain illustrated through shipwreck ordeals, starvation, and brutal punishment, as well as psychological anguish caused by enslavement, disregard for human rights, and helplessness. His descriptions of spiritual torment unveiled the disappointment of unfulfilled dreams, shattered plans, and a life falling short of expectations.

He truthfully depicted violence, particularly the atrocities committed in Portuguese colonial activities: mass killings, village burnings, prisoner enslavement, and torture of suspects. These acts were justified by religious beliefs, commercial interests, or the fact that Europeans possessed firearms while Asians did not until they learned to forge their own.

Mendes Pinto also revealed the essence of Asian civilizations. He debunked the notion that China was a backward empire awaiting European enlightenment, presenting it as an ancient civilization with sophisticated technologies and philosophies beyond European comprehension. Pinto's reflections align with Chinese philosopher Xunzi's assertion that 'humankind is profoundly influenced by what it sees and hears'; a quote reflecting China's intricate appreciation of music and ritual. Japan was portrayed accurately, possessing a refined culture that surpassed what Europeans knew; its customs appeared crude in comparison. In this regard, the words of Japanese tea master Sen no Rikyū resonate: 'Tea is nothing but this: first you boil water, then you make the tea and drink it properly. That is all you need to know.' Such perspectives illuminate the depth of cultural elegance Pinto encountered, challenging European preconceptions.

Moreover, he offered an honest self-portrait. Mendes Pinto confessed to his opportunistic nature, driven by survival and profit, which led him into unprincipled acts such as piracy and theft. However, he was also capable of empathy, curiosity, and personal growth. He learned to view Asian populations not as barbarous heathens but as human beings with their own valid ways of life.

This perspective was groundbreaking for his time. Most Portuguese accounts of Asia boasted about conquests and Christian conversions. Contrarily, Mendes Pinto's account expressed ambiguity, questioning the legitimacy of Portuguese actions in Asia and challenging the notion that forced conversion could be considered Christian. He recognized the hypocrisy displayed by those who invoked Christianity while committing atrocities.

This made him a threat to established narratives. Although his book was published and translated across Europe, it was dismissed as fiction because its content was too subversive. It was deemed more appropriate to classify it as entertainment rather than to take it seriously or accept it as legitimate testimony.

Yet, his book remained a significant testimony: an account of transformative times when the world began to shrink and isolated civilizations collided, leading to immense suffering and a lack of mutual understanding. It also bore witness to the potential of transcending cultural barriers, even if such understanding was always limited and fraught with challenges.

Mendes Pinto passed away, believing his life had been a failure. He sought fortune but only found survival; acquired wealth but lost it all; served crown and church without receiving meaningful recognition until it no longer mattered. He authored a book that few believed in.

However, the book endured, and through it Mendes Pinto's legacy survived not as the hero he may have wished to be or the faithful Christian he strove to become, but as something even more valuable: a witness who preserved crucial truths despite his fallibility, ensuring they would not be lost.

THE BEACH REMEMBERS

Tanegashima recalls a pivotal era. Museums stand as testaments to an event, Portuguese traders arriving with firearms, forever altering Japan's destiny. They don't name Fernão Mendes Pinto; Japanese records omit him, favoring others' accounts instead. Yet the story lingers in both Portuguese and Japanese memory.

Different cultures remember shared events differently. History is debated and multifaceted. What Portugal calls discovery, others term invasion. European civilization, to some, is Asian destruction. Triumph for one culture might be a tragedy to another.

Pinto grasped this duality perhaps more than he realized. His book doesn't laud Portuguese expansion; it examines, critiques, and questions it. He saw himself as both participant and observer, insider and outsider, a Portuguese man who spent most of his life in Asia, a Christian witnessing Buddhist devotion, a colonizer acknowledging the humanity of the colonized.

This ambivalence, this refusal to embrace simplistic triumphalism, makes "Peregrinação" valuable; it's an early work on transcultural understanding, recognizing no single culture owns truth. Different people perceive and understand the world differently, and they deserve mutual respect.

The Portuguese pun, "Fernão, Mentes? Minto!" has endured centuries but might be misinterpreted. Pinto may not have been saying "I lie," but "I tell stories." Stories aren't lies; they're shaped truths, selective emphases to make sense of chaos.

We are all liars in this context, our life stories being selective and slightly inaccurate. We focus on certain details, overlook others, and shape narratives for coherence, meaning, and purpose.

Pinto's "crime" was telling stories too well, so compelling that they were mistaken for false. His punishment: remembered mostly as a liar, his insights dismissed.

Yet truth resurfaces. Scholarship eventually vindicated him in part; today, Peregrinação is recognized as valuable despite its flaws. Pinto witnessed events worth recording and realities worth preserving.

His ambivalence, questioning the empire without reservation and critiquing one's own culture, was prescient. He anticipated critiques that would dismantle colonial superiority centuries later.

Forgotten by his court and unrecognized in his homeland, yet his words endured. In those preserved words, we hear a voice witnessing the world's transformation, participating in it, imperfectly aiming to tell truths about what he saw.

The storm driving him onto Tanegashima was just one of many life storms he faced, adrift on forces beyond his control and crashing onto unexpected shores. Yet he survived and bore witness. Just as hundreds of years ago, Pinto's journey resonates with the survival stories of today's refugees, caught in circumstances beyond their making, seeking safety and hope amidst chaos. As we reflect on his narrative, let this provoke an examination of our own role in these modern migrations. How do we foster understanding and empathy in the face of such human resilience and adversity?

That witness is crucial. Most voices from this time are lost. Slaves, sailors who drowned, merchants who never returned home, and Asian villagers caught in the Portuguese expansion's path remain mostly nameless statistics or erased histories.

Pinto's account survives because he wrote it down; within his story, glimpses of those other lives appear as supporting characters, though they had their own stories.

Testimony preserves not just facts but texture, events, and experiences. It allows future generations to feel the transformations that reshaped the world.

Tanegashima's beach remains, its storms persisting. Modern navigation prevents shipwrecks, yet the island commemorates that day of foreign sailors landing with history-changing weapons. Today's echoes linger; modern migrants face their own oceans and storms. Their survival tales, like Pinto's, subtly alter lives and histories.

He would inaccurately recount events or emphasize certain aspects but reveal truths unrecorded by others. Despite doubt, we should appreciate his unique perspective.

The storm abated. The man endured. A story unfolded.

Vasco Fernandes Coutinho

April 1535. Vila Velha, Captaincy of Espírito Santo, Brazil.

The screams emerged from the forest. Vasco Fernandes Coutinho lingered at the border of his clearing, heart pounding as smoke twisted skyward from soil his settlement was never intended to touch. He had arrived in Brazil two years earlier, accompanied by colonists, tools, seeds, livestock, and dreams of a tropical paradise mirroring Portugal. He poured his fortune into this vision. He risked everything on a future yet unseen, guided only by hope. Now, the Tupiniquim ignited his fields.

Vila Velha sat perched upon a hillside overlooking Guanabara Bay. Its rudimentary structures, referred to by colonists as houses, offered little protection from the elements. The stockade remained unfinished while the church stood roofless, a testament to their incomplete promise. Before violence stormed in and before fields meant to feed the colony could be fully cleared, Vasco, at forty-five years old, found himself steering destruction. He was a seasoned warrior of Morocco and India and had once been part of Portugal's finest knights. Yet now he faced a new kind of battle, a struggle for survival where each decision carried the weight of life or death.

This outcome was not what Vasco had expected. In 1534, King João III granted him the Captaincy of Espírito Santo. The king divided vast stretches of Brazilian coastline and all adjoining land into parcels for Portuguese nobles to settle. For Vasco, this was a chance at redemption. He had spent his wealth in service to the crown, fighting bravely yet gaining little. Despite his title and family name, he remained impoverished. This captaincy was his opportunity to build something enduring, create wealth for his family, and turn his service into prosperity.

Yet, Vasco found himself powerless, forced to observe as his dream crumbled to ash before him. A deep sense of helplessness pressed upon him like a weight too heavy to lift.

The colonial project in Brazil would one day flourish, transforming Portuguese America into a sugar empire that would reshape the Atlantic world. But in 1535, such a future seemed distant and elusive. Vasco Fernandes Coutinho lingered by his faltering settlement, eyes fixed on smoke curling along the horizon as dread coiled within him. He could not foresee that his desperate struggle would help decide whether Portugal's gamble in Brazil would endure amidst looming threats from French rivals lurking offshore.

THE WEIGHT OF EMPTY POCKETS

Vasco Fernandes Coutinho was born around 1490 during a time when Portugal stood at both the zenith and precipice of its power, beginning its long descent. His family, the da Coutinhos, were ancient nobility with roots tracing back to the Reconquista wars that slowly pushed Muslim rulers from the Iberian Peninsula. They held lands in Marialva, northern Portugal, providing modest income and a secure place in society. However, by the late fifteenth century, land alone no longer guaranteed prosperity.

The Portuguese nobility was divided into two distinct groups: the great nobles, dukes and counts, who dominated through vast estates, titles, and royal favor; and the lesser nobility, such as the Coutinhos, who possessed honor but lacked wealth. For these men, royal service was not just a duty but an economic necessity. Opportunities lay in military campaigns in Morocco, expeditions to India, and colonization efforts in the Atlantic islands, which presented alternatives to a stagnant homeland where land was distributed and wealth concentrated among merchants and great noble families.

From his youth, Vasco understood this path laid before him, though it was clear in broad strokes rather than in specific detail. He would receive an education suitable for his class. He mastered arms, horsemanship, Latin, mathematics, and the skills required to serve as an officer or administrator. Entering royal service, likely within the military ranks, he held onto hope. Through merit or fortune, he might secure a prosperous future for himself and his descendants.

By his twenties in the 1510s, Vasco served in Morocco, where Portugal maintained fortified trading posts and garrison towns along the North African coastlines. These possessions were costly to maintain yet always threatened by hostile Muslim populations. They operated as training grounds for Portuguese officers and symbols of their crusading mission. Fighting was brutal, with garrisons perpetually outnumbered and vulnerable to siege, relying on supplies from across the Strait of Gibraltar. Officers learned resilience in the face of chaos, making critical decisions amidst perpetual uncertainty.

Vasco's competence earned him recognition in Morocco. He persevered when many perished. He demonstrated qualities valued by Portuguese commanders: bravery, judgment, and the ability to function under duress. His reputation led him to his next assignment in India.

THE INDIAN EXPERIENCE

By the early 1520s, when Vasco arrived in India, the Portuguese Estado da Índia had moved beyond its initial, uncertain years under Vasco da Gama. It was now an established, if precarious, imperial system. Portuguese forts dotted the Indian Ocean's rim, from Mozambique to the Spice Islands. Armed ships patrolled the seas, enforcing a system where Asian merchants had to buy Portuguese safe-conduct passes and pay duties at Portuguese-held ports. Portuguese governors and captains blended diplomacy with coercion. They constantly projected naval strength. Vasco served under some of the most important Portuguese commanders of the period. He witnessed both the impressive operation of an empire at its height and its substantial costs. Portuguese power was maintained through a network of alliances with local rulers, control of strategic ports and fortifications, and monopolization of certain high-value trades. Disease claimed Portuguese soldiers and sailors at staggering rates. Great distances meant reinforcements and supplies were always uncertain. Local populations resented Portuguese dominance and sought opportunities to challenge it. Profits from Asian trade, though substantial, were not evenly distributed. This fact was crucial for Vasco's future. The Portuguese Asian empire made some men spectacularly wealthy. Governors and high officials could accumulate fortunes through legal and illegal means. Merchants who secured the proper licenses could profit enormously from trade in spices, textiles, and precious goods. Ordinary officers and soldiers, even those of noble birth like Vasco, often found that their salaries barely covered expenses. Opportunities for enrichment were limited by a lack of capital and connections. Vasco returned from India after several years of service with his honor intact, but his finances were precarious. He had served loyally and competently, but he had not grown wealthy. He was now in his late thirties, approaching middle age by sixteenth-century standards. He had a military reputation but lacked the resources to maintain his position in Portuguese society. He needed something more, an opportunity that would allow him to turn his service into lasting prosperity.

THE BRAZILIAN LOTTERY

That opportunity seemed to arrive in 1534. King João III announced his plan to colonize Brazil through a system of hereditary captaincies. The king would divide the Brazilian coast into strips running inland from the sea, each about fifty leagues along the coast and extending to the Line of Tordesillas, the treaty boundary between Spain and Portugal that, in theory, divided the entire non-Christian world between the two powers. Each captaincy would be granted to a Portuguese nobleman, the donatary, who had near-total authority and the task of colonizing his domain at his own expense. For the Portuguese crown, this system solved a problem. Portugal had claimed Brazil since Pedro Álvares Cabral's accidental discovery in 1500, but three decades later, the territory remained almost entirely uncolonized. Portuguese ships stopped along the Brazilian coast to take on water and provisions and to trade with indigenous peoples for brazilwood, the dyewood that had given the territory its name. But there were no permanent settlements, no infrastructure, no real assertion of Portuguese authority. Meanwhile, French ships had begun frequenting the Brazilian coast, trading with indigenous groups and threatening Portugal's claim to the territory. João III needed to colonize Brazil but lacked the resources to do so directly. Portugal's attention and treasury were focused on maintaining the Asian empire, which was profitable but expensive. A direct royal colonization effort in Brazil would require capital, ships, and administrative attention that Portugal could not spare. The captaincy system offered an alternative. By granting portions of Brazil to individual nobles and making them responsible for colonization, the crown could establish a Portuguese presence without direct royal expenditure. The donors would bear the costs and the risks. If they succeeded, the crown would benefit through taxes and through the enhancement of Portuguese power. If they failed, the crown would lose nothing but distant claims to territory it could not effectively control anyway. For men like Vasco Fernandes Coutinho, the captaincy system represented a gamble. The king offered each donatary extensive powers, the right to distribute land, to found settlements, to administer justice, to collect taxes, and to pass these rights to his heirs. The donatary would effectively be a king in miniature within his captaincy, constrained only by his obligation to respect Portuguese law and royal authority. This was an extraordinary grant of power, the kind of opportunity that lesser nobles rarely received. But the opportunity came with crushing burdens. The donor had to finance the colonization effort himself. Vasco needed to secure a substantial amount of money, far exceeding the resources of most lesser nobles. He borrowed 20,000 cruzados from local merchants, mortgaged the family estate in Marialva, and sold precious family heirlooms, including his grandmother's silver candelabras. With each sum borrowed and each item sold, the weight of his gamble grew heavier. He needed to recruit colonists, purchase ships, acquire tools, seeds, and livestock, hire soldiers to protect the settlement, and support the colony until it became self-sustaining. Call on every favor in Portugal, Vasco turned to distant cousins and even former rivals, hoping to gather whatever resources he could. And they would be betting everything on the hope that the colony would eventually produce enough wealth to repay their investment and generate profit. Vasco received the Captaincy of Espírito Santo, a fifty-league stretch of coastline north of Rio de Janeiro with excellent natural harbors and fertile land but no existing Portuguese presence. He immediately began the ruinous preparations for his expedition. He sold property. He borrowed from merchants and from wealthier relatives. He recruited colonists from among the landless and the desperate, men who saw Brazil as a chance to acquire land they could never hope to own in Portugal. He purchased ships and hired crews. He assembled the tools and supplies his colonists would need. And in 1535, having transformed his entire fortune into ships, supplies, and promises, he set sail for Brazil with sixty colonists and everything he owned.

LANDFALL IN AN ALIEN WORLD

Vasco's arrival along the Brazilian coast revealed a land both enchanting and formidable. The harbor he found was striking; it lay serene from the ocean's fury, deep enough for his ships and embraced by lush forests teeming with life. Climate bore warmth but not excess heat. The soil showed promise of richness, hinting at abundant harvests if tilled wisely. Initially, local Tupiniquim people appeared amenable to trade and even partnership with the newcomers.

Yet beneath this veneer of opportunity lay challenges that would test Vasco's resolve. His company was diverse, from those fleeing poverty and criminals banished overseas to some respectable souls seeking fortune. Among them was

João Silva, a former shoemaker from Lisbon with a history marred by minor thefts, choosing exile over punishment. Arriving with dreams of wealth or easy success, he clung to the hope of starting anew with his battered bag in hand. Yet, confronted by endless forests and demanding labor, many, including João, questioned their decision. The majority lacked the skills to cultivate a settlement from the wilderness. They came seeking riches but found, instead, the harsh reality that survival required effort; most were ill-prepared or unwilling to put in the effort.

Nature itself posed obstacles. Tropical heat weakened those accustomed to the gentler Portuguese climate. Disease swept through, incapacitating some and leaving others too weak to labor. Mosquitoes tormented them endlessly, spreading sickness further. Tasks such as clearing land, building shelters, and planting crops advanced at a painfully slow pace, far from the swiftness Vasco had anticipated.

But it was the relationship with the Tupiniquim that proved most daunting. Initial contact was peaceful, even amicable; they traded goods and labor for European wares. It seemed that peaceable coexistence could be maintained. However, as the settlement expanded and demands grew, tensions rose. The Tupiniquim had their own societal order, conflicts with neighboring tribes, and strategic considerations. They quickly discerned that the Portuguese presence shifted power dynamics in detrimental ways.

Within months, hostile encounters escalated. Raiders ambushed colonists working in fields, and remote farms were set ablaze. Those straying from fortified enclosures sometimes disappeared without a trace. No longer was Vasco forging a burgeoning colony; he now commanded a beleaguered garrison, relying on his North African and Indian campaigns to defend and counter-attack. Resources dwindled while hope faded.

By 1537, after only two years of Vila Velha's founding, Vasco faced ruinous financial prospects. He had invested everything in the colony yet reaped little return. The settlement barely sustained itself, let alone produced surplus for profit as expected. Promised sugar plantations remained unrealized due to land clearing and infrastructure needs outstripping available capital and labor. Colonists were exhausted, gripped by fear, increasingly convinced that Espírito Santo was a trap rather than an opportunity.

Vasco needed more resources, additional colonists to expand the settlement and provide an agricultural workforce, soldiers for defense against indigenous threats, tools, seeds, livestock, and time, but he had no financial means to procure any of these. He had already mortgaged future earnings that failed to materialize. His creditors in Portugal grew impatient. The crown, having granted him the captaincy without support, offered sympathy devoid of assistance.

This system was fundamentally flawed from its inception; it presumed that individual nobles possessed the necessary wealth to establish control over a challenging and unfamiliar environment, that colonization would progress swiftly enough to yield profits before initial resources were depleted, and that newly founded colonies would generate sufficient revenues to repay investments. While the Spanish encomienda model integrated local labor under direct royal military and administrative oversight, Portugal's reliance on private capitalization without sustained institutional support placed a heavier burden on individual donataries. As a result, out of the fifteen captaincies established under João III, only a few survived beyond their original founders, with just two, Pernambuco and São Vicente, achieving notable prosperity. This outcome illustrates how structural differences in colonial administration had lasting ramifications within the broader context of Iberian imperial expansion in the Atlantic world, with Portugal's decentralized model proving largely less effective than the Spanish approach in fostering stable, profitable settlements.

Vasco's Captaincy of Espírito Santo teetered between complete failure and spectacular success. The settlement clung to existence primarily through Vasco's unyielding perseverance as his fortune dissipated. But it never flourished; it remained a struggling outpost, perpetually deficient in supplies and perilous, sustained more by sheer resilience than any sign of true prosperity.

THE JESUIT INTERVENTION

In the 1540s, a new component emerged in Brazilian colonization that significantly impacted Espírito Santo and Vasco's circumstances. The Society of Jesus, an organization established by Ignatius Loyola and sanctioned by Pope Paul III in 1540, initiated the dispatch of missionaries to Brazil as part of their worldwide mission to propagate Catholicism and counteract the Protestant Reformation. These Jesuits adopted a distinct methodology when interacting with indigenous communities, which differed from most Portuguese colonizers' practices. Rather than seeking dominance or exploitation over these populations, the Jesuits aimed to convert them, assembling them in mission villages called aldeias, where they could receive instruction in Christianity and European customs while being shielded from the worst excesses of colonial exploitation.

The initial Jesuit contingent reached Brazil in 1549, accompanying Governor-General Tomé de Sousa, entrusted with integrating the diffuse captaincies and reinforcing royal jurisdiction over regions that individual donataries had struggled to colonize effectively. Several Jesuits arrived in Espírito Santo early in the 1550s, among them José de Anchieta, who would become renowned as one of Brazil's most distinguished missionaries.

For Vasco, the presence of the Jesuits entailed both benefits and complications. On the one hand, they mitigated violence between colonists and indigenous groups by mediating disputes and facilitating the Tupiniquim's integration into aldeias, thereby enhancing settlement security and enabling land cultivation. Conversely, the Jesuits advocated for shielding indigenous peoples from extreme forms of exploitation, notably enslavement, which clashed with colonists' aspirations to utilize indigenous labor for economic gains.

This conflict between exploiting labor and safeguarding indigenous populations shaped much of Brazilian colonial history. Colonists depended on labor for land clearing, crop planting, and infrastructure construction. Indigenous individuals represented a viable workforce. However, utilizing such labor necessitated either compensation unaffordable for colonists or coercion at odds with the Jesuits' moral stance. This conflict manifested as a three-way struggle between colonists seeking to exploit indigenous workers, Jesuit missionaries seeking to protect and convert indigenous peoples, and indigenous communities striving to maintain their sovereignty, leveraging divisions among colonists and missionaries.

Throughout the 1550s, Vasco grappled with this conflict. He required labor; he sought peace; and he aimed to uphold his role as donatary amid the growing presence of royal officials and Jesuit missionaries, each with their own agendas and sources of authority. His position shifted from ruling a domain to contending for power among multiple stakeholders, all vying for the colony's future while grappling with persistent resource shortages.

THE SUGAR GAMBLE

Despite his financial setbacks, Vasco recognized that the only path to prosperity for Espírito Santo lay in sugar production. As demand for this valuable commodity soared in the Atlantic world, surpassing supply, the Portuguese successfully set up sugar plantations in Madeira and São Tomé, proving that with adequate resources, tropical climates could efficiently cultivate it. Brazil's climate and soil were particularly well-suited for sugarcane.

Building a sugar plantation required significant upfront investment, not just land and sugarcane but also an intricate infrastructure. The engenho was central to this, transforming harvested cane into exportable sugar amidst the earthy scents of its operation. This machine demanded specialized equipment, skilled labor, and a reliable power source, such as water wheels or oxen's persistent pull. Proximity to both canefields and a port for shipping to European markets made strategic mill placement crucial. The entire venture relied heavily on labor for planting, tending, harvesting, and processing the cane.

Vasco lacked the capital to establish an engenho himself, but incentivized others by offering land and privileges within his captaincy in the 1540s and 1550s. Though modest compared to operations in Pernambuco and Bahia, these small sugar ventures marked progress for Espírito Santo. The colony started exporting sugar, generating customs revenue for Vasco as donatary and providing employment.

However, sugar production demanded an insatiable labor force. Depleted by disease and conflict, the indigenous population couldn't or wouldn't meet these demands. Colonists turned to importing enslaved Africans, setting in motion Brazil's eventual status as the largest destination for the transatlantic slave trade. This required additional capital, deepening planters' debts and creating a wealth gap between plantation owners and most colonists.

Vasco watched Espírito Santo evolve from an experiment to a modest sugar outpost, not the transformation he had envisioned. He remained in debt but found a bittersweet victory in the colony's survival against all odds.

THE PERSONAL TOLL

Occasionally, records provide glimpses into Vasco's personal struggles. Married before leaving for Brazil, but with his wife's name unrecorded, she bore him children and shared his poverty as their fortune vanished. His letters to the crown expressed frustration over Lisbon's lack of support and the unfairness of shouldering costs alone. Despite decades of service in Morocco, India, and Brazil, he sought financial aid and recognition of the flaws in the captaincy system.

Ignored by the crown, struggling with its own financial issues and maintaining an increasingly costly Asian empire, Vasco faced emotional turmoil. His dreams of transforming wilderness into prosperity were dashed as he presided over a faltering settlement, deep in debt, while other captaincies thrived or failed. Yet, he persisted, lowering expectations, learning from defeats, laying the groundwork for a colonial society through unglamorous daily management and problem-solving.

Vasco's journey from conquistador dreams to survival and administration showed that true colonial success lay not in grand gestures but in small victories: building homes, clearing fields, settling disputes, forging peace, and dispensing justice. These efforts transformed a military outpost into a living society.

THE JESUIT CHRONICLE

Our primary insights into Vasco's later years stem from Jesuit writings and records, notably those of José de Anchieta, who sojourned in Espírito Santo during the 1550s. Anchieta's accounts, safeguarded within Jesuit archives and subsequently published, depict an aged Vasco, still vigorous yet visibly fatigued by years of hardship. Anchieta portrays Vasco as devout, munificent toward the Jesuits, and earnestly concerned for indigenous welfare - at least from a contemporary viewpoint. However, he also notes Vasco's destitution and his resentment towards the crown's treatment.

The Jesuit narratives indicate that by the late 1550s, Vasco had reconciled with his circumstances. He would not attain wealth; his captaincy wouldn't be the crown jewel of Portuguese Brazil. Yet, it would endure, bestowing some inheritance upon his progeny, even if that legacy fell short of his initial aspirations. This acceptance of modest fortunes might signify Vasco's final transformation: from an aspiring knight envisioning fortune through Brazil, to a desperate colonizer striving for any gains, to a seasoned administrator measuring success merely by survival.

Anchieta also documents tensions between Vasco and certain settlers, particularly the affluent planters, who were hostile towards his governance and resented the restraints he imposed on their exploitation of indigenous labor. During one intense confrontation, a planter protested, "What authority do you claim to wield, Vasco, when your survival hinges on our success?" These disputes unveiled the inherent frailty of the donatory system. A donatary held legal jurisdiction over his captaincy, but that power was only as robust as his capacity to enforce it. A prosperous, influential donatary could dominate his colonists. But an impoverished donor like Vasco, dependent on those same settlers for taxes and economic activities that generated his meager income, had to bargain rather than dictate. The feudal concept of the captaincy system clashed with the emerging capitalist reality of colonial sugar production, leaving the donataries at a disadvantage.

DEATH AND SUCCESSION

Vasco Fernandes Coutinho died in 1560, likely in Vitória, at around 70 years old. The precise details surrounding his demise are not documented in extant records. He was interred in Espírito Santo - the land he had struggled to colonize for over two decades, a land that absorbed his wealth without granting him prosperity. His son, also named Vasco Fernandes Coutinho, succeeded him and pursued similar yet varied fortunes.

The younger Vasco faced similar challenges to his father: scarce resources, clashes with indigenous communities, struggles against settlers resentful of donatary authority, and the perpetual economic marginality of a captaincy that couldn't rival Pernambuco or Bahia's success. However, he inherited not only these issues but also a functioning colony - albeit modestly prosperous, with established settlements, fertile agricultural lands, and a populace growing albeit slowly. While less than ideal, the foundations laid by the elder Vasco provided a platform for the younger to build upon.

Over time, the Captaincy of Espírito Santo assimilated into Portugal's broader Brazilian administration as the crown increasingly assumed control over territories the donatory system had proven incapable of effectively cultivating. The Coutinho family retained titular authority and select privileges but saw real influence shift towards royal officials as Brazil became too integral to Portugal's economy to be overseen by uncertain nobility. By the 17th century, the captaincy system was practically obsolete, replaced by a more centralized colonial regime capable of overseeing sugar production, safeguarding against foreign threats, and facilitating the vast importation of enslaved Africans that propelled Brazil to the heart of Portugal's Atlantic commerce.

THE FORGOTTEN FOUNDATION

Vasco Fernandes Coutinho's enduring legacy may not be as widely known today, overshadowed by other colonizers and by the broader narrative of Brazil's history. He did not amass great wealth nor found a lasting dynasty; instead, he left behind a colony that struggled against all odds. Despite his lack of celebrated successes or stirring tales of conquest, Coutinho's efforts established Portuguese influence along the coast, laying the groundwork for future endeavors, even if they were less glamorous. The marginal successes and failures alike played crucial roles in mapping out Portuguese claims, fostering settlements, and learning from trials and errors.

Coutinho's perseverance amid hardship revealed the fundamental challenges faced by early modern colonizers, including substantial investments, labor shortages, complex relations with indigenous peoples, conflicting interests among various stakeholders, and personal sacrifices. His decision to bet everything on Brazil in 1535 led him into a prolonged struggle that would define his life and legacy.

Though he could not foresee the eventual rise of Vitória and Espírito Santo as thriving cities and states, his foundation laid the basis for their growth into the twenty-first century. Despite historical amnesia relegating his achievements to obscurity, his efforts marked the beginning of a much larger narrative of resilience, adaptation, and transformation.

Ultimately, what defined Coutinho's story was not grandeur or triumph but his unwavering commitment to transforming wilderness into community, persisting through adversity with quiet determination. His tale, untold in the annals of conquest and glory, whispers the power of perseverance amid the backdrop of emerging nations, where every city starts as a fragile settlement and every empire is born from individual choices and struggles against the odds.

In essence, Vasco Fernandes Coutinho was a builder among history's forgotten figures. By his own and contemporaries' standards, he may have fallen short, yet in the most crucial measure, he succeeded: enduring alongside his fledgling colony. The smoke of 1535 signaled not an end but the commencement of a saga of endurance and quiet metamorphosis. His persistent choice to rebuild and carry on, even as abandonment loomed as a viable option, became the bedrock upon which others could build, though he might never see the fruits of his labor.

Smoke ascended once more, the forest enclosing ever tighter, yet Vasco Fernandes Coutinho turned back to face the labor of survival with unyielding resolve.

THE THRONE AND THE TIDE: KINGS, COINS, AND THE COST OF DISCOVERY

THE GREAT WAGER

Lisbon, March 1419

The counting room's atmosphere was rich with the scent of damp parchment and flickering candle flames. Prince Henry studied João Gonçalves Zarco, an experienced squire whose demeanor reflected the perils of the deep ocean. Between them lay a formal agreement, outlining the essence of two caravels: salt pork, hearty ship's biscuits, pitch, sturdy sailcloth, and plentiful coils of rope that would entwine their fortunes together. The price was immense. Henry's treasurer had voiced objections, his words resonating with the depleted state of the royal treasury following the costly acquisition of Ceuta.

Yet a valuable lesson from the North African fortress eluded Henry's treasurer; it was something that Henry himself had witnessed from the ramparts, caravans making their way, camels laden with gold dust and ivory, while spices gleamed under the sun's gaze. This wealth traversed the Sahara under the watchful eyes of the Moors, who dictated prices and safeguarded access to Africa's abundance.

With determined resolve, Henry affixed his signature to the contract. Zarco would be at the helm, navigating westward into the Atlantic's vast, uncharted azure expanse, in pursuit of islands to conquer and lands to cultivate with wheat and sugarcane. He was bound to return with treasures or dissolve into folklore.

The prince met Zarco's gaze. "You comprehend all these terms?" Zarco acknowledged with a nod, understanding acutely that he himself was akin to cargo, his abilities assessed as any other asset within this venture. Triumph promised him land, wealth, and elevation to nobility, whereas defeat would leave behind only silence for his widow.

This marked the commencement of an agreement destined to reshape the world's geography. Not a saga born from pure valor but one that balanced risk with reward, sending forth multitudes into the unknown and securing empires in their clutches.

THE ROYAL ARCHITECTS OF EXPANSION

To comprehend the essence of the Portuguese Age of Discovery, one must recognize that exploration extended beyond mere adventure, becoming an integral economic strategy. It served as a sort of venture capital, fueling economic growth through thoughtful investments and calculated risk-taking. The Portuguese monarchs were not visionaries chasing fantasies but prudent financial minds who were well aware that their kingdom's sustenance hinged upon dismantling the Venetian and Muslim monopolies over trade in the East.

Dom João I (1385-1433): The Revolutionary Accountant

King João the First ascended to power in 1385 following a revolution marred by bloodshed. An illegitimate son of Pedro I, he garnered support from Lisbon's merchants and military orders, which granted him a unique perspective on governance, one that prioritized economic stability over divine right. Despite ruling a small kingdom bordered by the Atlantic and Castile with only around 1.5 million inhabitants, João recognized the indispensability of the sea, not as an option but as a necessity. He devised a primarily economic strategy centered on Ceuta, a strategic juncture for trans-Saharan gold caravans because of its location at the mouth of the Mediterranean. Gathering a fleet of approximately 200 ships at an enormous cost that consumed two years' worth of royal revenue, João was no gambler. To put this into perspective, imagine a modern government allocating nearly 20-25% of its annual GDP for a single military endeavor, reflecting the magnitude of such an investment. João had been informed by intelligence from Jewish merchants and renegade Moors about Ceuta's potential for capture and control.

Ceuta fell to the Portuguese in a single day on August 21, 1415, with Portuguese soldiers discovering warehouses laden with exotic goods. However, victory did not cement João's reign as Muslim merchants rerouted their caravans, transforming Ceuta into an expensive stronghold rather than a source of wealth. Through this costly lesson, João realized the futility of controlling a port without dominance over the supply routes.

This failure would shape Portugal's trajectory henceforth. Should land routes be blocked, Portugal must navigate by sea to reach their sources. Therefore, João sanctioned Henry to establish himself at Sagres on Portugal's southwestern tip and commence the methodical exploration along the African coast. Although João did not live to witness the fruition of his decision, passing in 1433, he had ignited an economic engine that would persist for centuries.

Dom Duarte (1433-1438): The Hesitant Inheritor

Duarte found himself ruling over a realm burdened by financial obligations. Specifically, the Ceuta garrison demanded a significant portion of royal funds just to maintain its presence. Facing Henry's proposal for broader exploration, Duarte was conflicted. As a scholar-king, his inclinations leaned more towards philosophical discourse than fiscal management. His court was rife with disagreement; conservatives were vehement in their belief that investments into Atlantic expeditions were ill-spent resources, akin to throwing good money out the window. They advocated for focusing on securing Portugal's territories in Morocco rather than embarking on quests for uncharted lands and speculative trade routes.

Nevertheless, Henry presented a compelling case for the ventures' potential profitability. He posited that these voyages could, over time, become self-sustaining endeavors. Newly discovered islands could be settled and turned into lucrative sugar plantations, leveraging expertise from Sicily. Furthermore, the slave trade promised immediate returns, while crossing the Sahara might yield rich gold deposits. Duarte, albeit with reservations, authorized limited funding for these endeavors. In doing so, he established a fiscal framework that was akin to early joint-stock companies - a system which would shape Portuguese exploration for generations to come. The Crown initiated expeditions with its capital support, while explorers were granted the autonomy of captaincies and monopolies over lands they discovered. This allowed them to tax trade, enslave indigenous populations, and extract resources, essentially acting as agents within a corporate venture framework, taking high risks in anticipation of substantial personal gains.

This fiscal arrangement laid the groundwork for later European investment models, demonstrating continuity in the financing methods of exploration. Unlike Renaissance Italy's patronage system, where affluent families funded artists for prestige, this was an instance of venture capital - strategic investments aimed at achieving tangible returns. Duarte's reign, unfortunately cut short by his untimely death from plague in 1438, did not live to see the outcomes of these endeavors. However, with the mechanisms now in place, the stage was set for what was to unfold.

Dom Afonso V (1438-1481): The African King

Afonso was six years old when he became king. During his prolonged minority, actual power remained in the hands of his uncle Dom Pedro, who served as regent. The exploration program continued apace, though it was Prince Henry who spearheaded it with unyielding passion. As the master of the Order of Christ, which succeeded the Knights Templar and controlled vast wealth from its lands, Henry enjoyed financial autonomy that enabled him to equip expeditions independently without continually petitioning the royal treasury. This fiscal independence enabled exploration to continue even during periods when the monarchy's finances were tight, embedding self-sustaining, profit-driven principles into Portugal's broader strategic framework. From 1434 to 1460, under Henry's guidance, his captains advanced steadily southward. Gil Eanes broke through the psychological barrier by rounding Cape Bojador in 1434, which had previously intimidated sailors from the Mediterranean. Each voyage yielded new intelligence; the desert coast gradually gave way to inhabited lands teeming with people who could be enslaved and sold. The slave trade gained momentum in the 1440s, driven by lucrative economic prospects that bolstered the Crown's commitment to further exploration. Captives arrived in Lagos, where they were traded like livestock, illustrating how maritime enterprise and human trafficking became integral to shaping Portugal's imperial economic policy.

The financial returns were clear from inception. Enslaved individuals captured on the African coast for mere trifles such as copper bracelets, glass beads, or red cloth could be sold in Portugal for twenty times their original cost. The profits were enormous. Portuguese raiders often returned with hundreds of captives at once. Between 1450 and 1500, approximately 140,000 African slaves were brought to Portugal and its Atlantic islands, generating immense wealth for captains, investors, and royal officials.

As Afonso matured, his obsession shifted towards Morocco, diverting resources from sea exploration. He dreamt of crusades and conquering Muslim territories in North Africa to achieve military glory. His failed attack on Tangier in 1437 was disastrous, with his brother Ferdinand captured and dying in Moorish hands. Despite this, Afonso persisted in launching campaigns that drained Portugal's funds and nobility. He referred to himself as "the African," adopted Moorish attire, and squandered money on fortifications that brought no income.

Even while Afonso fixated on Morocco, the maritime infrastructure Henry established continued to function effectively. The Casa da Guiné, founded in 1455, managed trade along the West African coast. Only licensed captains were permitted to sail south of Cape Bojador, with all goods required to pass through designated ports where royal tax collectors collected their share. Typically, the Crown claimed one-fifth of total profits under the practice known as the "royal fifth," which became standard throughout the Portuguese empire.

This bureaucratic shift was as significant as any navigational breakthrough. Portugal was constructing the frameworks to govern, regulate, and profit from a vast trading empire spanning thousands of miles. Letters of captaincy detailed how profits were distributed: usually, ten percent went to the captain, shares were divided among crew members based on rank, and the remainder was split between financiers and the Crown.

Prince Henry's death in 1460 left him deeply in debt despite the profitability of the slave trade. He had exhausted all his resources funding expeditions, shipyards, tools, and hiring pilots and cartographers. Nevertheless, his demise did not halt the system he created. Afonso continued to issue licenses for African voyages, while Portuguese merchants, having witnessed the profits, were motivated enough to carry on without royal inducement.

Dom João II (1481-1495): The Perfect Prince

If it were not for João the First setting the stage, we might never have heard of Henry initiating exploration. João the Second refined what was started, becoming king with a clear vision for Portugal's place in the world. At that time, Venice and Genoa controlled Mediterranean trade, while Muslim powers dominated the land route to India. If Portugal could just find an alternative sea route around Africa, it would break both monopolies. Every shipment of spices that arrived in Lisbon by sea meant lost profits for Venice or the Ottoman Empire. João understood something earlier kings had overlooked: information was as valuable as gold itself.

In a moment that would change the course of history, João's board of cosmographers, astronomers, and pilots unveiled an astonishing breakthrough: the refined astrolabe. This navigational marvel allowed sailors to determine their latitude with unprecedented accuracy, transforming the perils of sea voyages into calculated journeys. As the instrument was introduced, it was as if the fog of uncertainty was lifted from the vast ocean.

João also took direct control of expeditions, unlike Afonso, who allowed private adventurers more freedom. He required royal approval for every major exploration. He negotiated the Treaty of Alcáçovas with Castile in 1479, securing Portugal's monopoly over the African coast while Spain retained control of the Canary Islands. João was strategic in his diplomacy, ensuring exclusive rights to the route to India even before it was discovered. In 1482, he sent Diogo Cão south with clear orders to find the end of Africa. Cão reached the Congo River, marked the territory with stone monuments for Portugal, and brought back news of strong African kingdoms and lucrative trade opportunities.

João saw potential in this and sent embassies to the Kongo king along with missionaries and craftsmen to build lasting diplomatic ties. This was an empire-building approach combining military power, commerce, and culture. João's most daring move was sending Pêro de Covilhã overland to India in 1487. Disguised as a Muslim merchant, Covilhã traveled through Egypt, down the Red Sea, into India, and along Africa's eastern coast. He sent back critical information about Indian ports, Calicut's spice markets, and the monsoon winds affecting Indian Ocean trade. With this knowledge, João could now anticipate what his ships would encounter if they sailed around Africa.

The journey became a deliberate mission rather than just an expedition. When Bartolomeu Dias returned in December 1488 with news that he had sailed around Africa's southern tip, João knew Portugal's economic future was guaranteed. The route was real and could be utilized. All that was left was to make the most of it. The king was determined to safeguard this advantage. When Christopher Columbus presented a plan for a western route to Asia, João listened, consulted his advisors, and dismissed the idea as mathematically flawed. He likely made the right call since Columbus had grossly underestimated Earth's size. Furthermore, João perceived a threat; if Spain found a western route, Portugal's costly African endeavors could become obsolete. This was not an option.

João passed away in 1495 before any Portuguese ships reached India. But his work was far from over; he prepared everything meticulously. Navigation methods were refined, ships were constructed and tested, financing was arranged, and diplomatic ties were established. Vasco da Gama would be the one to complete the journey João had envisioned.

THE MACHINERY OF INVESTMENT

How a Common Sailor Became a Captain of Enterprise

For those outside the realm of noble lineage or abundant wealth, the exploration ventures stood as a rare beacon within European society: an opportunity for transformation through risk-taking into a proprietor of standing and influence. Accessing this doorway required navigating intricate webs of patronage, debt obligations, and royal benevolence that balanced meritocracy with rigid hierarchy.

Envision Martim, an amalgamation embodying the stories of numerous men who embarked on such a path. In 1490, at twenty-three years old, Martim was from Porto; his upbringing under the guidance of a master cooper surrounded by barrels and taverns near the port. He possesses literacy, fundamental arithmetic skills, and considerable experience aboard fishing ships. His aspiration? To helm a voyage to Guinea. The steps he must take are clear.

Step One: Acquiring Technical Credentials

Maritime endeavors were deemed prohibitively costly for those without prior experience. Martim had to demonstrate his worthiness before embarking on such ventures. To gain expertise in navigation and ship operations, he joined a trading fleet destined for Flanders. As an entry-level sailor, he received about 500 réis monthly, just enough to cover basic needs like food and housing. Over the span of three years, he mastered celestial navigation, dead-reckoning techniques, and leadership among his crewmates. His resilience was tested by surviving two violent Atlantic storms and narrowly avoiding capture by French privateers. What proved most significant in his journey were the bonds he formed along the way. He particularly gained favor with his captain, a certain nobleman named Álvaro Fernandes, who had influential ties to royalty.

Step Two: The Letter of Recommendation

Following his fourth voyage, Martim seeks out Fernandes in a Lisbon tavern, presenting his case: he claims to know the route to Arguim, the Portuguese fortress located on the Saharan coast. He has observed the slave trade and is aware of which West African rivers provide the best captives, as well as understanding what Africans desire in exchange for them through barter. Martim aims to lead his own expedition.

Fernandes remains skeptical due to Martim's common birth, lack of capital, and political connections. However, he acknowledges that the man possesses competence and genuinely likes him. What ultimately matters is Fernandes' recent financial loss from a failed voyage; he seeks capable captains willing to accept smaller shares in order to minimize his risk.

Ultimately, Fernandes agrees to write a letter of recommendation to Fernando Coutinho, a wealthy fidalgo involved in various trading ventures. This endorsement is vital; without noble support, Martim may spend years petitioning the crown without securing an audience.

Step Three: The Merchant Prince

Coutinho is bombarded by numerous requests. For the most part, he disregards them. However, Fernandes' name holds significant influence, and piques Coutinho's curiosity enough to arrange a meeting at his Lisbon residence. The encounter is short-lived and degrading. Coutinho, adorned in luxurious silks that exceed Martim's lifetime earnings, remains composed while Martim presents his proposition: embarking on a six-month journey to Guinea aboard two caravels with a crew of thirty men, estimated costs totaling 600,000 réis, anticipating profits threefold through the sale of slaves and merchandise.

Coutinho's inquiries are meticulous and unsettling. What percentage of sailors perish during such voyages? (Martim admits around fifteen percent.) How many slaves die en route across the Atlantic? (Between ten and twenty percent.) What would happen if one of the caravels sank? (The expedition would lose half its capacity and likely incur losses.)

These aren't poetic queries about triumphs and discoveries. Rather, they are a shrewd businessman's evaluation of risk-adjusted returns. Coutinho is evaluating Martim's proposal against other alternatives: agricultural landholdings, textile manufacturing, and financing the monarchy. The African commerce is lucrative yet unstable. Whole expeditions may disappear. Others return with fortunes that appear almost magical.

Coutinho agrees to review the proposal. He will consult his agent, assess current market prices for slaves and African goods, and scrutinize the monarch's current licensing requirements.

Step Four: The Syndicate

Three weeks later, Martim receives a call back into action. Coutinho determines to proceed but not alone; he gathers around him a syndicate of four investors: himself and two other merchants, alongside an emissary from the Order of Christ. Together, they will pool capital totaling 650,000 réis, slightly exceeding Martim's request, thereby allowing for unforeseen expenditures.

The terms are outlined in a contract drawn up by a scrivener, which grants Martim a Letter of Captaincy, endowing him with leadership of the expedition and authority over its crew. Compensation will not be in the form of a salary; rather, he will receive 10% of net profits after costs have been recovered. In sharp contrast, the syndicate, comprised of numerous investors, secures 70% of the net revenues. This dichotomy highlights the variance between risk and remuneration: Martim shouldering operational perils yet only entitled to a modest share, while investors reap a far greater share with minimal personal risk. The remaining twenty percent is allocated as taxes to the Crown. In the eventuality where the expedition incurs a loss, Martim receives nothing and finds himself indebted to the syndicate for service on a future voyage in order to repay their investment in him.

Further stipulations are delineated with an unyielding degree of exactitude: should Martim engage in private commerce for his own account utilizing resources from the expedition, he risks forfeiting his captaincy and faces prosecution by royal justice for embezzlement. Should he lose a vessel through carelessness or intoxication, he would be answerable for its replacement value. Additionally, if he abandons the expedition or mutinies against the syndicate's interests, his possessions and family are liable to seizure as reparation for the investors.

Martim signs. He has no alternative. This route is the singular option available to him.

Step Five: The Royal License

But without the royal authority's blessing, such agreements are void. Portuguese commerce with Africa is under the exclusive purview of the monarchy. No ship may venture past Cape Bojador unless it bears a permit from the Casa da Guiné. Martim seeks out Coutinho for aid in this matter. The request must pass through numerous hands, each demanding payment before reaching the sovereign's council.

The king assesses perhaps two dozen such pleas every moon. Most are granted if the candidates have paid the required amount and show substantial backing. However, the monarch's whims dictate otherwise at times. Licenses may be withheld to compel individuals into alliances with those favored by the crown. It serves as a means of political manipulation, rewarding loyalty and penalizing dissent.

Martim's plea succeeds, but not without stipulations: he must transport an emissary from the throne who will personally inspect all dealings on behalf of the royal family. He is forbidden from navigating certain waterways where exclusive rights have been granted to other commanders by the king. Any novelties found or intelligence gleaned from African rulers must be relayed back posthaste.

The license incurs a fifty-thousand réis fee, covered by Coutinho and added to Martim's voyage expenses.

Step Six: Outfitting the Vessels

As the arduous journey looms ahead, Martim embarks on a rigorous phase of readiness. His task is to equip two caravels for an expedition that could last 6 to 8 months. The principal vessel is approximately 70 feet long, with a displacement of 40 tons and a crew of 18. The secondary, slightly smaller ship, will accommodate fifteen crew members.

The financial burdens incurred are immense and documented with meticulous detail:

- Expenses for two caravels, either through purchase or lease: 280,000 réis
- Costs of sails, rope, pitch, and naval supplies: 80,000 réis
- Valuation of trade goods (copper bracelets, manillas, beads, red fabric, knives): 120,000 réis
- Provisioning (biscuits, salted pork, dried fish, olive oil, wine, water barrels): 90,000 réis
- Allocation for weaponry (crossbows, spears, blades, armors): 30,000 réis. Investment in navigation tools, maps, and miscellaneous items: 20,000 réis. Advance payment of wages to the crew prior to departure: 30,000 réis
- Aggregate expenditure: 650,000 réis, precisely in line with the predetermined budget.

Every acquisition is channeled through middlemen who reap their share of profit. The shipbuilders artificially inflate prices. The merchant of provisions sells biscuits infested with weevils. The purveyor of trade goods delivers copper and bracelets that corrode swiftly. Martim is aware he has been swindled, yet he has no bargaining power. He operates within a financial envelope not of his own, obligated to accept the terms presented to him.

Step Seven: Assembling a Crew

Discovering thirty-three men willing to hazard death in Africa proves surprisingly effortless. Lisbon's waterfront is bustling with jobless sailors, younger sons of impoverished families, criminals escaping debts or accusations. Martim simply needs to publicize his voyage at the dockworkers' recruitment fair. The candidates form a diverse mosaic of Portuguese seafaring society: seasoned voyagers with prior experience, young boys seeking excitement, and several African slaves belonging to Coutinho who will act as interpreters and liaisons. Martim picks his crew based on their expertise and physical condition. He steers clear of confirmed alcoholics and troublemakers; however, the pool of suitable candidates is limited. He ends up selecting several sailors whose reliability is questionable. The compensation package is individually negotiated. Seasoned sailors typically request shares in the profits rather than a fixed salary. Ordinary seamen usually receive a flat monthly fee ranging from 400 to 600 réis. The least skilled laborers, including the slaves, earn nothing but sustenance and the promise of spoils. Every crew member signs or marks an agreement delineating their responsibilities and remuneration. The notary's fees for these legal documents contribute another 5,000 réis to the expedition's budget.

Step Eight: Departure

On a morning in late April 1491, Martim's two caravels glide down the Tagus under a feeble sun. Coutinho and the other investors gaze from the shore, estimating when they might see returns on their investments. The crew, thirty-three souls aboard vessels not much larger than today's pleasure crafts, starts the daily tasks of ocean travel: manning watch duties, examining the rigging, conserving water supplies.

For Martim, gripping the tiller as Lisbon recedes into the distance, this is the fulfillment of years of endeavor and the start of perilous challenges. If the expedition fails, he reverts to being a common seafarer or worse, accountable for debts that cannot be repaid. Success may bring enough wealth to finance further voyages, gradually enhancing his reputation and capital until he can lead enterprises independently.

Such was the pattern year after year; it wasn't valor that propelled these journeys but the harsh arithmetic of profit return, men's ambition to exceed their current circumstances, and the readiness of investors to venture their funds on speculative endeavors if the promised gains were sufficiently vast. Ultimately, it was the relentless pursuit of profit that fueled the voyages, outweighing any measure of bravery or adventure.

THE COURT AND THE COUNTING HOUSE

The Nobility's Gamble

The distinguished Portuguese bloodlines approached exploration through an angle markedly distinct from that of the monarchy or the humble ship captains. For them, expeditions to Africa and Asia were not merely adventures; they were strategic instruments employed in the ceaseless struggle for royal patronage, advancement within their noble lineage, and the amassing of fortunes secure against the caprices of royalty.

The Braganza clan, preeminent among Portugal's ducal houses, exemplifies this calculated engagement. Fernando I, holder of the dukedom in its third generation, presided over vast territories in northern Portugal, held sway over thousands of vassals, and commanded resources independent of the Crown resources that rivaled those of the monarchy itself. Yet, despite such a commanding position, he invested heavily in exploration, not out of necessity but as part of a broader strategic plan. Aware that royal favor was transient, his father had been executed by João II in 1483 on questionable charges; the Braganza family's survival depended on becoming indispensable to the monarchy.

Fernando sponsored numerous expeditions to West Africa during the 1490s, forming alliances with merchant consortia yet ensuring his household members captained the vessels and controlled the division of profits. Upon these voyages' returns with human cargoes and gold dust, Fernando generously contributed a portion to the Crown, showcasing loyalty and service to the throne. The financial gain was secondary to the political benefits accrued.

Other noble factions played analogous games. The Almeida lineage, the Coutinhos, and the Castros each cultivated networks in maritime commerce, investing in ventures as a form of portfolio diversification and political safeguarding. When Dom Francisco de Almeida was appointed the inaugural Portuguese viceroy of India in 1505, it marked the culmination of years of familial investment in exploration endeavors. The honor was no act of benevolence but recognition of capital risk and political backing furnished.

The nobility also secured exclusive rights to specific trade routes or goods. The importation of pepper from India, the slave commerce from certain regions of Africa, and the cultivation of sugar on Atlantic archipelagos were each granted as royal franchises to families who undertook development in exchange for sole trading privileges. This system fostered an economic hybrid, a blend of capitalistic drive and feudal tradition, that integrated aristocratic privilege with entrepreneurial spirit in ways alien to more conservative courts across Europe.

The Church's Percentage

The Catholic Church played dual roles in the exploration era - both as a moral authority and an economic player. Papal bulls sanctioned the Portuguese rulers over conquered lands and allowed for the enslavement of non-believers under the guise of religious missions, all while the church benefited financially.

The Order of Christ, which succeeded the Portuguese Templars, amassed significant landholdings and was allotted a share of the profits from African trade by its leader, Prince Henry. This arrangement endured beyond his reign. Like any enterprise, this order funded expeditions, owned ships, and profited from its investments. It is worth noting that, while the religious narrative of spreading

Christianity through exploration may have been genuine to those involved, it also went hand in hand with financial gain.

Various dioceses and monasteries also financially supported these ventures, as they would with any profitable business. They invested in slave-trade activities, received shipments of African captives to work their properties, and advocated for advantageous positions within colonial territories. For instance, part of the funds used to build Lisbon's magnificent Jerónimos monastery came from the lucrative spice trade.

To contemporary participants, this was not a deceptive practice. The medieval belief system didn't perceive a conflict between religious goals and financial outcomes. Wealth was regarded as a divine boon, and the Church's acquisition of it was seen as fulfilling God's will. However, over time, especially among Protestant reformers, this interconnectedness came under scrutiny as proof of Catholic corruption.

The Jewish and New Christian Networks

Portugal's Jewish community contributed substantially to the financing of exploration endeavors, despite being officially vulnerable following the 1497 mandatory conversions. These Jewish merchants possessed extensive international networks, expertise in long-distance trade, and access to capital, rendering them indispensable for such expeditions. Before their expulsion and conversion, they openly invested in African expeditions; afterwards, under the guise of Christianity while maintaining private Jewish practices, they continued supporting ventures discreetly.

The community known as New Christians, forced converts, specialized in specific trades and dominated Madeira and São Tomé's sugar industry, controlled much of the slave trade financing and distribution, and served as factors and interpreters in African and Asian ports. Their linguistic skills and commercial connections were crucial to Portugal's success.

However, this engagement came with a high price. The Portuguese Inquisition, established in 1536, targeted New Christians accused of secretly practicing Judaism for centuries. Thousands faced arrest, property seizure, and business destruction. Despite relentless persecution, the community's role in trade remained too valuable to suppress entirely. Periodically, the Crown acknowledged its indispensability by issuing guarantees of safety to New Christian merchants, understanding that Portugal's empire relied on their expertise. This arrangement resulted in both immense wealth and great suffering, a cruel bargain with far-reaching consequences.

The Italian Connection

Portuguese rulers sought out connections within Italian financial circles, particularly with Genoese bankers who possessed the funds and knowledge required for long-distance trade expeditions. Notably, Columbus hailed from Genoa, as did various captains and financiers employed by Portugal. These individuals introduced Mediterranean trade practices to the Atlantic, including advanced strategies such as risk-sharing, maritime insurance, and joint-stock investments.

The presence of these Italian traders was not universally accepted. Portuguese merchants were envious of the competition posed by their Italian counterparts and campaigned for laws that curtailed foreign involvement in the African trade. However, the monarchs recognized the significant value of Italian capital and expertise and chose not to ban them entirely from participation. Consequently, a compromise was struck: Italians could engage in investment and partnerships but were barred from holding absolute control. They operated alongside Portuguese entities, contributing financial resources and sharing profits while acquiescing to a secondary role within the political sphere.

THE WAGES OF SUCCESS

Vasco da Gama's Return: The Accounting

Upon his return to Lisbon in September 1499, Vasco da Gama brought back only a fraction of his original fleet, commanding just two ships out of the initial four. Tragically, over half of his crew succumbed to scurvy, conflicts, or shipwreck, marking the voyage as a human catastrophe.

However, in terms of financial success, it was an unprecedented triumph for Portugal's commerce. These two surviving vessels were laden with approximately 300 tons of pepper, cinnamon, cloves, and other valuable spices, sourced either through purchase in Calicut or through the seizure of Arab ships they had captured. Though these goods were common commodities in the Indian Ocean, their value soared in Lisbon to exceed even that of gold. Calculations by the Casa da Índia valued the total cargo at over 2 million réis. Considering the costs incurred for vessels, provisions, and the compensation of the deceased crew members' wages, the total amounted to roughly 200,000 réis.

The profit margin reached an astounding 1000%.

King Manuel I recognized immediately that the European economic landscape had taken a dramatic turn. Future expeditions to India promised returns that would multiply their costs many times over. The once-unchallenged spice monopoly held by Venice crumbled under Portuguese competition, and the centuries-long Muslim trading hegemony in the Indian Ocean faced an aggressive European contender with superior naval capabilities, firepower, and a relentless drive for dominance.

In response, King Manuel heavily invested in constructing another fleet to follow da Gama's pioneering voyage. Pedro Álvares Cabral embarked on this new endeavor in March 1500 with 13 ships and 1,500 men at an enormous estimated cost of around 2 million réis, yet the anticipated profits made the investment worthwhile. His orders were clear: establish a trading post in Calicut by all means necessary, secure plentiful spices, and demonstrate Portuguese naval might to deter any opposition.

Cabral's expedition not only discovered Brazil but also fought in Calicut, destroying much of its Arab fleet. In terms of the cargo of spices brought back, he outdid da Gama himself. The financial influx from these expeditions was so staggering that it radically transformed Portugal's state finances. Unlike João II, who had grappled with budgetary strains, King Manuel found himself in a wealth beyond any previous monarch's imagination.

The Albuquerque System: Empire as Corporate Enterprise

Afonso de Albuquerque, who served as Portuguese governor in Asia from 1509 to 1515, was instrumental in developing an economic model that maintained Portugal's empire for more than a century. He understood that simple trading posts were insufficient to guarantee Portuguese supremacy. What was needed was full control of the Indian Ocean's critical bottlenecks and the methodical dismantling of rival commercial networks. Albuquerque strangled Goa in 1510, seized Malacca in 1511, and conquered Hormuz in 1515. Each conquest was carefully orchestrated for optimal economic impact. Goa dominated the western coast of India. Malacca monopolized the strait through which all trade between East Asia and the Indian Ocean passed. Hormuz controlled the Persian Gulf. Whoever commanded these three positions could dictate or disrupt nearly all Asian trade.

However, Albuquerque's vision extended beyond mere taxation; he aggressively sought to rewire trade itself. He mandated that all merchant vessels obtain cartazes, Portuguese permits, to navigate Asian waters. Any ship discovered without a cartaz was seized, its cargo confiscated, and its crew potentially enslaved or killed. This marked the first time a European power systematically enforced a maritime protection scheme across an entire ocean. The financial gains were staggering. The Crown collected taxes on all commerce passing through Portuguese-controlled ports. Individual governors, captains, and officials profited from their positions legally and otherwise. Within a generation, serving in the Estado da Índia, Portugal's Asian empire became synonymous with wealth for its nobility. Positions were bought and sold at court for astronomical sums. A three-year term as captain of a major fortress could secure one's fortune for life, provided one survived the perils of disease, violence, and distance that conspired against European aspirations in Asia.

The Human Price

The prosperity achieved was constructed upon the deaths of countless individuals. Mortality amongst Portuguese sailors venturing towards India was alarmingly high, surpassing thirty percent in the sixteenth century. To put it in perspective, this would be like losing 600 crew members from a modern aircraft carrier, highlighting the staggering scale of loss. A vessel setting sail from Lisbon with a crew of 200 might only arrive with approximately 120 souls remaining. Many perished while serving in Asia before they could complete their term and return home.

The mistreatment inflicted upon Asian populations was relentless and merciless. When the Zamorin of Calicut rejected Portuguese trade proposals, Albuquerque responded with an onslaught on the city, executing captured defenders from his ships' yardarms and annihilating the Arab mercantile community. Malacca's resistance led to its destruction by fire, the slaughter of thousands, and the enslavement of those who survived. This savagery was not merely wanton brutality but a calculated tactic intended to instill fear in other regional rulers and force them into submission.

In tandem, the transatlantic slave trade burgeoned significantly. By 1550, an estimated 10,000 African slaves were being annually transported to Portugal and its offshore islands. The economic ramifications for African societies were catastrophic. Large areas became sparsely populated due to ceaseless warfare, as kingdoms raided their neighbors in a bid to secure captives to sell to Portuguese traders. These consequences reverberated through generations.

Despite these grim realities, the notion of an era of abundance took hold in Portugal. Lisbon rapidly ascended to become one of Europe's most affluent cities. The Manueline architectural style, replete with its nautical symbols and opulent embellishments, epitomized the kingdom's burgeoning assurance and riches. The court attracted a plethora of artists, intellectuals, and explorers drawn by its allure. Portuguese galleons monopolized the trade route between Asia and Europe, solidifying Portugal's position as a maritime superpower. This tiny realm that had staked everything on naval conquests appeared to have secured an outcome surpassing even the most ambitious predictions.

THE FRAGILITY OF FORTUNE

The Hidden Weaknesses

However, despite its outward strength in the early sixteenth century, the Portuguese empire was built on shaky ground, as evidenced by its inability to sustain itself through various issues. To begin, Portugal faced a significant population deficit, lacking enough people to effectively control and maintain the territories it had claimed. With an estimated 1.5 million inhabitants in 1500, maintaining fortresses from Brazil to Japan proved beyond the nation's resources. As each fortress required military garrisons, administrative staff, and supporting populations, sustaining such a vast empire was impossible.

By 1530, the Crown was already struggling to find enough men willing to serve in Asia, despite the rising salaries and costs. Furthermore, the empire's naval superiority was tenuous at best; while Portuguese ships were well-built and well-armed compared to most Asian vessels in 1500, it did not take long for other powers, such as the Ottoman Empire, to catch up. Moreover, local rulers began purchasing European cannons and hiring renegades to train their sailors, ensuring that Portugal no longer held complete dominance over Asian waters by 1550.

Additionally, the Portuguese economy was fundamentally extractive rather than generative, meaning it did not create new wealth but merely redistributed existing Asian trade through channels under its control. While this generated immediate profits, it failed to build lasting infrastructure and relied on the continued cooperation of distant economies that produced the desired goods. Unlike contemporaries such as China and India, whose maritime economies were integrally connected to local production and commerce, fostering long-term growth and stability, Portugal's position was precarious at best. If these Asian economies chose to cease trading with Portugal, or if competing European powers established alternative routes, the Portuguese empire would crumble under the weight of its structural issues.

The Spanish Rivalry

The Treaty of Tordesillas in 1494 divided the world between Portugal and Spain, but this line of demarcation proved both ill-defined and unenforceable. With Spanish expeditions crossing into the Pacific from Mexico and claiming the Philippines by 1565, direct competition between these two Iberian empires arose within Asian waters.

Spain's superior resources included a larger population, greater territories, and the immense wealth generated by American silver mines. As Spanish silver arrived in Manila and entered Chinese trade networks, it offered Asian merchants an alternative to Portuguese intermediaries, gradually eroding Portugal's numerical advantage in Asian commerce.

Portugal's monarchs acknowledged this threat but struggled to counteract it effectively due to rising imperial maintenance costs. By 1580, when Philip II of Spain inherited the Portuguese throne through dynastic succession, Portugal's golden age had all but vanished. The union between the two crowns endured for 60 years until Portugal regained independence in 1640; however, its Asian empire had dwindled into a shadow of what Albuquerque had built, with the Dutch and English seizing most of the most lucrative territories during Spanish rule.

THE LEDGER'S LAST LINE

Lisbon, November 1515

As the funeral procession solemnly navigated through Lisbon's streets, it bore with it Afonso de Albuquerque's remains from India. This was a place where he had ended his life in disgrace after being dethroned as governor by an envious sovereign. Onlookers remained silent, reflecting on the figure he once was to Portugal - its most renowned conqueror in Asia and architect of the empire itself. However, others couldn't help but recall the bloodshed, burnt cities, and thousands enslaved under his command.

Meanwhile, back at Casa da Índia, accountants diligently recorded entries into their ledgers, as the fleet's return with full cargo was a yearly affair by now. The profitability of spices like pepper, cinnamon, and cloves had become so common that it no longer raised eyebrows.

King Manuel, an aging ruler brimming with suspicion, scrutinized these financial records within his palace walls. They were impressive indeed, serving as a testament to the shrewdness in sponsoring maritime explorations. Yet beneath this façade of strength, he sensed vulnerability: the expenses were mounting, competition was intensifying, and the victories led by figures such as Albuquerque couldn't be sustained indefinitely.

Regardless, the machinery of exploration churned on: ships departed and returned with their precious loads of gold and spices, investors cashed in on their shares, the Crown took its due fifth, and sailors fulfilled contracts before meeting their anonymous ends at sea. Slaves marched through Lisbon's markets in shackles, their acquisition and sale just another transaction itemized in grand accounts.

What romantic narratives choose to overlook is this: the Age of Discovery was primarily about a methodical endeavor by Portugal, a small kingdom at that, to wield maritime technology and ruthless trading strategies for siphoning global wealth into its hands. Records from Casa da Índia, such as Vasco da Gama's returns in 1499 boasting up to 1000% profit margins (Boxer, 1969; Godinho, 1963-1971), unveil how this translated into immense wealth for investors and nobles, documented in merchant records and royal contracts of Arquivo Nacional Torre do Tombo. Moreover, primary sources like the Crónica dos feitos notáveis que se passaram na Conquista da Guiné detail the origins of plantation systems and the Atlantic slave trade, which secondary studies (Russell-Wood, 1998; Saunders, 1982) argue laid foundational patterns for centuries of global history.

It all commenced in rooms like those where Prince Henry and João Gonçalves Zarco affixed their signatures in 1419. Here, men reduced grand dreams and suffering to mere figures on paper. The essence of the Portuguese Age of Discovery, turning hope and peril into quantifiable gains and losses, was epitomized by these calculations. This process dictated the destinies of both commoners and nobles and charted the course of global economic and imperial expansion, revealing that true discovery was propelled neither by valor nor by fate, but by the relentless pursuit of financial gain. As we look back, we must ponder: What ledgers are we still balancing in this way today?

The ships continued to set sail as long as the arithmetic favored them. Desperate souls kept coming because Portugal alone in Renaissance Europe offered a beacon of hope: a humble cooper's son from Porto might one day command fleets, amass estates, and elevate his lineage into nobility through wealth amassed in the blood-drenched markets of continents.

This was the toll of empire. This was the reward for discovery. Not glory but calculation; not heroism but return on investment; not destiny but tallying. The final entry in any ledger was always profit or loss, everything else being mere tales of valor.

BIBLIOGRAPHY

PRIMARY SOURCES

Chronicles and Contemporary Accounts

Albuquerque, Afonso de. *The Commentaries of the Great Afonso Dalboquerque.* Translated by Walter de Gray Birch. 4 vols. London: Hakluyt Society, 1875-1884.

Álvares, Francisco. *Verdadeira Informação das Terras do Preste João das Índias.* Lisbon, 1540. English edition: *The Prester John of the Indies,* edited by C.F. Beckingham and G.W.B. Huntingford. Cambridge: Cambridge University Press, 1961.

Barros, João de. *Décadas da Ásia.* Lisbon, 1552-1615.

Cadamosto, Alvise. *Navigazioni.* Mid-15th century. English edition edited by G.R. Crone. London: Hakluyt Society, 1937.

Caminha, Pero Vaz de. Letter to King Manuel I. 1500. In Greenlee, William Brooks, ed. *The Voyage of Pedro Álvares Cabral to Brazil and India.* London: Hakluyt Society, 1938.

Castanheda, Fernão Lopes de. *História do descobrimento e da conquista da Índia pelos portugueses.* Coimbra, 1552-1561.

Correia, Gaspar. *Lendas da Índia.* Written c. 1550s. Published Lisbon: Academia Real das Ciências, 1858-1866.

Fróis, Luís. *Historia de Japam.* Edited by José Wicki. 5 vols. Lisbon: Biblioteca Nacional de Lisboa, 1976-1984.

Frutuoso, Gaspar. *Saudades da Terra.* Written 1586-1590. 6 vols. Ponta Delgada: Instituto Cultural de Ponta Delgada, 1978-1983.

Góis, Damião de. *Crónica do Felicíssimo Rei D. Manuel.* Lisbon, 1566-1567.

Pacheco Pereira, Duarte. *Esmeraldo de Situ Orbis*. Written c. 1505-1508. English edition translated by George H.T. Kimble. London: Hakluyt Society, 1937.

Pigafetta, Antonio. *Relazione del primo viaggio intorno al mondo*. c. 1525. English edition: *Magellan's Voyage*, translated by R.A. Skelton. New Haven: Yale University Press, 1969.

Pina, Ruy de. *Crónica de El-Rei D. João II*. Written c. 1504.

Pinto, Fernão Mendes. *Peregrinação*. Lisbon: Pedro Crasbeeck, 1614. English edition: *The Travels of Mendes Pinto*, edited and translated by Rebecca D. Catz. Chicago: University of Chicago Press, 1989.

Pires, Tomé. *Suma Oriental*. Written c. 1512-1515. Translated by Armando Cortesão. London: Hakluyt Society, 1944.

Velho, Álvaro. *Journal of the First Voyage of Vasco da Gama to India, 1497-1499*. Translated by E.G. Ravenstein. London: Hakluyt Society, 1898.

Xavier, Francis. *The Letters and Instructions of Francis Xavier*. Translated by M. Joseph Costelloe. St. Louis: Institute of Jesuit Sources, 1992.

Zurara, Gomes Eanes de. *Crónica do Descobrimento e Conquista de Guiné*. Written c. 1453. English edition edited by C. Raymond Beazley and Edgar Prestage. London: Hakluyt Society, 1896-1897.

Letters and Documents

Anchieta, José de. *Cartas, correspondência ativa e passiva*. Edited by Hélio Abranches Viotti. São Paulo: Edições Loyola, 1984.

Anonymous. *Relação do Piloto Anónimo* (Anonymous Pilot's Account). c. 1500.

Cartas de Afonso de Albuquerque. Edited by Raymundo António de Bulhão Pato. 7 vols. Lisbon: Academia Real das Sciencias, 1884-1935.

Transylvanus, Maximilianus. *De Moluccis Insulis*. 1523.

Literary Sources

Camões, Luís de. *Os Lusíadas*. Lisbon, 1572.

SECONDARY SOURCES

Portuguese Maritime Expansion: General Works

Bethencourt, Francisco, and Diogo Ramada Curto, eds. *Portuguese Oceanic Expansion, 1400-1800*. Cambridge: Cambridge University Press, 2007.

Boxer, C.R. *The Portuguese Seaborne Empire, 1415-1825*. London: Hutchinson, 1969.

Crowley, Roger. *Conquerors: How Portugal Forged the First Global Empire*. New York: Random House, 2015.

Diffie, Bailey W., and George D. Winius. *Foundations of the Portuguese Empire, 1415-1580*. Minneapolis: University of Minnesota Press, 1977.

Disney, A.R. *A History of Portugal and the Portuguese Empire*. 2 vols. Cambridge: Cambridge University Press, 2009.

Godinho, Vitorino Magalhães. *Os Descobrimentos e a Economia Mundial*. 4 vols. Lisbon: Editorial Presença, 1981-1983.

Newitt, Malyn. *A History of Portuguese Overseas Expansion, 1400-1668*. London: Routledge, 2005.

Prestage, Edgar. *The Portuguese Pioneers*. London: A. & C. Black, 1933.

Russell-Wood, A.J.R. *The Portuguese Empire, 1415-1808: A World on the Move*. Baltimore: Johns Hopkins University Press, 1998.

Subrahmanyam, Sanjay. *The Portuguese Empire in Asia, 1500-1700: A Political and Economic History*. London: Longman, 1993.

Thomaz, Luís Filipe F.R. *De Ceuta a Timor*. Lisbon: Difel, 1994.

Biographical and Expedition Studies

Ames, Glenn J. *Vasco da Gama: Renaissance Crusader*. New York: Pearson Longman, 2009.

Axelson, Eric. *Congo to Cape: Early Portuguese Explorers*. London: Faber and Faber, 1973.

Bergreen, Laurence. *Over the Edge of the World: Magellan's Terrifying Circumnavigation of the Globe*. New York: William Morrow, 2003.

Biggar, H.P., ed. *The Precursors of Jacques Cartier, 1497-1534*. Ottawa: Government Printing Bureau, 1911.

Braga, J.M. *China Landfall 1513: Jorge Álvares' Voyage to China*. Macau, 1955.

Fernández-Armesto, Felipe. *Straits: Beyond the Myth of Magellan*. London: Bloomsbury, 2022.

Greenlee, William Brooks, ed. *The Voyage of Pedro Álvares Cabral to Brazil and India*. London: Hakluyt Society, 1938.

Morison, Samuel Eliot. *The European Discovery of America: The Northern Voyages, A.D. 500-1600*. New York: Oxford University Press, 1971.

Rogers, Francis M. *The Quest for Eastern Christians: Travels and Rumor in the Age of Discovery*. Minneapolis: University of Minnesota Press, 1962.

Russell, Peter. *Prince Henry 'the Navigator': A Life*. New Haven: Yale University Press, 2000.

Subrahmanyam, Sanjay. *The Career and Legend of Vasco da Gama*. Cambridge: Cambridge University Press, 1997.

Regional Studies: Asia

Andaya, Barbara Watson, and Leonard Y. Andaya. *A History of Malaysia*. London: Macmillan, 1982.

Boxer, C.R. *The Christian Century in Japan, 1549-1650*. Berkeley: University of California Press, 1951.

Brook, Timothy. *The Confusions of Pleasure: Commerce and Culture in Ming China*. Berkeley: University of California Press, 1998.

Chaudhuri, K.N. *Trade and Civilisation in the Indian Ocean: An Economic History from the Rise of Islam to 1750*. Cambridge: Cambridge University Press, 1985.

Fairbank, John K. *The Chinese World Order: Traditional China's Foreign Relations*. Cambridge: Harvard University Press, 1968.

Flores, Jorge. "The Origins of Portuguese Macau." In *Portuguese Asia*. London: Routledge, 2015.

Lach, Donald F. *Asia in the Making of Europe. Vol. I: The Century of Discovery*. Chicago: University of Chicago Press, 1965.

Pearson, M.N. *The Portuguese in India*. Cambridge: Cambridge University Press, 1987.

Reid, Anthony. *Southeast Asia in the Age of Commerce, 1450-1680*. 2 vols. New Haven: Yale University Press, 1988-1993.

Scott, William Henry. *Barangay: Sixteenth-Century Philippine Culture and Society*. Quezon City: Ateneo de Manila University Press, 1994.

Regional Studies: Indian Ocean

Chaudhuri, K.N. *Trade and Civilisation in the Indian Ocean: An Economic History from the Rise of Islam to 1750*. Cambridge: Cambridge University Press, 1985.

Das Gupta, Ashin, and M.N. Pearson, eds. *India and the Indian Ocean, 1500-1800*. Calcutta: Oxford University Press, 1987.

McPherson, Kenneth. *The Indian Ocean: A History of People and the Sea*. Delhi: Oxford University Press, 1993.

Pearson, Michael N. *The Indian Ocean*. London: Routledge, 2003.

Regional Studies: Africa

Hilton, Anne. *The Kingdom of Kongo*. Oxford: Clarendon Press, 1985.

Saunders, A.C. de C.M. *A Social History of Black Slaves and Freedmen in Portugal, 1441-1555*. Cambridge: Cambridge University Press, 1982.

Thornton, John K. *Africa and Africans in the Making of the Atlantic World, 1400-1800*. 2nd ed. Cambridge: Cambridge University Press, 1998.

Regional Studies: Brazil and the Atlantic

Bethell, Leslie, ed. *Colonial Brazil.* Cambridge: Cambridge University Press, 1987.

Hemming, John. *Red Gold: The Conquest of the Brazilian Indians*. Cambridge: Harvard University Press, 1978.

Marshall, Ingeborg. *A History and Ethnography of the Beothuk*. Montreal: McGill-Queen's University Press, 1996.

Metcalf, Alida C. *Go-Betweens and the Colonization of Brazil, 1500-1600*. Austin: University of Texas Press, 2005.

Schwartz, Stuart B. *Sugar Plantations in the Formation of Brazilian Society: Bahia, 1550-1835*. Cambridge: Cambridge University Press, 1985.

Schwartz, Stuart B., ed. *Tropical Babylons: Sugar and the Making of the Atlantic World, 1450-1680*. Chapel Hill: University of North Carolina Press, 2004.

Vieira, Alberto. *Os Escravos no Arquipélago da Madeira: Séculos XV a XVII*. Funchal: CEHA-Madeira, 1991.

Navigation, Cartography, and Maritime Technology

Cortesão, Jaime. *A Política de Sigilo nos Descobrimentos*. Lisbon: Imprensa Nacional, 1960.

Domingues, Francisco Contente. *Os Navios do Mar Oceano: Teoria e Empiria na Arquitectura Naval Portuguesa dos Séculos XVI e XVII*. Lisbon: Centro de Estudos de História e Cartografia Antiga, 2004.

Randles, W.G.L. "The Emergence of Nautical Astronomy in Portugal in the XVth Century." *Journal of Navigation* 51, no. 1 (1998): 46-57.

Unger, Richard W. *The Ship in the Medieval Economy, 600-1600*. London: Croom Helm, 1980.

Waters, David W. "Science and the Techniques of Navigation in the Renaissance." *Maritime Monographs and Reports*, No. 19. Greenwich: National Maritime Museum, 1976.

Military and Naval History

Monteiro, Saturnino. *Batalhas e Combates da Marinha Portuguesa, Vol. I (1139-1521)*. Lisbon: Livraria Sá da Costa Editora, 1989.

Comparative and Contextual Studies

Kamen, Henry. *Spain's Road to Empire: The Making of a World Power, 1492-1763*. London: Allen Lane, 2002.

Parry, J.H. *The Age of Reconnaissance: Discovery, Exploration and Settlement, 1450-1650*. London: Weidenfeld and Nicolson, 1963.

Articles

Catz, Rebecca D. Various articles on Fernão Mendes Pinto. *Portuguese Studies* and *Bulletin of Hispanic Studies*, 1980s-1990s.

Godinho, Vitorino Magalhães. "As incógnitas de Tristão da Cunha." *Revista da Faculdade de Letras de Lisboa*, 3rd series, 8 (1964): 60-83.

Johnson, Harold B. "The Donatary Captaincy in Perspective: Portuguese Backgrounds to the Settlement of Brazil." *Hispanic American Historical Review* 52, no. 2 (1972): 203-214.

Thornton, John K. "Early Kongo-Portuguese Relations: A New Interpretation." *History in Africa* 8 (1981): 183-204.